Conflicts in Conservation

Navigating Towards Solutions

Conflicts over the conservation of biodiversity are incr
conservation objectives and people's lives and livelihoods. Changing patterns in land use, over-exploitation, pollution, climate change and the threat posed by invasive species all challenge the way we currently manage biodiversity – from the local management of single species to the international management of resources. The solutions are as much about people as they are about biodiversity.

Presenting approaches from different academic disciplines and practitioners, this volume offers new insights and approaches to deal with conflicts. Ground-breaking strategies for conservation are analysed, and a large section of the book is devoted to exploring case studies of conflict from around the world.

Aimed primarily at academics, researchers and students from disciplines relating to conservation, ecology, natural resources management and environmental governance, this book will be equally valuable to conservation NGOs and practitioners, and to the policy community at national and international levels.

STEPHEN M. REDPATH is a conservation scientist at the University of Aberdeen. He started his career as an ecologist, working on birds of prey and red grouse. Much of his work now focuses on understanding and searching for sustainable solutions to a wide range of conservation conflicts.

R. J. GUTIÉRREZ is a wildlife ecologist at the University of Minnesota, St. Paul, where he holds the Gordon Gullion Endowed Chair in Forest Wildlife Research and Education. He has studied spotted owls and the conflict between forestry and conservation interests. Recently, his focus has expanded from science and management to the roots of this conflict.

KEVIN A. WOOD is a researcher at Bournemouth University interested in conservation and ecology. His interest in the natural world began with a childhood chasing insects, climbing trees and trying to catch fish in the Oxfordshire countryside where he grew up. His research aims to predict how organisms will respond to environmental change, and how species can be managed to prevent population declines.

JULIETTE C. YOUNG is a political ecologist at the NERC Centre for Ecology and Hydrology, where she has been working since 2002. Her portfolio of national and international research is focused on interdisciplinary approaches to understand and address conservation conflicts, attitudes towards biodiversity and its conservation, and science–policy interfaces.

Ecological Reviews

Ecological Reviews publishes books at the cutting edge of modern ecology, providing a forum for volumes that discuss topics that are focal points of current activity and likely long-term importance to the progress of the field. The series is an invaluable source of ideas and inspiration for ecologists at all levels from graduate students to more established researchers and professionals. The series has been developed jointly by the British Ecological Society and Cambridge University Press and encompasses the Society's Symposia as appropriate.

Biotic Interactions in the Tropics: Their Role in the Maintenance of Species Diversity
Edited by David F. R. P. Burslem, Michelle A. Pinard and Sue E. Hartley

Biological Diversity and Function in Soils
Edited by Richard Bardgett, Michael Usher and David Hopkins

Island Colonization: The Origin and Development of Island Communities
By Ian Thornton
Edited by Tim New

Scaling Biodiversity
Edited by David Storch, Pablo Margnet and James Brown

Body Size: The Structure and Function of Aquatic Ecosystems
Edited by Alan G. Hildrew, David G. Raffaelli and Ronni Edmonds-Brown

Speciation and Patterns of Diversity
Edited by Roger Butlin, Jon Bridle and Dolph Schluter

Ecology of Industrial Pollution
Edited by Lesley C. Batty and Kevin B. Hallberg

Ecosystem Ecology: A New Synthesis
Edited by David G. Raffaelli and Christopher L. J. Frid

Urban Ecology
Edited by Kevin J. Gaston

The Ecology of Plant Secondary Metabolites: From Genes to Global Processes
Edited by Glenn R. Iason, Marcel Dicke and Susan E. Hartley

Birds and Habitat: Relationships in Changing Landscapes
Edited by Robert J. Fuller

Trait-Mediated Indirect Interactions: Ecological and Evolutionary Perspectives
Edited by Takayuki Ohgushi, Oswald Schmitz and Robert D. Holt

Forests and Global Change
Edited by David A. Coomes, David F. R. P. Burslem and William D. Simonson

Trophic Ecology: Bottom-Up and Top-Down Interactions across Aquatic and Terrestrial Systems
Edited by Torrance C. Hanley and Kimberly J. La Pierre

Conflicts in Conservation

Navigating Towards Solutions

Edited by

STEPHEN M. REDPATH
University of Aberdeen, UK

R. J. GUTIÉRREZ
University of Minnesota, USA

KEVIN A. WOOD
Bournemouth University, UK

JULIETTE C. YOUNG
Centre for Ecology & Hydrology, UK

CAMBRIDGE
UNIVERSITY PRESS

CAMBRIDGE
UNIVERSITY PRESS

University Printing House, Cambridge CB2 8BS, United Kingdom

Cambridge University Press is part of the University of Cambridge.

It furthers the University's mission by disseminating knowledge in the pursuit of education, learning and research at the highest international levels of excellence.

www.cambridge.org
Information on this title: www.cambridge.org/9781107017696

First published 2015

Printed in the United Kingdom by TJ International Ltd. Padstow Cornwall

A catalogue record for this publication is available from the British Library

Library of Congress Cataloguing in Publication data
Conflicts in conservation : navigating towards solutions / edited by Stephen M. Redpath,
University of Aberdeen, U.K. [and three others].
pages cm
Includes bibliographical references and index.
ISBN 978-1-107-01769-6 (hardback: alk. paper) 1. Conservation biology. 2. Conservation of natural resources. I. Redpath, S. M. (Steve M.)
QH75.C625 2015
333.95′16 – dc23 2015005505

ISBN 978-1-107-01769-6 Hardback

For Simon Thirgood
(1962–2009)

Contents

The colour plates are located between pages 110 and 111

Contributors

WILLIAM M. ADAMS
Department of Geography, University of Cambridge, UK

ARJUN AMAR
Percy FitzPatrick Institute, DST-NRF Centre of Excellence, University of Cape Town, South Africa

DENNIS R. BECKER
Department of Forest Resources, University of Minnesota, St Paul, Minnesota, USA

HERBERT H. BLUMBERG
Department of Psychology, Goldsmiths, University of London, UK

ESTHER CARMEN
NERC Centre for Ecology and Hydrology, Edinburgh, UK

SCOTT CASHEN
California, USA

ANTONY S. CHENG
Department of Forest & Rangeland Stewardship, Colorado State University, Fort Collins, Colorado, USA

RUTH CROMIE
Wildfowl & Wetlands Trust, Slimbridge, Gloucestershire, UK

FRANCIS DAUNT
NERC Centre for Ecology and Hydrology, Edinburgh, UK

AMY J. DICKMAN
WildCRU, Zoology, University of Oxford, The Recanati-Kaplan Centre, Tubney, UK

NIGEL DOWER
School of Divinity, History and Philosophy, University of Aberdeen, UK

IOAN FAZEY
School of Environment/CECHR University of Dundee, Perth Road, Dundee, UK

JENS FRANK
Department of Ecology, Swedish University of Agricultural Sciences, Sweden

DAVID GANZ
USAID LEAF, Bangkok, Thailand

ISLA M. GRAHAM
Lighthouse Field Station, Institute of Biological & Environmental Sciences, University of Aberdeen, Cromarty, UK

JOHN GUNN
Spatial Informatics Group, Cumberland, Maine, USA

R. J. GUTIÉRREZ
Department of Fisheries, Wildlife, and Conservation Biology, University of Minnesota, St. Paul, Minnesota, USA

NICK HANLEY
Department of Geography and Sustainable Development, University of St Andrews, UK

JOHANNES P. M. HEINONEN
The Institute of Biological and Environmental Sciences, University of Aberdeen, UK

ALAN HOLLAND
Department of Politics, Philosophy and Religion, Lancaster University, UK

ANDREW JORDAN
School of Environmental Sciences, University of East Anglia, UK

NIELS KANSTRUP
Danish Academy of Hunting, Rønde, Denmark

ISIDORA KATARA
Instituto Português do Mar e da Atmosfera, Portugal

JASPER O. KENTER
Aberdeen Centre for Environmental Sustainability, University of Aberdeen, UK

ALISON A. KOCK
Department of Biological Science, University of Cape Town, Rondebosch, South Africa

ROBERT A. LAMBERT
Department of History, University of Nottingham, UK

JOHN D. C. LINNELL
Norwegian Institute for Nature Research, Trondheim, Norway

MICHAEL LIQUORI
Soundwatershed, Alameda, California, USA

FRANCINE MADDEN
Human–Wildlife Conflict Collaboration, USA

MYSORE D. MADHUSUDAN
Nature Conservation Foundation, Mysore, India

JESPER MADSEN
Department of Bioscience, Aarhus University, Denmark

MARIELLA MARZANO
Forest Research, Northern Research Station, Roslin, Midlothian, UK

ALY McCLUSKIE
RSPB Centre for Conservation Science, Edinburgh, UK

BRIAN McQUINN
Institute of Cognitive and Evolutionary Anthropology, University of Oxford, UK

L. DAVID MECH
U.S. Geological Survey, Northern Prairie Wildlife Research Center, Jamestown, ND, USA

AMY MERRILL
Stillwater Sciences, Berkeley, California, USA

CHARUDUTT MISHRA
Snow Leopard Trust and Nature Conservation Foundation, Mysore, India

JULIA NEWTH
Wildfowl & Wetlands Trust, Slimbridge, Gloucestershire, UK

MATTHEW T. O'HARE
Centre for Ecology and Hydrology, Edinburgh, UK

M. JUSTIN O'RIAIN
Department of Biological Science, University of Cape Town, Rondebosch, South Africa

CRISTINA PITA
Centre for Environmental and Marine Studies, University of Aveiro, Portugal

DIANA POUND
Dialogue Matters, UK

WILLIAM PRICE
Gifford Pinchot Institute for Conservation, Washington, DC, USA

STEPHEN M. REDPATH
Institute of Biological & Environmental Sciences, University of Aberdeen, Aberdeen, UK

MARK S. REED
Knowledge ExCHANGE Research Centre, The Birmingham School of the Built Environment, Birmingham City University, UK

JOANNA L. ROBINSON
Glendon College, York University, Toronto, Ontario, Canada

PAUL ROGERS
School of Social and International Studies, University of Bradford, UK

DAVID S. SAAH
Department of Environmental Science, University of San Francisco, San Francisco, California, USA

ROGER SIDAWAY
Independent Facilitator, Edinburgh, UK

JULIAN SIDOLI DEL CENO
Knowledge ExCHANGE Research Centre, The Birmingham School of the Built Environment, Birmingham City University, UK

PETER SIMMONS
School of Environmental Sciences, University of East Anglia, UK

RICHARD A. STILLMAN
Bournemouth University, Poole, UK

MARK C. J. STODDART
Department of Sociology, Memorial University of Newfoundland, Canada

KULBHUSHANSINGH R. SURYAWANSHI
Snow Leopard Trust and Nature Conservation Foundation, Mysore, India

WILLIAM J. SUTHERLAND
Department of Zoology, University of Cambridge, UK

DAVID B. TINDALL
Department of Sociology, The University of British Columbia, Vancouver, Canada

JUSTIN M. J. TRAVIS
The Institute of Biological and Environmental Sciences, University of Aberdeen, UK

ARIE TROUWBORST
Tilburg Law School, Department of European and International Public Law, Tilburg University, The Netherlands

ALLAN WATT
NERC Centre for Ecology and Hydrology, Edinburgh, UK

ANDREW WHITEHOUSE
The Department of Anthropology, University of Aberdeen, UK

KEVIN A. WOOD
Bournemouth University, Poole, UK

JULIETTE C. YOUNG
NERC Centre for Ecology and Hydrology, Edinburgh, UK

Foreword

As pressures on the environment from ever-increasing numbers of people with more needs and demands for natural resources grow, so must the number of conservation problems. Examining conservation problems solely from the perspective of the species or habitat that is put under pressure simply leads to a rapidly escalating list of ever-more-difficult problems. Solutions are hard to find, especially solutions that will be effective over longer time periods. In fact, conservation is almost never this straightforward anyway. It is much more complex and usually there are many different winners and losers in different sectors of society, and often over time and in different places as well. No wonder, then, that progress to meet conservation targets has been so challenging. As the authors in this book show over and over again, the apparent conflicts between people and the environment are better tackled by appreciating that the conflicts are actually between different groups of people. While sometimes the conflicts are very clear, especially to conservationists working on the front line who encounter stakeholders with different perspectives, this is rarely the case, and it is even rarer that the issues can be simply or neatly circumscribed.

A key advance, then, is the recognition that conservation is an interdisciplinary endeavour and that biologists acting alone can never hope to make lasting progress. Instead, recognising the many linkages between nature and society, and building on the different kinds of values that people have and hold for nature will allow much more fundamental and ultimately sustainable progress to be made. Conservation will never work when it is either apparently or actually at odds with the needs and wishes of those who are nature's stewards.

So, this is a very important book, focussing as it does very clearly at the interface between conservation and livelihoods, and scrutinising many different strategies for enhancing both conservation success and social outcomes in mutually supporting relationships. Of course there are few easy solutions, but the many different perspectives and case studies presented here provide a remarkable resource for lasting solutions. I commend Steve Redpath and the other editors for the extraordinary effort to include such a wide range of different disciplinary experts and conflict situations in such broad geographical and political contexts. Among a plethora of conservation manuals and text books, this stands out for its breadth and depth, and the potential it offers for so many hard conservation problems.

Georgina Mace
University College London

Acknowledgements

The initial idea for this book grew out of a conference on human–wildlife conflicts in London in 2002, partly organised by our friend and colleague, Simon Thirgood. This conference ultimately led to the excellent book *People and wildlife: conflict or coexistence*, by Woodroffe *et al.* (2005). The focus of that conference was on human–animal interactions. Yet, as recognised by contributors at that conference, conservation is fundamentally about personal and cultural values and the choices people make, and we therefore need to understand the people as much as we do the human–animal interactions. To start exploring these complex problems, we held a conference, 'Conflicts in conservation: strategies for coping with a changing world', in August 2011 that was organised by the Aberdeen Centre for Environmental Sustainability (ACES). The discussions from this conference led to a review paper on conservation conflicts (Redpath *et al.*, 2013), and ultimately this book.

This book is the result of collaboration with over 60 colleagues studying conflicts from diverse disciplines around the world. We thank them all for their wisdom, their insight and their wonderful contributions, as well as their patience in seeing this book to completion. Thanks to Catherine Young, Adam Vanbergen and Viktor Bahktin for their beautiful illustrations. We also want to thank Georgina Mace for her foreword to the book.

We are grateful to The British Ecological Society, especially, and Aberdeen University for supporting initial meetings to discuss these ideas. We are particularly indebted to Mark Reed and Anna Evely for their energy in getting this book off the ground. We would also like to thank colleagues, notably those attending the Society for Conservation Biology meeting in Aukland in 2011, and the Human Dimensions of Wildlife conference in 2012, who helped shape the ideas in this book.

Thanks to our colleagues who have argued and discussed and criticised our ideas and helped us strengthen them and craft the book, especially Allan Watt, Mike Manfredo, Justin Irvine, Nils Bunnefeld, Freya St. John, EJ Milner-Gulland, Charudutt Mishra, Kulbhushansingh Suryawanshi, Phil Hulme, Camilla Sandstrom, Adrian Treves, Martin Sharman and Tomas Willebrand. Our thanks also go to Catherine Hill and Hefin Jones at the British Ecological Society, and Ilaria Tassistro, Dominic Lewis, Kate Harrison, Renee Duncan-Mestel, Sarah Payne and all their colleagues at Cambridge University Press for their help and support in the writing of this book.

Steve Redpath was supported by the University of Aberdeen; R. J. Gutiérrez by the University of Minnesota (UMN) Agriculture Experiment Station Project MIN-41-036, the UMN Department of Fisheries, Wildlife, and Conservation Biology and the Gordon Gullion Endowed Chair; Kevin A. Wood by Bournemouth University's School of Applied Sciences and Juliette Young by the UK Natural Environment Research Council Centre for Ecology and Hydrology.

References

Redpath, S. M., *et al.* (2013). Understanding and managing conservation conflicts. *Trends Ecol. Evol.*, 28, 100–109.

Woodroffe, R., Thirgood, S. and Rabinowitz, A. (eds) (2005). *People and Wildlife: Conflict or Coexistence?* Cambridge: Cambridge University Press.

PART I

Introduction to conservation and conflict

CHAPTER ONE

An introduction to conservation conflicts

STEPHEN M. REDPATH
University of Aberdeen
R. J. GUTIÉRREZ
University of Minnesota
KEVIN A. WOOD
Bournemouth University
ROGER SIDAWAY
Independent Facilitator, Edinburgh, Scotland
and
JULIETTE C. YOUNG
NERC Centre for Ecology & Hydrology

The conservation of biodiversity is an increasingly challenging endeavour. Current pressures from a growing human population have led to concerns of a sixth mass extinction event, bringing mounting pressure to find effective ways of conserving biodiversity (Barnosky *et al.*, 2011). However, our ability to meet this challenge is affected by the fact that not everyone supports conservation objectives. People naturally have different interests and priorities, some of which may be diametrically opposed to conservation objectives. In some cases, these differences lead to damaging and costly conflicts that we see emerging across the world and which present major challenges to modern conservation (MacDonald and Service, 2007).

At a cursory glance, the conflicts that surface around conservation often appear to be about impact: the impact of carnivores on livestock; the impact of wind farms on birds; or the impact of protected areas on livelihoods. Consequently, a common approach to these problems has been to build robust science and develop an evidence base to understand these impacts and find ways of reducing them, often through technical solutions. This approach, however, rarely works for the simple reason that many of these conflicts are about much more than impact. So even if we can develop the science to quantify impacts and show how they can be reduced, the conflicts can stubbornly persist. Indeed, beneath the surface of any of the conflicts discussed in this book is a complex layering of diverse issues related to different world views, issues of trust, power

Conflicts in Conservation: Navigating Towards Solutions, ed. S. M. Redpath, R. J. Gutiérrez, K. A. Wood and J. C. Young. Published by Cambridge University Press.

imbalances or latent historical issues – issues that lie well outside the sphere of the natural sciences. So, if we really want to understand and tackle these thorny problems, we need insights from other disciplines as well as from the practitioners specialising in resolving conflicts.

The growing recognition of the complexity within conflicts has led many authors to suggest more cross-disciplinary approaches, especially through better integration of ecological and social science (Manfredo and Dayer, 2004; Sillero-Zubiri *et al.*, 2007; Treves, 2009; Dickman, 2010; White and Ward, 2010). This book seeks to build and expand on these ideas. One of the difficulties of doing this lies in understanding what the many different disciplines can offer to the understanding and management of conflicts, and which disciplines and conflict-management practitioners we should be working with. In this book we begin to address this challenge. Let's start, however, by defining what we mean by conservation conflict.

Defining conservation conflict

Conflicts are generally defined as 'a state of opposition or hostilities' or 'a clashing of opposed principles' (*Oxford Concise Dictionary*). The term therefore implies action rather than simply a passive reflection on differences and disagreements. So, in conservation, conflicts occur when parties clash over differences about conservation objectives and when one party asserts, or at least is perceived to assert, its interests at the expense of another (Redpath *et al.*, 2013). For example, conservation conflicts emerge when people kill predators or destroy important habitats that other people want conserved. Similarly, conflicts emerge when conservation results in the protection of predators or new protected areas that threaten the livelihoods or well-being of other people. These examples are clearly very different, occur at different scales and involve a range of different people, but the principles are the same – conflicts are about clashes in priorities and world views and the imposition of one value system on another.

In the literature, much of the research on conservation and conflicts has been focused on human–wildlife conflict, which has been defined as occurring whenever actions by either humans or wildlife have an adverse effect on the other (Conover, 2002). Despite being widely used, this term is problematic, in part because it falsely suggests that wildlife species are knowingly antagonists in conflict (Peterson *et al.*, 2010). In addition, the term focuses attention on the interactions between humans and wildlife. Yet, as highlighted above, this is just one of the components inherent in these conflicts – so the term obscures the underpinning conflicts that occur between those humans who are affected by the wildlife and those humans who are defending pro-wildlife objectives (Young *et al.*, 2010). Thus, we consider that it is critical to partition human–wildlife and human–human dimensions, because they highlight very different elements that we need to understand and address. This distinction also helps clarify the main adversaries involved in conflict, and by doing so it opens up the space and expertise to search for sustainable solutions (Redpath *et al.*, 2014). In light of these terminological problems, we have used the term 'conservation conflict' throughout this book to help disentangle the two components. In reality, of course, the overwhelming majority of so-called human–wildlife conflicts can be considered conservation conflicts because the wildlife species involved are typically of conservation concern (Redpath *et al.*, 2014).

To illustrate our point, consider two very similar situations in the UK. In the first, introduced mink *Neovison vison* have had large impacts on species of conservation concern, and considerable effort has been expended trying to manage this problem (Bonesi and Palazon, 2007). There has been widespread recognition of the problem and broad agreement about the need to reduce the mink population. In contrast, hedgehogs *Erinaceus europaeus* were introduced into the western islands of Scotland where they have also had large impacts on birds of conservation concern (Jackson *et al.*, 2004). Like the mink, considerable effort has been expended on removing hedgehogs. In this instance, however, there were strong disagreements about lethal control and a conflict erupted between conservationists and animal welfare organisations (Warwick, 2012). Both of these issues would be considered to be human–wildlife conflicts. However, while mink, like many other pest and invasive species, certainly present some challenging management problems to reduce impacts, they are not at the centre of a conservation conflict, in a way that the hedgehog is.

The complexity of conservation conflicts

Grey wolves *Canis lupus* and humans have long had an uneasy relationship. Throughout human history, wolves have been viewed as a threat to human life and livelihoods. This in turn has led to the species being extirpated from large parts of its global range (Delibes, 1990; Woodroffe, 2000). More recently, however, other voices have been heard in favour of wolves, either because of recognition of the role they play in ecosystems or because of their inherent right to exist (Mech, 2011). We have moved over time from a 'vermin' control problem to a conservation conflict where groups argue for and against wolves. Recently, these carnivores have returned to parts of their former range, either by spreading naturally, or with the help of reintroduction schemes (Wabakken, 2001; Mech, 2011). These changes and the response to them vividly highlight the complexity of conflict (Nie, 2003; Box 8; Box 15).

There is concern and disagreement about the impact of wolves on livelihoods; there are arguments and uncertainties about the positive benefits of wolves for the ecosystems; there are very deeply held values on both sides of the conflict and anger and passionate arguments for and against wolves both from individuals and from specifically formed groups; there are strong cultural, ethical and moral dimensions that underpin these values and arguments and there are ongoing legislative battles about how wolves should be managed. In addition, there are researchers, some of whom remain neutral while others become advocates, there are the state and federal authorities who must manage this difficult conflict, and there is a diverse variety of information that influences the managers' judgement, the politicians' stance and the public's opinion – knowledge from experiences of those who live and work in these ecosystems, evidence from scientific studies and articles from the media.

This brief glimpse into this world highlights the bewildering dimensions involved in conflicts. We not only need to understand the impact that wolves have on livestock and ecosystems, including the uncertainties inherent in such research, we need to quantify how negative impacts can be mitigated. More than that, however, we also need to understand the world views, values, beliefs and attitudes of those people involved in the conflict; the moral and ethical arguments involved; why people behave in the way that they do towards wolves; how effective different forms of knowledge and communication are at altering attitudes and human behaviour; how history and economic arguments influence perspectives; what position scientists should take when engaged in the conflict; how decisions should be made about wolf management; if 'top-down' approaches imposed on people will lead to better or worse social and ecological outcomes than 'bottom-up' approaches that involve mediation or dialogue; how people should be encouraged to engage in the search for solutions; how an effective dialogue process should be designed; and what role government, mediators and independent facilitators should play in tackling these problems. These are just some of the complex dimensions that demand a multi-pronged approach and that we seek to cover in this book.

Types of conflict

Throughout the book, a diversity of conflicts is depicted by authors. Each is unique and context-dependent, but at the same time there are similar issues that run through them. Six broad, often overlapping, categories of conflict have been identified (Sidaway, 2005; Young *et al.*, 2010; see also Moore, 1996). This typology partitions the different dimensions of a conflict and helps identify key themes. This, in turn, can highlight what types of information and approaches might be useful in helping our understanding of the specific conflict, and ways of managing it. These categories are as follows.

- Conflicts of interest – two groups want different things from the same habitat or species. This is well demonstrated in the chapters and boxes in this book. A typical example of a conflict of interest is over forest resources, where some groups want to harvest trees, and other groups want to preserve the forest as a habitat for a specific species (e.g. owls; see Box 19).

- Conflicts over beliefs and values – differences exist over normative perceptions, such as perceptions of what human activities should be allowed, or what species should be conserved. These types of values are explored in more detail in Chapters 7 and 10. One example of such a conflict may be related to species reintroductions. For example, stakeholders may have strong beliefs and disagree about whether species such as sea eagles *Haliaeetus albicilla* that can kill lambs should be reintroduced into Scotland or Ireland (O'Rourke, 2014). Deep-set beliefs can be non-negotiable and where they clash can lead to conflicts difficult to resolve.
- Conflicts over process – different approaches to decision-making and fairness taken by different people, groups, or agencies. These conflicts can often be stand-alone, or part of the above two conflicts. An example is when two groups involved in a conflict have very different cultures of how to manage conflict. Another example would be where one group favours a participatory, consensus approach to searching for solutions, while the other group favours a more authoritarian approach.
- Conflicts over information – situations where information is lacking, misunderstood, or perceived in different ways by different stakeholders. As with conflicts over process, often these types of conflict will be embedded in broader conflicts over values or over interest. Perhaps one of the most common conflicts in this category occurs when scientific knowledge is not in line with knowledge held by local stakeholders. In such cases, one form of knowledge is refuted and challenged and can hinder understanding between parties and generate mistrust.
- Structural conflicts – relate to social, legal, economic and cultural arrangements. These conflicts are often latent, involving inequalities between parties, and only becoming apparent once a conflict has become more visible. An example could be a conflict in which a large multi-national corporation has many more resources than a small grass-roots organisation and can exercise power, in terms of legal, political or economic leverage, over them.
- Interpersonal conflicts – relate to personality differences between individuals or groups, including issues of communication and trust. Interpersonal conflicts are inherent not only to conservation conflicts, but to all aspects of society. Interpersonal conflicts can be linked to perceptions of groups and individuals, before such individuals and groups ever actually meet face to face. Such conflicts therefore need to be recognised, and acknowledged as integral to conservation conflicts.

This typology highlights the value of spending time at the outset identifying the types of conflicts involved and the approaches required to find solutions. For example, if a conflict is primarily about beliefs or interpersonal issues, a detailed scientific study of impact may not help greatly in finding solutions. If the conflicts are over information, there may be merit in bringing stakeholders together to co-produce knowledge and thereby increase shared understanding (Hage *et al.*, 2010).

The search for solutions to conservation conflicts

Academic disciplines are essential in helping us understand and tease apart the dimensions inherent in conflict. However, they are often of less help when it comes to searching for solutions. The issue of how we address conservation conflicts raises several important points. First, do we actually want to solve the conflict? In some cases, conflicts can be constructive because they stimulate change (Wittmer *et al.*, 2006). However, more commonly they are damaging to people's lives and livelihoods, to relationships between individuals and institutions and to biodiversity (Treves and Karanth, 2003; Woodroffe *et al.*, 2005; Box 1). So, assuming we do wish to reduce the negative consequences of conflict, the challenge then becomes finding effective ways of moving away from a damaging, destructive situation to one that improves conservation and livelihood outcomes.

Other issues to consider are what a solution should look like and whose solution we are seeking. Consider the example in the UK of a long-running conflict between conservation organisations and game managers over the illegal killing of birds of prey (Thirgood and Redpath, 2008; Box 3). To put it simply, conservation organisations want more predators in the environment and game managers want more game for their hunting clients to shoot. To many on the conservation side the best solution would be more predators being imposed through enforcement and legislation, possibly without regard to the monetary or social costs to managers and hunters. In contrast, game managers might be more interested in a solution that involves the continued removal of predators from large areas of land managed to sustain their current levels of game. These are the types of adversarial positions we commonly see in conflict. For a variety of very good reasons there is often little attempt from either side to understand and reduce the human–human conflicts that underpin such issues.

The broader idea of conflict resolution is to recognise that conflicts represent shared problems, in this case that both hunting and conservation are legitimate activities, and to determine if parties can be moved from their original adversarial positions to ones of shared agreement. So the solution might involve effective dialogue between both sides, leading to agreement that encourages management activities to eliminate illegal killing, reduces the level of social conflict and allows predators and game interests to coexist. There is good evidence that dialogue improves trust and reduces social conflict, but it is less clear how this relates to ecological outcomes (Young *et al.*, 2013). From a conservation perspective, the question then becomes: is it better to hold the adversarial position and seek an outcome that would impose high numbers of predators onto an unwilling party, recognising that there will be continued tension and conflicts; or is it better to engage and search for a shared solution, with the recognition that there will realistically need to be some compromise in the numbers of predators? In other words, under what conditions is it better for individuals and organisations to fight an adversary and escalate conflict, rather than seek an acceptable solution through dialogue and discussion? While many practitioners engaged in long-running conflict are likely to seek solutions that reduce the level of conflict, their willingness and ability to do so will depend on issues such as anger, a lack of trust, their underlying values, power imbalance and the leadership of those in authority.

Any resolution process operates within political and legal realities, in addition to the scientific, ethical and practical considerations. These factors can limit the options available. In the case above, for example, birds of prey are legally protected, so a solution that involves any form of lethal control would be difficult to implement unless previously hard-won laws are modified, which is something that conservationists are inevitably going to feel uncomfortable about. So, while legislation can force change (see Chapter 19; Box 4), it can also provide a barrier to change, as stakeholders may take strong positions behind the legislation rather than focusing on solutions and seeking to adapt the legislation accordingly (Heydon *et al.*, 2011).

Engagement and the potential for compromise may often be seen as a weakness by parties engaged in conflict and this may therefore limit the options for dialogue. However, in any dialogue, stakeholders will have aspects that are negotiable and aspects that are non-negotiable. For example, species survival will be a non-negotiable aspect for those from the conservation side. Such aspects may require fundamental a priori agreement among parties. The challenge then becomes seeing if there is sufficient flexibility between the positions of the two parties to find a solution. By engaging in dialogue on raptors in the UK uplands we might not expect hunters to accept high densities of raptors and therefore lose their shooting interests. Similarly, conservationists will not accept illegal killing, or extinction. Here the question would be: is there enough flexibility in the two positions to find a solution that would allow some grouse shooting and some breeding raptors?

In any conflict management process, we must also understand and incorporate the uncertainties involved. There are several different types of uncertainty that need to be considered, including those that relate to the natural systems, or the willingness of people to implement management decisions (Milner-Gulland and Rowcliffe, 2007). The incorporation of an adaptive management framework

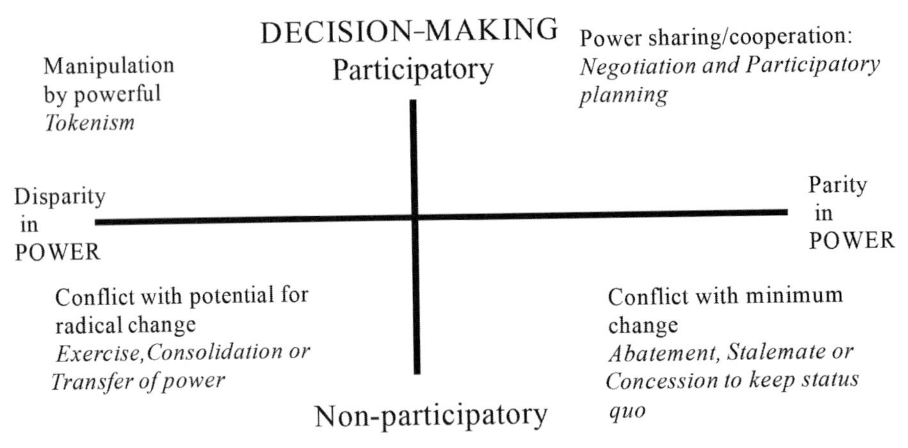

Figure 1.1 Alternative approaches to managing conflicts based on differences in power and decision-making processes (reprinted from Sidaway, 2005. © Sidaway, 2005): the extent to which decision-making is participatory (i.e. inclusive of different stakeholders) is set on the vertical axis. The horizontal axis conveys the power relationships between the stakeholder groups to assess the relative parity or disparity of power among stakeholders.

is therefore an integral part in the development of effective solutions (Holling, 1978).

Given all this complexity, it is worth pointing out that 'silver-bullet' solutions are unlikely ever to be available. We are dealing with multi-layered conflicts that cannot be readily resolved. However, that does not of course mean that it is impossible to reduce and manage the level of conflict in a way that allows coexistence. The complexity means that conflicts can re-emerge unexpectedly and so require long-term management and trust-building.

Conflicts and power

The distribution of power is a central element in conflicts that we have already touched on. A pre-requisite for conflict management lies in understanding the relative distribution of power in the decision-making processes, while understanding that power is dynamic and can shift. Although complete parity is unrealistic, in the context of conflict management, settings can be created that are more or less conducive to the sharing of power between stakeholders. By treating disparities in power and decision-making as independent dimensions, it is possible to represent alternative outcomes of conflicts diagrammatically (see Fig. 1.1).

Arguably, all outcomes on the right-hand side of Fig. 1.1 are stable as long as a relative parity of power between the parties is maintained, while outcomes on the left-hand side of the diagram are potentially unstable, because of relative disparities of power. In the bottom left quadrant, disparity allows the more powerful stakeholders to *exercise or consolidate their power* and they are assisted by the adversarial, non-participatory system of decision-making. However, the injustices resulting from the disparity of power mean that, eventually, opposing groups may emerge and there is the potential for radical change by the *transfer of power*. During the course of the power struggle, disparities in power may be reduced and the conflict moves towards the bottom right quadrant. Concessions may be made to maintain the status quo, the conflict may subside (*abatement*) or *stalemate* may be reached and the conflict remains in the bottom right quadrant.

At the point of stalemate, the prospect of dialogue becomes more attractive, thus there is an opportunity to change the system of decision-making and, if this is taken, it is possible for the disputing parties to work together. Conflict gives way to *cooperation or collaboration* in the top right quadrant. This sequence of events lends support to the argument that a period of stalemate may be necessary before collaboration is feasible. In other words, the politics of collaboration only become a viable possibility when the politics of power have been exhausted or substantively reduced (Amy, 1987; see also Chapter 19). Collaboration implies that decisions are made collaboratively by stakeholders, or their representatives, including in some cases policy makers, government representatives and legislators, who voluntarily work towards finding a mutually acceptable outcome to the conflict. When the interested parties are prepared to reach an outcome, the stakeholders may collaborate among themselves or they may seek assistance from a neutral third party. This could be an arbitrator who listens to their evidence and recommends a solution in their best interests, or a mediator who helps them seek agreement and create their own solution.

If a system is more participatory without equalisation of power (e.g. see Chapter 19), the more powerful stakeholders can manipulate the situation (top left quadrant) as they are not committed to respond to the needs and wishes of other participants. Involving interest groups in decision-making runs counter to many bureaucratic, legal and political cultures. The powerful can be reluctant to begin, or feel threatened by, negotiating with stakeholders. The powerful may see this as the best strategy for maintaining their power, but the injustice perceived by the powerless may act as a stimulus for their politicisation. It is in these situations that consultative planning exercises can be perceived to be *tokenistic* – 'tell us what you think, although we have already made up our minds'.

The book

There have been literally thousands of academic papers and several books that provide analysis and discussion about a diversity of conservation conflicts. So why, then, does the world need another book? Our motivations were threefold. The first was a personal one. All four editors have been involved in conservation conflicts of one sort or another in our work and were driven by a desire to make sense of these 'wicked' problems and make a positive contribution to finding some way through the often seemingly intractable challenges. When you have been in the midst of conflict they can seem overwhelming, so this book has provided a great opportunity to reflect on the many interwoven strands inherent in conflict. For the first time we have attempted to bring those strands together in one place, to give those interested in or struggling with conflict a place where multiple perspectives would be available.

Second, we seek to build on previous work by developing the concept of interdisciplinary approaches, to allow others to reflect on conservation conflicts from their different disciplinary viewpoints and to attempt to draw together the lessons. Hence we include chapters from a wide range of disciplines, written by experts in their field. Each of these disciplines has their own worlds, their subdivisions, their schools, their specific journals and their textbooks. We cannot hope to capture all of that in this book, but we seek to provide a small insight into each world that will hopefully allow a way in for those interested. Conservation conflicts involve a variety of philosophical and ethical considerations (Chapters 2 and 10). Of course, these conflicts occur because there is an ecological imperative present (the conservation of a species, biological community or ecosystem) and so ecology is a central discipline of relevance to understanding and resolving conservation conflicts (Chapter 3). These issues, however, are at their root about people, so we explore anthropological, sociological and psychological perspectives (Chapters 7, 9 and 11). We also explore the historical aspects and the crucial legal, political and policy dimensions to conflicts (Chapters 4, 5, 8 and 13). Conflicts often affect people's livelihoods so this means there are economic ramifications to conservation conflicts and attendant socio-economic issues (Chapter 6). Although we have selected this particular set of disciplines to represent, they are not the only ones relevant to conservation conflicts. In addition to the disciplinary perspectives, we also explore the relevance of models and scale to their understanding and opportunities for offering solutions (Chapters 14 and 15). Peace studies research has a long and relevant history of disentangling the seemingly endless reasons for conflict among people and is of relevance to the conflicts discussed here (Chapter 12). We have tried to minimise jargon, but have allowed the authors the intellectual freedom to express what they considered to be the salient ideas, approaches and contributions of their field. Consequently, we as editors may or may not agree with all ideas expressed by them, as the authors may not agree with our conclusions.

Third, we wanted to produce a book that was not just an academic treatise, but one that examined the practical challenges involved with finding a way through these difficult issues and that therefore dealt with some of the important transdisciplinary issues involved with working with stakeholders. The latter chapters of the book therefore explore approaches to managing conflict, including the use of mediation (Chapter 16). Two chapters are written by practitioners in

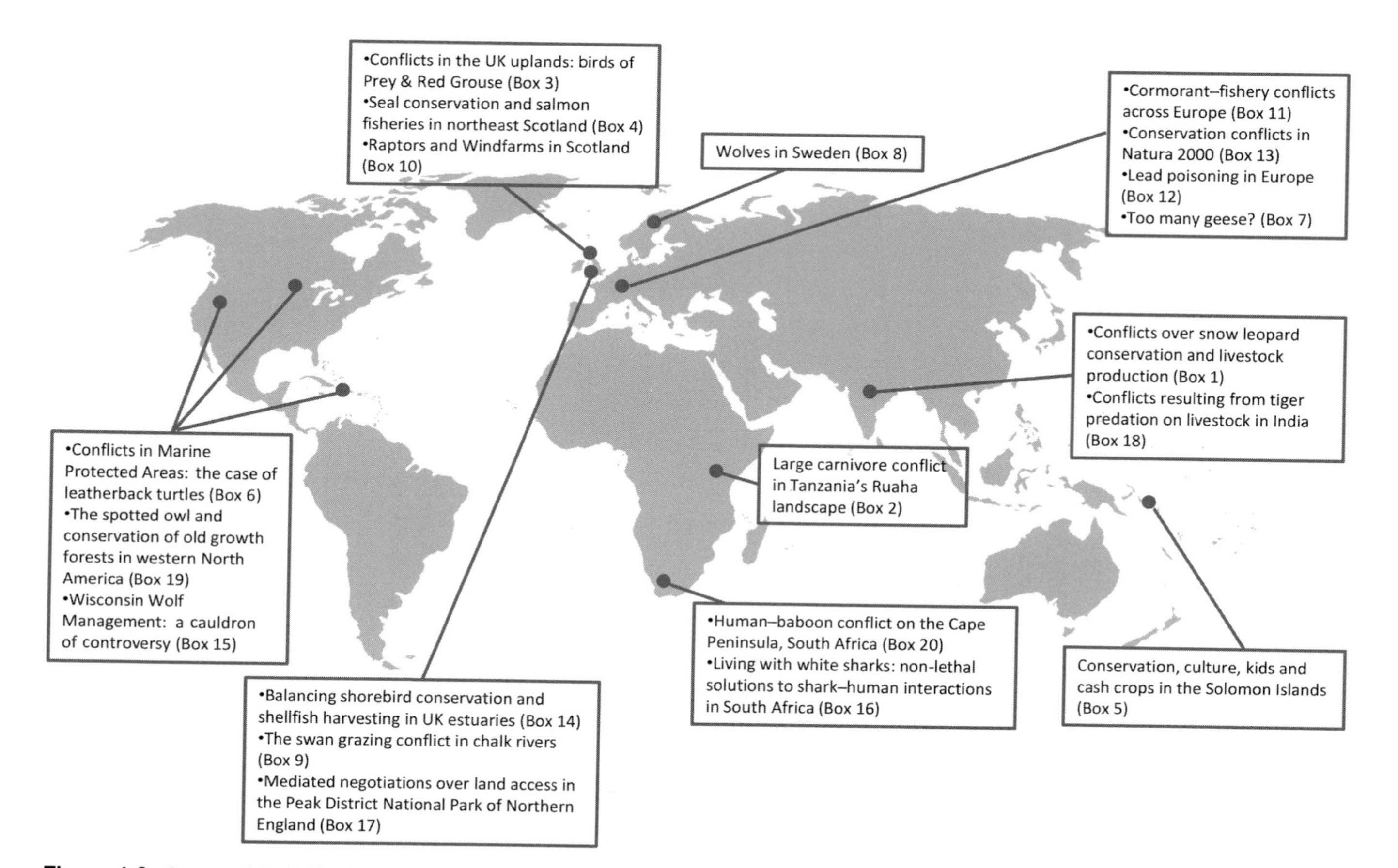

Figure 1.2 Geographical distribution of conservation conflicts described in short boxes or vignettes throughout the book (map reproduced from Wikimedia Commons).

conflict resolution – people at the frontline who support stakeholders in finding solutions to some difficult conflicts (Chapters 17 and 18). Finally, we describe the complexities and lessons learned from a hybrid conflict resolution process of 'legislated collaboration' in California (Chapter 19). At the end of the book we attempt to draw what we consider to be the key messages from this book into a final synthesis (Chapter 20).

Throughout the book, we also provide a series of 20 vignettes of conservation conflicts from around the world (Fig. 1.2). These examples depict conflicts associated with charismatic species such as snow leopards, tigers, lions, raptors and great white sharks, less charismatic but equally problematic species such as baboons and cormorants, and land-use conflicts such as those related to windfarms and protected areas. Some are slightly depressing and reflect seemingly perpetual conflict, whereas others are more positive stories that reflect successful change and coexistence. The idea behind these examples is to bring the subject of this book to life. We are dealing with big, live challenges in conservation that demand effective engagement and management. It is inevitable that such issues are only going to become more prevalent as human impacts on the planet increase. Our hope is that this book will help those currently engaged in conflicts find a way forward and stimulate others to get involved in helping develop solutions that will ultimately benefit conservation through recognising the underlying causes and the shared humanity of those involved on either side.

References

Amy, D. (1987). *The Politics of Environmental Mediation*. New York: Columbia University Press.

Barnosky, A. D., *et al.* (2011). Has the Earth's sixth mass extinction already arrived? *Nature*, 471, 51–57.

Bonesi, L. and Palazon, S. (2007). The American mink in Europe: status, impacts, and control. *Biol. Conserv.*, 134, 470–483.

Conover, M. (2002). *Resolving Human–Wildlife Conflicts: The Science of Wildlife Damage Management*. Boca Raton: CRC Press.

Delibes, M. (1990). *Status and conservation needs of the wolf (*Canis lupus*) in the Council of Europe member states (No. 47)*. Council of Europe.

Dickman, A. J. (2010). Complexities of conflict: the importance of considering social factors for effectively resolving human–wildlife conflict. *Anim. Conserv.*, 13, 458–466.

Hage, M., Leroy, P. and Petersen, A. C. (2010). Stakeholder participation in environmental knowledge production. *Futures*, 42, 254–264.

Heydon, M. J., Wilson, C. J. and Tew, T. (2011). Wildlife conflict resolution: a review of problems, solutions and regulation in England. *Wildl. Res.*, 37, 731–748.

Holling, C. S. (1978). *Adaptive Environmental Assessment and Management*. New York: John Wiley & Sons.

Jackson, D. B., Fuller, R. J. and Campbell, S. T. (2004). Long-term population changes among breeding shorebirds in the Outer Hebrides, Scotland, in relation to introduced hedgehogs (*Erinaceus europeus*). *Biol. Conserv.*, 117, 151–166.

MacDonald, D. W. and Service, K. (2007). *Key Topics in Conservation Biology*. Oxford: Blackwell Publishing.

Manfredo, M. J. and Dayer, A. A. (2004). Concepts for exploring the social aspects of human–wildlife conflict in a global context. *Hum. Dimens. Wildl.*, 9, 1–20.

Mech, D. L. (2011). Is science in danger of sanctifying the wolf? *Biol. Conserv.*, 150, 143–149.

Milner-Gulland E. J. and Rowcliffe J. M. (2007). *Conservation and Sustainable Use: A Handbook of Techniques*. Oxford: Oxford University Press.

Moore, C. W. (1996). *The Mediation Process: Practical Strategies for Resolving Conflict*. Second edition. San Francisco: Jossey-Bass.

Nie, M. (2003). Drivers of natural resource-based conflict. *Policy Sci.*, 36, 307–341.

O'Rourke, E. (2014). The reintroduction of the white-tailed sea eagle to Ireland: people and wildlife. *Land Use Pol.*, 38, 129–137.

Peterson, M. N., Birckhead J. L., Keong, K., Peterson, M. J. and Peterson, T. R. (2010). Rearticulating the myth of human–wildlife conflict. *Conserv. Lett.*, 3, 74–82.

Redpath, S. M., *et al.* (2013). Understanding and managing conservation conflicts. *Trends Ecol. Evol.*, 28, 100–109.

Redpath, S. M., Bhatia, S. and Young, J. (2014). Tilting at wildlife: reconsidering human–wildlife conflict. *Oryx*, doi: 1001017/S0030605314000799.

Sidaway, R. (2005). *Resolving Environmental Disputes*. London: Earthscan.

Sillero-Zubiri, C., Sukumar, R. and Treves, A. (2007). Living with wildlife: the roots of conflict and the solutions. In *Key Topics in Conservation Biology*, eds. D. W. MacDonald and K. Service, pp. 266–272. Oxford: Blackwell Publishing.

Thirgood, S. and Redpath, S. (2008). Hen harriers and red grouse: science, politics and human–wildlife conflict. *J. Appl. Ecol.*, 45, 1550–1554.

Treves, A. (2009). The human dimensions of conflicts with wildlife around protected areas. In *Wildlife and Society: The Science of Human Dimensions*, eds. M. J. Manfredo, J. J. Vaske, P. J. Brown, D. J. Decker and E. A. Duke, pp. 214–228. Washington, DC: Island Press.

Treves, A. and Karanth, K. U. (2003). Human–carnivore conflict and perspectives on carnivore management worldwide. *Conserv. Biol.*, 17, 1491–1499.

Wabakken, P., Sand, H., Liberg, O. and Bjärvall, A. (2001). The recovery, distribution, and population dynamics of wolves on the Scandinavian peninsula, 1978–1998. *Can. J. Zool.*, 79, 710–725.

Warwick, H. (2012). Comment: Uist hedgehogs – lessons learnt in wildlife management. *Br. Wildl.*, 24, 111–116.

White, P. C. L. and Ward, A.I. (2010). Interdisciplinary approaches for the management of existing and emerging human–wildlife conflicts. *Wildl. Res.*, 37, 623–629.

Wittmer, H., Rauschmayer, F. and Klauer, B. (2006). How to select instruments for the resolution of environmental conflicts? *Land Use Pol.*, 23, 1–9.

Woodroffe, R. (2000). Predators and people: using human densities to interpret declines of large carnivores. *Anim. Conserv.*, 3, 165–173.

Woodroffe, R., Thirgood, S. and Rabinowitz, A. (eds) (2005). *People and Wildlife: Conflict or Coexistence?* Cambridge: Cambridge University Press.

Young, J. C., *et al.* (2010). The emergence of biodiversity conflicts from biodiversity impacts: characteristics and management strategies. *Biodivers. Conserv.*, 19, 3973–3990.

Young, J., *et al.* (2013). Does stakeholder involvement really benefit biodiversity conservation? *Biol. Conserv.*, 158, 359–370.

Box 1

Conflicts over snow leopard conservation and livestock production

Charudutt Mishra and Kulbhushansingh R. Suryawanshi

Snow Leopard Trust and Nature Conservation Foundation, 3076/5, 4th Cross, Gokulam Park, Mysore 570002, India

Snow leopards *Panthera uncia* continue to coexist alongside livestock in Asian mountains over an area of 2 million km^2 (Mishra *et al.*, 2010). In the past, people killed snow leopards in response to livestock predation, and perhaps some livestock loss was acceptable. This has changed because livestock have become a global economic asset (Berger *et al.*, 2013). Concomitantly, snow leopards have been provided the highest protection status in all their 12 range countries, rendering their killing a serious crime. Monetary losses and punitive action when retaliatory killings are discovered lead to frustration and reduced tolerance of the snow leopard.

Livestock populations in snow leopard habitats are increasing (Berger *et al.*, 2013). Wild prey populations are declining because of competition with livestock for forage, disturbance from herders and dogs, and hunting. Currently, wild herbivores on these rangelands probably contribute less than 5% of the total herbivore biomass, with the bulk being livestock (Berger *et al.*, 2013). Studies report 3–18% livestock losses to snow leopards and other predators, translating to monetary losses per family of up to half their mean annual per capita income (Mishra, 1997).

Early research, based on interviewing local people, described the extent of livestock predation and people's responses to these losses. People's perceptions of livestock predation extended well beyond actual losses. More recent research has distinguished between perception and reality of livestock damage by examining their correlates (Suryawanshi *et al.*, 2013). These authors reported that socio-economic factors such as stocking density of larger and more expensive livestock increased people's perception of threat from snow leopards. In contrast, actual depredation by snow leopards was higher in areas of greater abundance of both wild herbivores and snow leopards.

While snow leopards were included in national conservation strategies and their persecution became illegal, nations generally failed to assist local communities in mitigating livestock predation losses. In a few countries, compensation programmes became the earliest mitigation tool. However, these state-sponsored efforts were fraught with poor implementation (Mishra, 1997). Establishment of the international non-governmental organisation (NGO) The Snow Leopard Trust three decades ago catalysed more bottom-up approaches to management. Since then, conflict mitigation has included community-based livestock insurance programmes aiming to share economic losses; the Snow Leopard Enterprises programme that enhances the livelihood of local women thorough handicrafts, and improves the communities' tolerance towards predators; livestock vaccination programmes that try to improve people's tolerance through better veterinary services; predator-proofing of corrals that reduce instances of multiple livestock kills; and Himalayan Homestays that bring tourism revenue to local people as an incentive for snow leopard conservation (Mishra *et al.*, 2003; Jackson and Wangchuk, 2004).

Although more effective compared to earlier compensation systems, one of the key weaknesses of these efforts is their single-initiative focus. Conflicts are complex, and there are multiple dimensions to the issue of livestock depredation. Conflict management efforts therefore must be multipronged, and aim to simultaneously address at least three aspects: (1) reduce livestock losses through better livestock protection, (2) share and offset economic losses when livestock depredation does take place, and (3) improve the social carrying capacity for the predators through conservation-linked livelihood enhancement and sustained awareness. In our experience, multi-pronged programmes involving long-term and respectful community engagement have been successful in gaining community support for conservation.

Implementation of community-based, multi-pronged conflict-management programmes at the scale of large landscapes, as is necessary for wide-ranging species such as the snow leopard, is resource-intensive. NGOs have demonstrated model programmes, but their effective application at landscape scales depends on the participation of governments. As part of a recent global initiative led by the President of Kyrgyzstan, all 12 governments have recognised a strong role for local communities in snow leopard conservation. However, actual devolution of conservation responsibility and authority to local communities has yet to take place.

Conflict over livestock depredation is one important issue in the effective co-management of Asian rangelands for local livelihoods and biodiversity conservation. Even as conservationists develop mitigation strategies, new conflicts are emerging. Mining for minerals in snow leopard habitats has become a land-use conflict between herding communities, mining companies, governments and conservationists in countries such as Mongolia and Kyrgyzstan. Large hydroelectric dams have become sources of conflict between local communities, conservationists, the state and hydropower companies in Himalayan countries. Like most conflicts, they are challenging but not insurmountable. Governments need to bring greater transparency and equity in how land-use decisions are made and in according a legally stronger role for local communities in those decisions. All involved parties must also recognise that these are shared problems and that the solutions lie in respectful dialogue, compromise and reconciliation.

References

Berger, J., Buuveibaatar, B. and Mishra, C. (2013). Globalization of the cashmere market and the decline of large mammals in Central Asia. *Conserv. Biol.*, 27, 679–689.

Jackson, R. M. and Wangchuk, R. (2004). A community-based approach to mitigating livestock depredation by snow leopards. *Hum. Dimens. Wildl.*, 9, 1–16.

Mishra, C. (1997). Livestock depredation by large carnivores in the Indian Trans-Himalaya: conflict perceptions and conservation prospects. *Environ. Conserv.*, 24, 338–343.

Mishra, C., Allen, P., McCarthy, T., Madhusudan, M. D., Bayarjargal, A. and Prins, H. H. (2003). The role of incentive programs in conserving the snow leopard. *Conserv. Biol.*, 17, 1512–1520.

Mishra, C., Bagchi, S., Namgail, T. and Bhatnagar, Y. V. (2010). Multiple use of trans-Himalayan rangelands: reconciling human livelihoods with wildlife conservation. In *Wild Rangelands: Conserving Wildlife while Maintaining Livestock in Semi-Arid Ecosystems*, eds. J. Du Toit, R. Kock and J. Deutsch, pp. 291–311. Oxford: Blackwell Publishing.

Suryawanshi, K. R., Bhatnagar, Y. V., Redpath, S. and Mishra, C. (2013). People, predators and perceptions: patterns of livestock depredation by snow leopards and wolves. *J. Appl. Ecol.*, 50, 550–560.

CHAPTER TWO

Philosophy, conflict and conservation

ALAN HOLLAND
Lancaster University

Philosophy has been described as 'the conversation that Plato began'. In some remarks about 'the role of philosophy' in relation to environmental matters, Bernard Williams observed that "there is no special way in which philosophical considerations join the [. . .] discussion" (Williams, 1995: 233). Nor, in my opinion, can philosophy be defined by reference to any set of doctrines. It is with some trepidation, therefore, that I attempt to provide a more informative characterisation of philosophy and of how it might help us to understand and manage conservation conflicts.

Philosophy and its methods

The reference to Plato is more helpful than it seems at first sight, provided we know something about Socrates, the 'hero' of Plato's dialogues. Socrates is depicted, typically, as conducting an 'elenchos' or refutation, in which someone's claim to knowledge is shown to be baseless, not on empirical grounds but because it can be shown by a process of sound reasoning that it either harbours a contradiction or leads to some other claim that is unacceptable. Because sound reasoning depends crucially on understanding the concepts through which any reasoning proceeds, Socrates undertakes to dissect the relevant concepts. Such 'conceptual analysis' has been central to philosophy ever since. Early philosophers thought of themselves as 'lovers of truth' or 'lovers of wisdom' (from the Greek, '*philosophoi*') and hence as engaged primarily in a search for truth. The gap between those who sought truth using empirical methods (scientists) and those who continued with conceptual explorations (philosophers) widened gradually, although they can never be completely separate; Aristotle, for example, belongs to the canon of both science and philosophy.

Among philosophers, a division soon emerged between those who thought the point of philosophy was to *find* the truth, the so-called 'dogmatists', and those who thought the point of philosophy was to *keep on searching* for it, the so-called 'skeptics'. Socrates clearly belonged to the 'searching' school and his trademark question, 'what is X?', where X is a concept such as justice, knowledge, beauty or love, laid the foundation for much subsequent philosophy and the

Conflicts in Conservation: Navigating Towards Solutions, ed. S. M. Redpath, R. J. Gutiérrez, K. A. Wood and J. C. Young. Published by Cambridge University Press.

building up of bodies of theory such as the theory of knowledge, theories of ethics and aesthetics. These 'bodies of theory' are precisely not sets of doctrines, but tried and tested ways of approaching most of the central questions of human existence.

Logic was among the first bodies of theory to emerge. In its infancy, philosophy was like an infant plaything. Philosophers, or 'sophists' as they were often called, enjoyed themselves hugely devising paradoxes, such as the paradox of the liar (take the statement 'I am a liar'. If this is true, then it is false; and if it is false, then it is true); or the paradox of the heap – the so-called Sorites paradox (taking a grain from a heap of sand makes no significant difference; do this often enough and there is no heap left). They also enjoyed themselves in creating playful fallacies (e.g. 'this is your dog; this dog is a father; so, this dog is your father'). However, there was a serious purpose behind these paradoxes and fallacies. If we think of sound reasoning as the safe navigation from one truth to another, then the paradoxes and fallacies drew attention to problems that beset any such attempts. Logic, first systematised by Aristotle in the fourth century B.C., can be informally defined as a codification of the principles of sound navigation (from one truth to another).

Equally precocious were theories of ethics and knowledge. It was Plato who laid the foundations for the theory of knowledge in his *Theaetetus* (1973), and his definition of knowledge as 'justified true belief' remains either the default position or the starting point for any new investigation of the topic. Aristotle's *Nicomachean Ethics*, the first systematic treatise on value, is a response to the central question of Plato's Republic: 'how should we live our lives?' If Aristotle's science has been superseded, his ethics has not. It remains a constant source of reference for contemporary philosophers.

The relevance of philosophy for conservation conflicts

If we reflect on my brief account of some pre-occupations of philosophers, it is apparent that these pre-occupations are relevant to addressing conservation conflicts. In what follows, I assume that such conflicts are being addressed in a context that is open, honest and fair rather than a context, for example, with a power imbalance.

I give two disclaimers before proceeding. Given that philosophy does not ally with any specific empirical domain, and given that conservation conflicts will often involve disputes over matters of fact, the relevance of philosophy will often be of a complementary or ancillary nature. That said, and given the complex nature of conservation conflicts, it would be rare to find a conflict that was the domain of just a single discipline, and there is increasing recognition of the need for cross-disciplinary approaches (see Chapter 1). The second disclaimer is that these pre-occupations of philosophy are hardly unique to this discipline.

Everyone, one might hope, has an interest in, and aspires to be a practitioner of, sound reasoning. The difference is that for philosophers the pursuit of sound reasoning, and everything relevant to it, is an end in itself, rather than a means to some other end; it is their *raison d'être*.

The first question that a philosopher will ask about a conflict situation is: is there *really* a conflict? In other words, are there really *grounds* for conflict? Effectively, this means questioning the account of the conflict given by the participants themselves. In examining this question, further questions will naturally follow, such as:

- what are the presuppositions made by the parties to the dispute, and how convincing are they;
- how convincing are the justifications that are offered; and
- what goals are being pursued?

These questions in turn will spawn further questions. For example, the investigation of the goals pursued by each side will inevitably lead to questions about the value systems that are being endorsed. Any value system is capable of being challenged. The same is true of the justifications that are offered and knowledge claims that are made. In each case, alternative avenues might be opened up that clarify whether there are really grounds for conflict. Essentially, then, the role of philosophy in the understanding of conservation conflicts is to assist in the discovery of where there might be room for manoeuvre.

Room for manoeuvre

Presuppositions

An example of the importance of questioning presuppositions is to be found in a study conducted to evaluate the benefits of a coastal defence scheme at Aldeburgh in Suffolk, UK (Turner *et al.*, 1992). A conclusion of the study was that the benefits of the scheme outweighed the costs. The results were accepted by the regional water authority and the development subsequently funded by the Ministry for Agriculture, Fisheries and Food. However, as an anonymous commentator on the study pointed out, no equivalent study was undertaken of the 'do nothing' option, an option that would have led to a breach of the existing sea defences. Because a breach would almost certainly have meant the loss of the Royal Society for the Protection of Birds reserve at Orfordness, the presupposition was that developing the sea defences was the only way of protecting the conservation interest. However, one independent prediction was that, had a breach been allowed, 'an avocet nesting area three times the present one could be created' (anonymous commentator on Turner *et al.*, 1992: 102). The truth of such alternative projections is not the issue here, but rather it is the importance of identifying, and possibly questioning, the presuppositions of each side. A change of presupposition could make the difference between whether or not there is a conservation conflict.

Justification and consequences

To justify a belief, action or policy is to cite considerations that are thought to tell in its favour. When it comes to justifying actions or policies, this frequently takes the form of citing the good or bad consequences resulting from those actions or policies. Indeed, those philosophers who call themselves 'consequentialists' take the view that, ultimately, this is the *only* way in which actions and policies can be justified. What these same philosophers will point out, however, is that there are many ways of accounting for consequences.

To begin with, we need to distinguish between 'rule consequentialism' and 'act consequentialism'. A rule-consequentialist holds that the consequences relevant for the justifying of an action or policy are those that would result if the action or policy were to become a regular practice. An act-consequentialist, on the other hand, holds that we need only consider the consequences of the particular act or policy. Therefore, adopting a rule-consequentialist position can lead to different actions or policies than would be justified if we took an act-consequentialist approach. For example, the rescuing of a wild animal that has been injured, the feeding of wild birds, or interacting with an individual mountain gorilla *Gorilla beringei beringei* might each be considered innocent or even commendable if we consider only the consequences of that particular act. However, the consequences of such acts becoming regular practice – a 'rule' to be followed – are not obviously commendable and perhaps not commendable at all. In the case of the mountain gorilla, for example, it leads to their becoming vulnerable to human disease (Palacios *et al.*, 2011). A classic illustration of the problem is afforded by Garrett Hardin's 'tragedy of the commons' (Hardin, 1968). Here, each individual act of adding an extra grazing animal to the pasture is thought to have good consequences. Taken collectively, however, the consequences are disastrous. The lesson that rule-consequentialists take from this is that such an individual act is unjustified, even if we do not know whether other people will behave likewise, and even if we know that they will not.

A further distinction that can make a difference for whether actions or policies are considered justified is that between maximising and satisficing consequentialism (Slote, 1985). The question is whether we should attempt to bring about the greatest amount of good (the 'maximising' version) or, simply, a sufficient amount of good (the 'satisficing' version). It is a distinction that might well have application in discussions of biodiversity (assuming biodiversity to be a 'good'), especially when the 'enhancing' of biodiversity is a declared goal of international governments, including the British government (UK Biodiversity Action Plan Steering Group, 1995: 9–10). The goal of enhancing biodiversity, because it assumes that 'more is better', is, or implies, a maximising objective. However, it can be questioned whether the 'enhancing' of biodiversity is a desirable, or even an intelligible, objective, as opposed to, say, the protecting of an 'appropriate' biodiversity where the demand for a more than appropriate (or 'sufficient') biodiversity makes little sense. Some philosophers indeed, such as Stocker (1990: 342), find the maximising agenda a total anathema: 'Maximisation', he writes, 'is mistaken, irrelevant and parasitic', mainly, as he explains, because evaluation is most often conducted in terms of the *concrete* value of the items being evaluated rather than in the essentially comparative terms of whether they are 'best', or even in terms of how good they are.

Even if we settle for a 'satisficing' form of consequentialism, however, complications still remain. Six different versions of satisficing consequentialism have already been distinguished, each capable of justifying significantly different policies or actions (Bradley, 2006). Discussion of the value of these various forms of consequentialism is ongoing. Thus far there is no settled agreement on which version is the most satisfactory, and perhaps there never will be. Nevertheless, the point is this: each side to a dispute is bound to offer justifications for their respective positions, but if so prevalent a form of justification as the appeal to consequences is open to the wide variety of interpretations that have been described, then the structure of conflicts is never impossible to change.

Values

Values, understood to cover a variety of motivating factors, ranging from desires, through needs and interests to ideals and aspirations, are central to all forms of conflict; however, not all conflicts necessarily involve conflicts of value. Given that philosophy does not ally with any specific empirical domain, it tends not to generate what are normally thought of as 'case studies'. However, in the context of a volume devoted to conservation conflicts understood as a 'clash over conservation objectives' (Redpath *et al.*, 2013), I think it highly appropriate to illustrate the practice of philosophical analysis by examining the concept of 'the conservation interest', which is frequently how the interest of one or another of the conflicting parties is referred to.

One influential articulation of the conservation interest occurs in the foreword to Aldo Leopold's *Sand County Almanac*, where he writes: 'these essays are the delights and dilemmas of one who cannot live without wild things' (Leopold, 1949: vii). It is plausible to take him as saying that 'wild things' are an essential ingredient of what he considers a worthwhile life, and thus as making a 'normative', that is, an action-guiding claim. I assume here only that the conservation interest is at least mildly 'visceral', in that it expresses in some form or other what we might term a 'commitment to nature'. Yet, on further examination, 'the conservation interest' turns out to be a highly complex and even puzzling concept.

The complexity arises from the fact that conservation can hardly be reduced to some single and unique 'interest'. For among conservationists there are some deep divisions, with practical implications. One such division is between those who believe that the primary focus of attention should be on individuals and those who believe that the primary focus of attention should be on 'wholes' such as populations, species and ecosystems. Another division is between those who place considerable importance on the nature and degree of sentience found in non-humans, and those who believe that in an environmental context, at any rate, the nature and degree of sentience is a relatively unimportant matter or, at least, less important than human need. A third division is between those who are, and those who are not, prepared to tolerate the incursion of so-called 'invasive' or non-indigenous species, especially where these pose a threat to one or more of the indigenous species or even ecosystems, and where the invasion is, at least in part, a consequence of human activity.

The point of drawing attention to these divisions is that different policy recommendations can follow from different sides of each division. I give two examples to illustrate this. 'Conservationists' are said to be opposed to the trial badger cull recently undertaken in the UK (see Chapter 3). However, because the cull can hardly be depicted as a threat to the badger species as such, it is not clear that 'holists' can have any grounds for objecting to the cull. Or, if we consider certain 're-wilding' projects, such as the reintroduction of beavers into Scotland, even if it can be shown that such projects 'enrich biodiversity' (whatever that means), it is not clear that the effect on the welfare of the individual animals involved is even considered. Conservationists who are 'holists' might approve such projects, where 'individualists' might find grounds for objection.

The puzzling feature arises from the fact that it is difficult to see how the conservation interest can be said to be an 'interest' at all. Many environmental philosophers (Katz, 1987; Elliot, 1992; McShane, 2007) would hold that to reduce the conservation interest, merely, to the interests of conservationists, would be at best to undervalue nature and at worst a grievous travesty. The conservation interest can only represent nature, they say, if it views nature as, in some sense, having value in its own right, a value that transcends the potentially ephemeral interests of particular conservationists. The interest of a particular conservationist in scuba-diving or bird-watching may well coincide with the 'conservation interest', but could never be identified with it. The interest of the scuba-diver could be superseded by an interest in sky-diving because he or she finds the latter even more exciting; the interest of the bird-watcher could be superseded by an interest in stamp-collecting, because he or she finds it less energetic. As Katz (1987: 242–243) puts it: 'an environmental ethic cannot be based on human interests because of the contingent relationship between human interests and the welfare of the natural environment.'

Yet, perhaps conservationists can be said to represent 'the interests' of nature because they often claim to be the 'voice' of nature, or to 'speak on behalf of' nature. This political stance has its appeal. However, the problem is simply that nature cannot be said to have 'interests', at least not in any normal sense of the term. Varner (1998: 8) goes further. 'It makes no sense', he writes, 'to speak of what is in nature's interests', claiming that the term has application only to 'individual living organisms'. In fact, he probably overstates his case because entities such as football teams and universities are not individual living organisms but can be said to have interests. Although not organisms, they are, however, organisations, which implies an organised structure; they do, therefore, have features in common with organisms that entities such as species and ecosystems lack.

To infer that the conservation 'interest' is on a par, and could be compared with, say, an agricultural or fisheries interest is potentially misleading. This is because the conservation interest is understood by many to encompass essentially ethical considerations (Leopold, 1949). Although interests are often relevant for ethics, they cannot determine what we should do. Tyrants and paedophiles, for example, have interests that we should probably discount. The ethical considerations underlying the conservation interest have been expressed in several ways. As I have already remarked, a leading contention among environmental philosophers is that nature has 'intrinsic value', and is therefore a source of obligation. 'Here I argue that wild nature has intrinsic value', writes Elliot (1992: 138), 'which gives rise to obligations both to preserve it and restore it'. There are different accounts of the precise source of obligation. Elliot himself finds naturalness itself a source of added value. Others, such as Taylor (1986), see a value in every single living organism that demands our respect, while Rolston (1990) finds such value in the properties of ecosystems. An alternative idea is expressed by Leopold (1949: viii), who urges that we should see 'land' as a 'community to which we belong' rather than as a 'commodity belonging to us'. In this case, the obligations that we have with regard to the natural world are assimilated to the kinds of obligations we recognise as existing towards the 'communities' to which we belong – family, school, organisation, society.

There are many, however, both philosophers and non-philosophers, who would resist Leopold's suggestion, and fail to see nature as a source of ethical obligation, in itself. For example, in his devastating critique of the Stoic doctrine that we should 'follow nature', Mill (1874: 28–29) famously remarks:

'In sober truth, nearly all the things which men are hanged or imprisoned for doing to one another are nature's every day performances.' Besides killing, which 'nature does once to every being that lives', he proceeds to list torture, starvation, disease and a host of ways in which nature brings premature death to 'all but a small percentage of lives'. His criticism is twofold. The first criticism is that if these are indeed nature's everyday performances, then we can hardly be ethically obliged to emulate them. The second criticism is that if these are nature's everyday performances, then we can hardly be under an ethical obligation to uphold such a system. It is likely enough, however, as current concerns over climate change demonstrate, that those who fail to see nature as an intrinsic source of obligation would recognise the crucial *instrumental* importance of the natural world so far as human well-being is concerned. On that assumption, the notion of 'the conservation interest' does indeed gain sense – it is an important aspect of 'the human interest' in general.

Even if the conservation interest is indeed, at bottom, ethical, it is not obvious that ethical considerations are always more important than 'mere' utilitarian interests. In the opinion of some, it is true that ethical considerations trump all others, but this claim is readily challenged. It would seem excessively harsh, for example, to insist that a person who is starving is ethically obliged to desist from stealing, if that is the only way in which she can feed herself and her family. Perhaps all one can say is that if ethical considerations are present, they cannot be silenced because they will always leave a residue, say of guilt or remorse. When interests are overridden, on the other hand, the residue is more likely to be frustration or anger. More important, however, is the fact that interests themselves are likely enough at some level to encompass questions of need and of livelihoods, which in turn might properly constitute ethical demands.

The fundamental point is that conflicts of interest are very different from ethical conflicts, in two respects: first, as regards the basis of the case that each party brings to the dispute and second, as regards what kinds of accommodation among the parties is going to be considered appropriate. The point is highly relevant, therefore, both for the understanding of conflicts to which the conservation interest is party, and for identifying potential means of resolving those conflicts.

Philosophy and the management of conflict

It seems clear that there will be some relation between the kinds of measures that might mitigate or even resolve particular situations of conflict and the kinds of passion, interest or ideal that motivated the conflict in the first place. If, for example, the conflict is truly a conflict of interests, then negotiation, trade-off and compromise are usually the most appropriate means for resolving such disputes. In these kinds of conflict in particular, notions of fairness have a crucial role to play.

Philosophical theory draws our attention to the existence of at least three different kinds of fairness, any of which might have relevance for the conflict in question. There is fairness of treatment, as discussed by Williams in 'The Idea of Equality' (1973). This, roughly, enjoins differential treatment only on the grounds of relevant differences. Equally important are the notions of procedural fairness (roughly, fairness of process) and substantive fairness (roughly, fairness in the allocation of resources). Provided that the compromise can be represented as fair according to at least one of these senses of the term, then honour will be satisfied. A further proviso is, of course, that each party has sufficient incentive to resolve the dispute. Even if both parties come away from the deal feeling aggrieved (for one cannot control how people will feel), the important question is whether they have a right to feel aggrieved. If the deal was fair in one or more of the senses indicated, they have no such right.

However, the situation is very different if we are dealing with conflicting values or conflicts between, say, interests and needs, or between needs and ideals. Especially if ideals or values are involved, then the possibility of resolution must always be problematic because ideals and values are precisely the kinds of thing that do not normally allow negotiation or compromise. In conflicts of this kind, backing down entails what Lukes (1997: 188) refers to as 'sacrifice', further remarking that 'Trade-off suggests that we compute the value of the alternative goods on whatever scale is at hand, whether cardinal or ordinal, precise or rough and ready. Sacrifice suggests precisely that we abstain from doing so.'

Here too, as often as not, there are also what we might call 'meta-conflicts' at work. These are disagreements of a very fundamental kind, differences of 'worldview', as we might call them, exemplified by Leopold's (1949) distinction between those for whom land is a commodity and those for whom land is a community. In fact, many of our social, political and legal institutions are, effectively, ways of pre-empting conflict of this kind. The system of property rights, for example, that establishes your ownership of a piece of land, automatically removes various avenues that might otherwise be open for me to challenge your use of this land, whatever our 'worldviews' happen to be. However, supposing that the conflict is beyond the scope of our institutions to settle in this way, then two resorts remain. First, it is in fact rather rare for interests, goals, values and the like to be intrinsically conflicting. Conflict arises, rather, from the means chosen to fulfil the respective goals or interests. Hence, a question worth raising is whether either or both parties can see alternative ways of satisfying their interests, meeting their goals or adhering to their principles. Second, as Schmidtz (2000) remarks, some apparent conflicts of values are not conflicts of values at all but conflicts of priorities. The parties might in fact share values, although one or other party is simply not in a position to uphold those values. Thus, referring to the conflict between conservationists and subsistence farmers over the killing of elephants in Africa, Schmidtz (2000: 418) writes: 'subsistence farmers for whom getting enough food is a day by day proposition can have priorities that differ from [those of conservationists] not because their values are different but precisely because their values are the same'. Here mediation, for example, may prove irrelevant. It is rather a matter of trying to bring about social, political or economic change such that 'If we care about wildlife, we need to accept that wildlife will survive to the extent that people who have to live with it are better off taking care of it. It is roughly that simple!' (Schmidtz, 2000: 423).

Finally, there are those who urge that the stance known as 'environmental pragmatism' has a great deal to offer in the field of conflict resolution. Pragmatists tend to believe in 'value pluralism' (i.e. the existence of a diversity of values); they tend also to believe that this diversity should be respected. Accordingly, they recommend that we set aside particular substantive conservation values and focus instead on immediate, practical environmental problem-solving. Mediation should aim simply to discover common ground, to find out 'what works'. Thus, summarising the version of environmental pragmatism defended by Andrew Light and Eric Katz, Eckersley writes: 'Deliberation, creative conflict mediation and social learning thus replace any quest for ethical perfection' (Eckersley, 2002: 141–142). Yet Eckersley (2002) herself urges caution on three grounds. First, environmental pragmatism runs the risk of being 'too accommodating of the existing constellation of social forces' (Eckersley, 2002: 142). Second, it runs the risk of being too instrumentalist, in closing off dialogue where the 'pressure of practical imperatives is lifted' (Eckersley, 2002: 145), and where mutual trust and understanding may be generated. Third, it might be seen as not pluralist enough in that it 'ultimately comes to rest on the basic (monistic?) liberal humanistic principle of respect for individuals' (Eckersley, 2002: 146) and hence fails to address specifically environmental concerns. I have merely touched here on what is a lively and ongoing debate.

Conclusion

I have tried to show how philosophy can help us understand the structure of conservation conflicts by identifying the presuppositions, the supporting justifications and the values that are involved, and how, at the same time, by drawing attention to distinctions that might otherwise go unrecognised, it can help us to discern what Reynolds (2009: 40) calls 'a wealth of opportunities for seeing and doing things differently'. It is when no other means for the securing of interests or the achieving of goals can be identified, other than those which bring the two parties into dispute, that we have conflict that is unavoidable and, perhaps, irresolvable. While recognising that irresolvable conflicts are not always to be lamented and might indeed be welcomed as evidence of democratic vigour, I have tried to show also the various ways in which philosophy can assist in at least minimising the occasions when these occur.

References

Aristotle (1999). *Nicomachean Ethics*, tr. T. Irwin. Indianapolis: Hackett.

Bradley, B. (2006). Against satisficing consequentialism. *Utilitas*, 18, 97–108.

Eckersley, R. (2002). Environmental pragmatism. In *Democracy and the Claims of Nature*, eds. B. A. Minteer and B. P. Taylor, pp. 49–69. Washington, DC: Rowman & Littlefield.

Elliot, R. (1992). Intrinsic value, environmental obligation and naturalness. *The Monist*, 75, 138–160.

Hardin, G. (1968). The tragedy of the commons. *Science*, 162, 1243–1248.

Katz, E. (1987). Organism, community and the substitution problem. *Environ. Ethics*, 7, 241–257.

Leopold, A. (1949). *A Sand County Almanac*. Oxford: Oxford University Press.

Lukes, S. (1997). Comparing the incomparable: trade-offs and sacrifices. In *Incommensurability, Incomparability and Practical Reason*, ed. R. Chang, pp. 184–195. Cambridge, MA: Harvard University Press.

McShane, K. (2007). Why environmental ethics shouldn't give up on intrinsic value. *Environ. Ethics*, 29, 43–61.

Mill, J. S. (1874). *Three Essays on Religion*. London: Longmans, Green, Reader, and Dyer.

Palacios, G., *et al.* (2011). Human metapneumovirus infection in wild mountain gorillas, Rwanda. *Emerg. Infect. Dis.*, 17, 711–713.

Plato (1973). *Theaetetus (Clarendon Plato Series)*, tr. J. McDowell. Oxford: Oxford University Press.

Plato (1993). *Republic*, tr. R. Waterfield. Oxford: Oxford University Press.

Redpath, S. M., *et al.* (2013). Understanding and managing conservation conflicts. *Trends Ecol. Evol.*, 28, 100–109.

Reynolds, M. (2009). Environmental ethics. In *The Environmental Responsibility Reader*, eds. M. Reynolds, C. Blackmore and M. J. Smith, pp. 40–51. London: Zed Books.

Rolston, H. (1990). Duties to ecosystems. In *Companion to A Sand County Almanac*, ed. J. B. Callicott, pp. 246–274. Wisconsin: University of Wisconsin Press.

Schmidtz, D. (2000). Natural enemies: an anatomy of environmental conflict. *Environ. Ethics*, 22, 379–408.

Slote, M. (1985). *Common-sense Morality and Consequentialism*. London: Routledge & Kegan Paul.

Stocker, M. (1990). *Plural and Conflicting Values*. Oxford: Clarendon Press.

Taylor, P. (1986). *Respect for Nature: A Theory of Environmental Ethics*. Princeton: Princeton University Press.

Turner, R. K., Bateman, I. and Brooke, J. S. (1992). Valuing the benefits of coastal defence: a case study of the Aldeburgh sea-defence scheme. In *Valuing the Environment*, eds. A. Coker and C. Richards, pp. 77–100. London: Belhaven Press.

UK Biodiversity Action Plan Steering Group (1995). *Biodiversity: The UK Steering Group report volume 1 – Meeting the Rio Challenge*. London: HMSO.

Varner, G. (1998). *In Nature's Interests? Interests, Animal Rights and Environmental Ethics*. Oxford: Oxford University Press.

Williams, B. (1973). *Problems of the Self*. Cambridge: Cambridge University Press.

Williams, B. (1995). *Making Sense of Humanity*. Cambridge: Cambridge University Press.

Box 2

Large carnivores and conflict in Tanzania's Ruaha landscape

Amy J. Dickman
WildCRU, Zoology, University of Oxford, The Recanati-Kaplan Centre, Tubney House, Abingdon Road, Tubney OX13 5QL, UK

Large carnivores often generate intense conflict among people – they often rely heavily upon prey living in human-dominated lands so local communities bear the costs of coexistence (Dickman *et al.*, 2011). Yet, they are highly valued by people outside of these local areas, both for their intrinsic value and for tourism (Dickman *et al.*, 2011). The costs associated with large carnivore predation can be high; research in areas surrounding Tanzania's Ruaha National Park revealed that livestock depredation by large carnivores cost people 18% of their annual income. Although carnivores generate significant amounts of tourism revenue nationally (Lindsey *et al.*, 2012), villagers usually receive few or no benefits from carnivore presence to offset their losses.

Economic loss because of carnivore predation is the most commonly cited driver of conflict (Sillero-Zubiri and Laurenson, 2001), but the conflict around Ruaha is not purely driven by economics. People are often antagonistic towards the Park, because they see few personal benefits from its presence, feel it restricts their activities, and complain about being unable to use the Park while wildlife from the Park roam freely across village land. Antagonism towards protected areas and Park authorities often includes wildlife because people feel that their needs are secondary to the needs of outsiders and that wildlife conservation is imposed rather than voluntary (Naughton-Treves and Treves, 2005).

Conflict can also have deep-seated cultural drivers (Hazzah, 2011). Around Ruaha, there was evidence that intertribal tension was manifesting itself within the broader conflict. Even though different tribes interact and communicate, for instance during Village Assembly meetings, there is still uneasy coexistence among them. Depredation incidents can ignite conflicts, and there is suspicion among some tribes that neighbours from rival tribes can 'bewitch' hyaenas, send them to kill livestock and drag it back to their household, so that a villager can eat meat without having to kill his own stock (Dickman, 2010). Therefore, the hyaena is seen as an agent of another tribe, which intensifies local antagonism towards hyaenas. There are also beliefs that some people can conjure 'spirit lions', which they send to perform evil deeds (like killing people) without implicating themselves (West, 2001). These beliefs intensify antagonism towards carnivores and have more to do with conflict with other humans than the actual damage caused. For example, lion attacks on people have led to the murder of village witchdoctors because people from the affected tribe believed that their rivals used witchdoctors to bewitch the lion that caused the damage.

This conflict has led to substantial killing of carnivores around Ruaha (e.g. over 35 large carnivores were killed in

18 months in just 3 villages), so it is of conservation significance because it probably occurs in many other villages as well. This interaction between carnivores and people has probably always existed, but people now have more efficient methods for killing carnivores, such as guns and poison. Moreover, although depredation is often the precipitating factor for killing carnivores, there is also a cultural element. Pastoralists such as the Barabaig have traditionally killed lions or other dangerous wildlife to prove their bravery. A man may be rewarded with up to 20 cattle by spearing a lion and in an area of intense poverty such as Ruaha this represents an important way of securing wealth. It is legal for someone to kill a carnivore to protect themselves or their stock and much of this killing is done in remote areas without government oversight.

Oxford University's Ruaha Carnivore Project (RCP) has been seeking solutions to enhance carnivore conservation by reducing impacts to villagers since 2009, using a bottom-up approach. RCP is working with villagers to protect livestock from attacks, increase their interaction with Park staff, and encourage open discussion among tribes about carnivores. In all study villages, RCP encouraged people to vote at Village Assembly meetings on those benefits wanted in exchange for tolerating carnivores. The top three choices were healthcare, education and veterinary medicines, so RCP worked with the villagers to implement relevant benefit schemes. Importantly, the project has used the Kenyan Lion Guardians model to employ young men in lion conservation (Hazzah, 2011), so they gain wealth and status from lion presence rather than from killing them. These approaches have resulted in an 80% decline in carnivore killing in the core study area, but it has been a slow process because the Barabaig are suspicious of outsiders. Maintaining this lowered killing level will be challenging because this community-based approach is time-consuming and intensive, so it will be hard to replicate this success over all 22 local villages. However, the approach has demonstrated that although the conflict around Ruaha is complex and multi-faceted, a community-based approach involving damage limitation, benefit provision and the inclusion of the entire community can significantly improve the situation.

References

Dickman, A. (2010). Complexities of conflict: the importance of considering social factors for effectively resolving human–wildlife conflict. *Anim. Conserv.*, 13, 458–466.

Dickman, A. J., Macdonald, E. A. and Macdonald, D. W. (2011). A review of financial instruments to pay for predator conservation and encourage human–carnivore coexistence. *Proc. Natl Acad. Sci. USA*, 108, 13937–13944.

Hazzah, L. (2011). Exploring attitudes, behaviors, and potential solutions to lion (*Panthera leo*) killing in Maasailand, Kenya. PhD Dissertation, University of Wisconsin, Madison.

Lindsey, P. A., Balme, G. A., Booth, V. R. and Midlane, N. (2012). The significance of African lions for the financial viability of trophy hunting and the maintenance of wild land. *PLoS ONE*, 7, e29332.

Naughton-Treves, L. and Treves, A. (2005). Socio-ecological factors shaping local support for wildlife: crop-raiding by elephants and other wildlife in Africa. In *People and Wildlife: Conflict or Coexistence?* eds. R. Woodroffe, S. Thirgood and A. Rabinowitz, pp. 252–277. Cambridge: Cambridge University Press.

Sillero-Zubiri, C. and Laurenson, M. K. (2001). Interactions between carnivores and local communities: conflict or co-existence? In *Carnivore Conservation*, eds. J. L. Gittleman, S. M. Funk, D. W. Macdonald and R. K. Wayne, pp. 282–312. Cambridge: Cambridge University Press.

West, H. (2001). Sorcery of construction and socialist modernisation: ways of understanding power in postcolonial Mozambique. *Am. Ethnol.*, 28, 119–150.

PART II

Contrasting disciplinary approaches to the study of conflict in conservation

CHAPTER THREE

The value of ecological information in conservation conflict

STEPHEN M. REDPATH
University of Aberdeen
and
WILLIAM J. SUTHERLAND
University of Cambridge

The British countryside is renowned for its pastoral beauty: a rich mosaic of farmland, woodlands, hedgerows and winding lanes where biodiversity flourishes. Yet linger within this apparently serene landscape and you are likely to discover an equally rich mosaic of conflict, which can sometimes be bitter and acrimonious. We see conflicts emerge over a wide range of issues, such as the culling of badgers *Meles meles* to control disease in cattle, the impact of intensive farming techniques on biodiversity or the illegal killing of predators for the benefit of game species. As we see elsewhere in this book, such conflicts are not restricted to the UK; they occur worldwide. Conflicts differ in details and participants, but they are often similar in challenges and strategies for resolution.

Ecological arguments invariably are part of conservation conflicts. We need to understand what impacts our activities have on species and ecosystems. Typically, in conflicts, the information deficit model is followed (Burgess *et al.*, 1998). This model holds that more expert knowledge (in our case ecological data) and better communication are needed to help raise awareness, develop effective policies and change people's behaviour. Ecology is therefore often seen as providing the necessary objective evidence to enable decisions to be made to address conservation conflict. Consider, for example, the situation of conflicts involving predators. Those whose livelihoods depend on prey species commonly perceive predators as threats. So, ecologists commonly ask if perceptions of impact match ecological data (Sillero-Zubiri *et al.*, 2007; Dickman, 2010). To make a decision about predator management, we need to quantify predation levels, understand how predation varies in time and space and how predators impact prey populations. Once we understand these ecological interactions and impacts, we can then make predictions about when and where different interests may become incompatible and to initiate management decisions. Frequently there may be

Conflicts in Conservation: Navigating Towards Solutions, ed. S. M. Redpath, R. J. Gutiérrez, K. A. Wood and J. C. Young. Published by Cambridge University Press.

several different techniques available for reducing impact so we need to know the relative effectiveness of each (Smith *et al.*, 2014). Ecological data and analysis can help us distinguish between the effectiveness of alternative management strategies (see Box 9).

Questions of impact and management are therefore central to the objectives of ecologists. Here we explore the value of ecology for understanding and dealing with conservation conflicts. We first describe the discipline, and then explore some key concepts before illustrating our points with two conflicts.

What is ecology?

Ecology is a method for understanding how the natural world works, by studying the interrelationships between organisms and their surroundings. Its roots are in natural history; indeed in 1927, Charles Elton described ecology as 'scientific natural history'. Nowadays, ecologists work at multiple levels of organisation, from genes to ecosystems, with the discipline expanded to incorporate ideas from many associated fields including behaviour, genetics, physiology and evolution (Krebs, 2001). Humans, although part of these natural systems, are traditionally excluded from pure ecological study; this typically is the realm of the social sciences and humanities. However, applied ecology encompasses the study of the impacts of humans on the environment and the challenges of countering and resolving these problems. Ecology is therefore about explaining the patterns and interactions seen in nature. We make predictions based on hypotheses of how we think the system works, often using mathematical models to help specify those predictions. Ecology is not only concerned with the factors that explain current patterns, such as the impact of predators on their prey, but also with the underlying evolutionary explanations about why individuals behave the way they do.

Although ecology was initially descriptive, it acquired a strong theoretical underpinning. It has also had close links to applied problems. Population ecology, for example, grew out of a need either to understand why some species erupted to become pests or to design sustainable harvesting strategies. This led to advancements in our theoretical understanding of population dynamics and the application of that theory to these applied problems.

In the first half of the twentieth century the field of wildlife management grew out of the work of Americans such as Gifford Pinchot and Aldo Leopold. This field is essentially applied ecology as applied to wildlife. Leopold argued for the use of ecology and technology in the management of wildlife populations, and considered the tools of management to be the 'axe, plow, cow, fire and gun' (Leopold, 1933: vii). In the United States, wildlife management is placed under the Public Trust Doctrine and, unlike in other parts of the world such as the UK, considerable funding is available through taxes and licences on hunting to support this endeavour.

In the 1980s, conservation biology grew out of the recognition that the integrity of natural systems, not just wildlife species, was threatened by the impact of humans and its aim has been to understand the nature of threats and the means to mitigate them. As a consequence, conservation biology, like wildlife management, has been open to integration with social sciences. Both conservation biology and wildlife management are targeted towards a goal, but the main distinction between them is that conservation biology specifically advocates the conservation of biodiversity (Soulé, 1985).

Much of the focus on conflicts in the ecological literature has been on issues surrounding individual species. Ecology in its various forms has helped us understand the impacts of wildlife on people, the impacts of people on wildlife, the ecological relationships within systems where impacts occur, and the relative effectiveness of alternative mitigation strategies. We consider each of these in turn.

Quantifying impact

The attitudes and behaviour of participants in conflicts are often based on their perceptions of how systems work, rather than based on an assessment of independently collected ecological data. Perceptions and data may not tell the same story. A good illustration comes from US Great Plains and northern Mexico, where black-tailed prairie dogs *Cynomys ludovicianus* used to flourish on prairies. These small herbivores eat vegetation and therefore ranchers consider them to be competitors with their cattle (Reading *et al.*, 2005). The data, however, tell a different story. Prairie dogs avoid many plant species eaten by cattle, they may actually promote grazing-tolerant plant species and their impact on livestock varies spatially and appears to be a function of rainfall (O'Meilia *et al.*, 1982; Augustine and Springer, 2013).

When perceptions differ, ecological data may help move a debate forward. Consider the arguments around the impact of raptors on red grouse in the uplands of the UK (Box 3). For decades there has been conflict between those wanting more raptors and those wanting more grouse. Arguments, largely in the absence of independent data, centred on the magnitude of impact on grouse populations. Considerable ecological research effort was expended to quantify the impact. This research found that under certain conditions raptor predation was such that it could make intensive forms of management uneconomic (Redpath and Thirgood, 1997). The findings of this research moved the arguments on from ones of impact to ones of the most appropriate management strategy (Thirgood and Redpath, 2008). However, while the science has changed the arguments, it has not reduced the conflict (Thirgood and Redpath, 2008).

In conflicts, the impact of a species on a resource of interest to humans is not the only impact that is relevant; we may also be interested in the impact by people on the species, often carried out in retaliation (see Woodroffe *et al.*, 2005).

The perception of prairie dogs as pests led to their eradication from large areas of their original range over a 60-year period through poisoning programmes and habitat conversion (Dunlap, 1988), which ultimately led to the near extinction of the black-footed ferret *Mustela nigripes*, an obligate prairie dog predator (Miller *et al.*, 1996). It is clearly essential to have ecological studies that estimate the impact of humans and wildlife on each other as well as the consequences of those impacts for the whole ecosystem.

Understanding the mechanisms

Predicting impact is not always as straightforward as it might seem. For example, it seems intuitive that each animal killed by a predator means one less to harvest by humans. If true, we can simply determine the overall impact of predators by estimating the rate that prey are killed and multiplying that by the number of predators. However, such simplistic calculations are usually wrong and are often over- or underestimates of real impact.

The main reason why the impact of predation may be overestimated is that compensation (also known as density dependence) in the system means that predation losses might be offset by either improved survival or productivity of the remaining individuals. For example, Sandercock *et al.* (2011) examined the consequences of ptarmigan *Lagopus lagopus* hunting. When populations were unexploited, 38% of the birds died during the winter. However, after they harvested 15% of the population in the autumn, the subsequent survival rate of the remaining birds increased by 13% because of compensation.

In contrast, the impacts of predators can be underestimated when non-lethal effects are not considered (Lima, 1998). It is increasingly recognised that predators can create so-called landscapes of fear, where their presence may affect prey behaviour, habitat use, life-history strategies and other aspects. These non-lethal impacts can be as important as lethal effects (Cresswell, 2008). So, although the number of prey killed may be small, the overall impact on populations may be much greater. These examples show that predation effects are complex and require decision-makers to have a more sophisticated understanding of the ecological mechanisms when assessing impacts.

Our understanding of impact is often hindered by uncertainty. Because we cannot measure everything, there will be gaps and uncertainties in our knowledge. The level of uncertainty is important, not only because it affects our confidence in estimates, but also because it can be used as either an excuse for inaction or as an argument for or against a given position. Fortunately, new developments in analysis, such as Bayesian Belief Networks (Marcot *et al.*, 2006) and Management Strategy Evaluation (Smith *et al.*, 2008; Bunnefeld *et al.*, 2011) enable us to evaluate uncertainty and to explore management scenarios in collaboration with stakeholders.

Deciding among alternative management strategies

In the study of conservation conflicts, there has been considerable ecological research exploring the effectiveness of a wide range of techniques aimed at reducing impacts of predation, including electric fences, loud noises, chemical aversion therapy, supplemental feeding, translocation, and lethal control (e.g. Fischer and Lindenmayer, 2000; Nyhus *et al.*, 2005; Treves and Naughton-Treves, 2005; Box 16; Box 20). When deciding among strategies, it is important to consider the range of available options. The number of alternatives can be large. For example, Williams *et al.* (2013) list 28 means of reducing predation on bird populations, 24 means of reducing by-catch of birds during fishing, and 7 means of reducing the impact of artificial lights on birds. Providing an independent review of the evidence, as Williams *et al.* (2013) do for bird management, may help to reduce disagreements over the evidence base and reach agreement over the most appropriate strategy.

Two examples – ecology and conflict

We illustrate the approach taken by ecologists studying conflicts by briefly considering jaguars *Panthera onca* in Latin America and badgers in the UK. We use jaguars to illustrate the problems of living with large carnivores and show how ecologists are increasingly asking questions related to social and economic sciences to understand and manage conflict. The badger example illustrates how ecology has been used in a conflict where there is considerable investment and political pressure to find a solution. Our aim is not to provide a detailed review of evidence; this has been done elsewhere (see Cavalcanti *et al.*, 2010; Godfray *et al.*, 2013). Rather, we seek instead to illustrate the types of questions ecologists ask in conflict situations.

Jaguars

The jaguar occurs from the southwestern United States to northern Argentina and is classified as Near Threatened by the International Union for Conservation of Nature (Caso *et al.*, 2008). Jaguars are perceived as a threat to cattle across their range and retaliatory killing by ranchers is thought to be a serious threat to jaguar populations (Zeller, 2007). There is increasing recognition of the need to find solutions to the conflicts between those supporting jaguar conservation and those seeking livelihoods from livestock (Caso *et al.*, 2008).

In these types of situations, ecologists address questions about predation and management. Do the ecological data suggest that jaguars are important predators of cattle? What factors explain how many cattle are killed? What effect does this predation have on livestock numbers? How can jaguar impact be minimised? What is the most effective approach to minimising levels of retaliatory killing?

In the Pantanal of Brazil, 82% of ranchers considered jaguars a threat to their cattle, and 38% considered them to be the key problem on their ranch (Zimmermann *et al.*, 2005). Reported losses of cattle by jaguars averaged 2.3% of cattle holdings per year (Zimmermann *et al.*, 2005). In a fragmented forest in southern Brazil, reported losses varied from 0.3% to 1.2% of cattle over 2 years with a financial cost to ranchers during that period of between $290 and $1771 depending on the size of herds. Ecological data from radio-telemetry found that, at least in some regions, cattle were an important part of their prey, especially in terms of relative biomass consumed (Crawshaw and Quigley, 2002; Azevedo, 2008). However, large-scale, independent estimates of rates of cattle loss on ranches are now needed because estimates of impact made by ranchers may be imprecise (Cavalcanti *et al.*, 2010).

There is considerable temporal and spatial variation in estimates of the numbers of cattle killed (Azevedo and Murray, 2007). In many cases calves were the primary prey with calf mortality tending to be highest during the peak calving season (Michalski *et al.*, 2006). Cattle were vulnerable to jaguar in forested areas, especially when they left the pastures and entered forests, during periods of flooding and when wild prey were scarce (Rabinowitz, 1986; Hoogesteijn *et al.*, 1993; Polisar *et al.*, 2003; Michalski *et al.*, 2006; Azevedo and Murray, 2007; Palmeira *et al.*, 2008). An understanding of these forms of variation is helpful in highlighting potential solutions. Polisar *et al.* (2003) suggested various management recommendations to allow the coexistence of jaguars and cattle, including reducing risks of predation by moving cattle to safer areas, using fences to prevent them from entering forests, reducing the length of the calving season, protecting populations of wild prey and ensuring monitoring that allows ranchers to measure the impact of their management decisions. Other, more controversial suggestions include shooting or translocation of problem individuals (Rabinowitz, 1986; Isasi-Catalá, 2010).

Ecologists now recognise that conservation and livelihoods are inextricably linked, so socio-economic questions are addressed in an attempt to improve understanding of stakeholder positions and develop appropriate mitigation strategies. This is apparent in the jaguar example because rancher attitudes have been explored (Conforti and Azevedo, 2003; Zimmermann *et al.*, 2005) to understand their motivation for killing jaguars (Marchini and MacDonald, 2012).

Badgers

Bovine tuberculosis (bTB) is a respiratory infection of cattle caused by the bacterium *Mycobacterium bovis*. In the UK, this disease is widespread across the west of England and Wales and much of Northern Ireland. It is transmitted directly via aerosol droplets and indirectly through contamination of food, water and pasture. In the UK and Ireland, badgers are the main wildlife host for this disease. Governments aim to eradicate the disease and a major tool to deliver this is cattle testing and slaughter of infected cattle. In 2010, approximately 25,000 cattle were slaughtered in England as part of TB control measures, at a cost to the taxpayer of £90 million (DEFRA, 2011). Recently, however, two pilot badger culls have also been initiated, to test whether badger culling by farmers at a larger scale could be safe, humane and effective. Culling badgers has led to conflicts among those who farm cattle and those interested in the welfare and conservation of badgers (Krebs *et al.*, 1998; Delahay *et al.*, 2009).

Considerable ecological work has focused on the links between cattle, badgers and bTB (summarised in Godfray *et al.*, 2013). This body of evidence shows that badgers can act as reservoirs of bTB and that infection can be transmitted between cattle, between badgers, between cattle and badgers and vice versa. When it was first recognised that badgers carried bTB and therefore were a potential threat to cattle, they were killed in problem areas, initially using gassing, trapping and shooting. Such controls were not designed as experiments so their effectiveness could not be readily evaluated. This led Krebs (1997) to recommend a 10-year Randomised Badger Culling Trial (RBCT) in 30 areas each of approximately 100 km^2. Results of this major study showed that proactive badger culling was associated with an estimated 23% reduction in bTB incidence in cattle herds inside control areas (Bourne *et al.*, 2007). However, just outside these areas, the same culling was associated with a ~25% increase in the incidence of bTB, thought to be related to disruptive effects of culling on badger social structures and movements. The Bourne report concluded that 'badger culling can make no meaningful contribution to cattle TB control in Britain'. Culling is only likely to have a major impact on disease incidence if it is carried out on large numbers of badgers over very large geographical areas (Godfray *et al.*, 2013).

Both cattle and badgers can be vaccinated against bTB (Godfray *et al.*, 2013). Such vaccines, however, are not licensed for cattle in the UK, because of EU regulations and the difficulties of distinguishing infected cattle from vaccinated ones. Vaccines for both species are known to reduce symptoms of the disease and, therefore, presumably the prevalence and spread of *M. bovis* infection (Godfray *et al.*, 2013).

With cattle vaccines currently unavailable, the mitigation toolbox is limited to culling, improving testing, vaccinating badgers, reducing movement of cattle between farms and farming measures that reduce contact between cattle and badgers. Much of the current focus is still on culling of cattle and badgers, although trials to develop new cattle vaccines and test effectiveness of badger vaccines and farm management are ongoing. In 2013, two pilot areas of 256 km^2 and 311 km^2 were established in southwest England with the object of reducing the badger population by at least 70% over a 4-year period as a test of the efficacy of larger-scale culling on bTB in the cattle.

Because of the impacts of TB on human health and farmer livelihoods, there have been huge investments in research on this system. The RBCT alone cost an estimated £49 million. While there have been many insights from this rich ecological and epidemiological study, it is clear that research has currently not provided a solution nor reduced conflict. Surprisingly, little research has been done on the human dimension, despite that fact that this conflict is clearly between people advocating different viewpoints. Whatever the outcome of the trial, it seems inevitable that large-scale culling will be controversial and opposition vociferous.

Concluding remarks

These examples illustrate the value and limitations of ecological data when addressing conflicts. Ecological research allows us to quantify direct and indirect impacts, estimate the uncertainties in the system and test the effectiveness of specific mitigation techniques. Consequently, ecology can transform arguments based on belief systems, anecdotes and perceived impacts to ones based on empirical data.

Following the revolutionary impact of evidence-based medicine, there has been interest in establishing a similar movement in conservation (Sutherland *et al.*, 2004). Such an approach allows others to collate ecological evidence to provide independent assessments of the science to inform decision-making. For example, Stewart *et al.* (2007) show that systematic review and critical appraisal of evidence aided assessments of impact of windfarms on bird populations. Such reviews are increasingly becoming available to stakeholders and policy makers, for example through the website ConservationEvidence.com. Although ecological data are valuable in understanding how a system works, the importance of the human dimension is striking. People engaged in conflict have different views about the extent of the problem, how the problem is framed, the validity of different arguments and the merits of different options. Addressing this human dimension is fundamental to understanding and managing conflicts (Dickman, 2010; Redpath *et al.*, 2013). However, this recognition raises the question of whether the information deficit approach, as employed in the badger example, is always appropriate, or whether a different approach would be more effective. We could hypothesise that a bottom-up process of engagement and dialogue among main groups aimed at reducing conflict would (i) increase trust, (ii) indicate levels of stakeholder support for different approaches, (iii) refocus attempts to find a technical solution, (iv) enable the science to be better developed and understood, and (v) enable selection of a management solution with less controversy, reduced costs and less long-term resentment.

Another lesson from the examples is that ecological research at an appropriate scale can be extremely expensive and time-consuming. This can be frustrating to those engaged in conflict who seek rapid, robust answers and are reluctant to wait years for experimental results. Decision makers can sometimes use this fact to procrastinate (Thirgood and Redpath, 2008). Government agencies have been keen to avoid controversy so they establish working groups, commission desk studies and promote a 'cascade' approach to mitigation where the least intrusive techniques are tested in detail first, before other more intrusive techniques are considered. The net effect is a slowing of the process and 'kicking the can further down the road'. Such tactics frustrate all sides who seek to find a solution to conflict.

Acknowledgements

We are grateful to Fernanda Michalski and Robbie McDonald for their helpful comments on this chapter.

References

Augustine, D. J. and Springer, T. L. (2013). Competition and facilitation between prairie dogs and livestock. *Bull. Ecol. Soc. Am.*, 94, 177–179.

Azevedo, F. C. C. (2008). Food habits and livestock depredation of sympatric jaguars and pumas in the Iguacu National Park area, south Brazil. *Biotropica*, 40, 494–500.

Azevedo, F. C. C. and Murray, D. L. (2007). Evaluation of potential factors predisposing livestock to predation by jaguars. *J. Wildl. Manage.*, 71, 2379–2386.

Bourne, F. J., *et al.* (2007). *Bovine TB: The Scientific Evidence. Final report of the independent scientific group on cattle TB.* London: Defra.

Bunnefeld, N., Hoshino, E. and Milner-Gulland, E. J. (2011). Management strategy evaluation: a powerful tool for conservation? *Trends Ecol. Evol.*, 26, 441–447.

Burgess, J. Harrison, C. and Filius, P. (1998). Environmental communication and the cultural politics of environmental citizenship. *Envir. Plann. A*, 30, 1445–1460.

Caso, A., *et al.* (2008). *Panthera onca*. IUCN Red List of Threatened Species.

Cavalcanti, S. M. C., Marchini, S., Zirnmerrnann, A., Gese E. M. and Macdonald, D. W. (2010). Jaguars, livestock and people in Brazil: realities and perceptions behind the conflict. In *The Biology and Conservation of Wild Felids*, eds. D. Macdonald and A. Loveridge, pp. 383–402. Oxford: Oxford University Press.

Conforti, V. A. and Azevedo, F. C. C. (2003). Local perceptions of jaguars (*Panthera onca*) and pumas (*Puma concolor*) in the Iguacu National Park, south Brazil. *Biol. Conserv.*, 111, 215–221.

Crawshaw, P. G. Jr. and Quigley, H. B. (2002). Habitos alimentarios del jaguar y el puma en el Pantanal, Brasil, con implicaciones para su manejo y conservacion. In *El Jaguar en el Nuevo Milenio*, eds. R. A. Medellin, *et al.*, pp. 223–236. Mexico: Ediciones Cientificas Universitarias.

Cresswell, W. (2008). Non-lethal effects of predation in birds. *Ibis*, 150, 3–17.

DEFRA (2011). *Measures to address bovine TB in badgers: Impact Assessment*. London: Defra.

Delahay, R. J., *et al.* (2009). Managing conflict between humans and wildlife: trends in licensed operations to resolve problems with badgers *Meles meles* in England. *Mammal Rev.*, 39, 53–66.

Dickman, A. J. (2010). Complexities of conflict: the importance of considering social factors for effectively resolving human–wildlife conflict. *Anim. Conserv.*, 13, 458–466.

Dunlap, T. R. (1988). *Saving America's Wildlife*. Princeton, NJ: Princeton University Press.

Elton, C. S. (1927). *Animal Ecology*. London: Sidgwick & Jackson.

Fischer, J. and Lindenmayer, D. B. (2000). An assessment of the published results of animal relocations. *Biol. Conserv.*, 96, 1–11.

Godfray, H. C., *et al.* (2013). A restatement of the natural science evidence base relevant to the control of bovine tuberculosis in Great Britain. *Proc. R. Soc. B*, 280, 20131634.

Isasi-Catalá, E. (2010). Is translocation of problematic jaguars (*Panthera onca*) an effective strategy to resolve human-predator conflicts? *CEE review*, 08–018 (SR55).

Krebs, C. J. (2001). *Ecology: The Experimental Analysis of Distribution and Abundance*. New York: Harper-Collins College Publishers.

Krebs, J. R. (1997). *Bovine Tuberculosis in Cattle and Badgers*. London: Ministry of Agriculture, Fisheries and Food.

Krebs, J. R., *et al.* (1998). Badgers and bovine TB: conflicts between conservation and health. *Science*, 279, 817–818.

Leopold A. (1933). *Game Management*. New York: Charles Scribner's Sons.

Lima, S. L. (1998). Nonlethal effects in the ecology of predator–prey interactions. *Bioscience*, 48, 25–34.

Linnell, J. D., Odden, J. and Mertens, A. (2012). Mitigation methods for conflicts associated with carnivore depredation on livestock. In *Carnivore Ecology and Conservation*, eds. L. Boitani and R. A. Powell, pp. 314–332. Oxford: Oxford University Press.

Marchini, S. (2003). *Pantanal: opinião pública local sobre meio ambiente edesenvolvimento*. Rio de Janeiro: Wildlife Conservation Society and Mamiraua Institute.

Marcot, B. G., Steventon, J. D., Sutherland, G. D. and McCann, R. K. (2006). Guidelines for developing and updating Bayesian belief networks applied to ecological modeling and conservation. *Can. J. Forest Res.*, 36, 3063–3074.

Michalski, F., Boulhosa, R. L. P., Faria, A. and Peres, C. A. (2006). Human–wildlife conflicts in a fragmented Amazonian forest landscape: determinants of large felid depredation on livestock. *Anim. Conserv.*, 9, 179–188.

Miller, B., Reading, R. P. and Forest, S. (1996). *Prairie Night: Black Footed Ferrets and the Recovery of Endangered Species*. Washington, DC: Smithsonian Institution Press.

Nyhus, P. J., Osofsky, S. A., Ferraro, P., Fischer, H. and Madden, F. (2005). Bearing the costs of human–wildlife conflict: the challenges of compensation schemes. In *People and Wildlife: Conflict or Coexistence?* eds. R. Woodroffe, S. Thirgood and A. Rabinowitz, pp. 107–121. Cambridge: Cambridge University Press.

O'Meilia, M. E., Knopf, F. L. and Lewis, J. C. (1982). Some consequences of competition between prairie dogs and beef cattle. *J. Range Manage.*, 35, 580–585.

Palmeira, F. B., Crawshaw, P. G. Jr, Haddad, C. M., Ferraz, K. M. P. and Verdade, L. M. (2008). Cattle depredation by puma (*Puma concolor*) and jaguar (*Panthera onca*) in central-western Brazil. *Biol. Conserv.*, 141, 118–125.

Polisar, J., Maxit, I. and Scognamillo, D. (2003). Jaguars, pumas, their prey base and cattle ranching: ecological interpretation of a management problem. *Biol. Conserv.*, 109, 297–310.

Rabinowitz, A. R. (1986). Jaguar predation on domestic livestock in Belize. *Wildlife Soc. B*, 14, 170–174.

Reading, R. P., Mccain, L., Clark, T. W. and Miller, B. J. (2005). Understanding and resolving the black-tailed prairie dog conservation challenge. In *People and Wildlife: Conflict or Coexistence?* eds. R. Woodroffe, S. Thirgood and A. Rabinowitz, pp. 209–223. Cambridge: Cambridge University Press.

Redpath, S. M. and Thirgood, S. J. (1997). *Birds of Prey and Red Grouse*. London: Stationery Office.

Redpath, S. M., *et al.* (2013). Understanding and managing conservation conflicts. *Trends Ecol. Evol.*, 28, 100–109.

Sandercock, B. K., Nilsen, E. B., Brøseth, H. and Pedersen, H. C. (2011). Is hunting mortality additive or compensatory to natural mortality? Effects of experimental harvest on the survival and cause-specific mortality of willow ptarmigan. *J. Anim. Ecol.*, 80, 244–252.

Sillero-Zubiri, C., Sukumar, R. and Treves, A. (2007). Living with wildlife: the roots of conflict and the solutions. In *Key Topics in Conservation Biology*, eds. D. W. MacDonald and K. Service, pp. 266–272. Oxford: Blackwell.

Smith, A. D., *et al.* (2008). Experience in implementing harvest strategies in Australia's south-eastern fisheries. *Fish. Res.*, 94, 373–379.

Smith, R. K., Mitchell· R., Dicks, L. V. and Sutherland, W. J. (2014). Comparing interventions: the missing link in effective conservation practice. *Conserv. Evidence*, 11, 2–6.

Soulé, M. E. (1985). What is conservation biology? *Bioscience*, 35, 727–734.

Stewart, G. B., Pullin, A. S. and Coles, C. F. (2007). Poor evidence-base for assessment of windfarm impacts on birds. *Environ. Conserv.*, 34, 1–11.

Sutherland, W. J., Pullin, A. S., Dolman, P. M. and Knight, T. M. (2004). The need for evidence-based conservation. *Trends Ecol. Evol.*, 19, 305–308.

Thirgood, S. and Redpath, S. (2008). Hen harriers and red grouse: science, politics and human–wildlife conflict. *J. Appl. Ecol.*, 45, 1550–1554.

Treves, A. and Naughton-Treves, L. (2005). Evaluating lethal control in the management of human–wildlife conflict. In *People and Wildlife: Conflict or Coexistence?* eds. R. Woodroffe, S. Thirgood and A. Rabinowitz, pp. 86–106. Cambridge: Cambridge University Press.

Williams, D. R., *et al.* (2013). *Bird Conservation: Global Evidence for the Effects of Interventions.* Exeter: Pelagic Publishing.

Woodroffe, R., Thirgood, S. and Rabinowitz, A. (2005). The impact of human–wildlife conflict on natural systems. In *People and Wildlife: Conflict or Coexistence?* eds. R. Woodroffe, S. Thirgood and A. Rabinowitz, pp. 1–12. Cambridge: Cambridge University Press.

Zeller, K. (2007). *Jaguars in the New Millennium Data Set Update: The State of the Jaguar in 2006*. New York: Wildlife Conservation Society.

Zimmermann, A., Walpole, M. J. and Leader-Williams, N. (2005). Cattle ranchers' attitudes to conflicts with jaguar *Panthera onca* in the Pantanal of Brazil. *Oryx*, 39, 406–412.

Box 3

Conflicts in the UK uplands: birds of prey and red grouse

Arjun Amar[1] and Stephen M. Redpath[2]

[1] *Percy FitzPatrick Institute, DST-NRF Centre of Excellence, University of Cape Town, Rondebosch 7701, South Africa*

[2] *Institute of Biological & Environmental Sciences, Aberdeen University, 23 St Machar Drive, Aberdeen AB24 2TZ, UK*

In the USA and Europe, birds from the order *Galliformes*, such as grouse, pheasants and quail, are commonly hunted. In some places, gamebird management can be intensive and involve large-scale bird releases, habitat manipulation, predator control, the provision of supplementary food and water and disease control. Since the mid nineteenth century, large areas of moorland in northern England and east and central Scotland have been intensively managed for red grouse *Lagopus l. scoticus*. On these estates the aim is to maximise the number of grouse available for shooting between 12 August and 10 December. Grouse management can benefit other species, preserve moorland habitats, create jobs and augment local economies, but is detrimental for other reasons such as the illegal killing of predators. Notably there is an ongoing, contentious conflict between grouse managers and conservationists over illegal killing of raptors on grouse moors. The main species involved are hen harrier *Circus cyaneus*, peregrine falcon *Falco peregrinus* and golden eagle *Aquila chrysaetos*. This illegal killing has reduced breeding success, survival and population size of these and other raptors and has contributed to the virtual disappearance of harriers and eagles as breeding birds in England and on many grouse moors in Scotland (Etheridge *et al.*, 1997; Whitfield *et al.*, 2007; Amar *et al.*, 2012).

This conflict is complex and polarised. On one hand, landowners fear that uncontrolled birds of prey will make shooting economically unviable, resulting in reduced income and job losses. On the other hand, raptor conservationists seek to enhance the status of raptors through protection. The conflict also reflects differences between groups about politics, land ownership and decision making. Some conservationists resent that most of upland Britain is divided into large private estates, whose owners may flaunt environmental laws and regulations. In contrast, landowners do not understand why conservationists are more concerned about birds of prey than they are for human welfare such as community income and jobs, especially when those species have secure populations elsewhere in the world.

The main protagonists have been conservation NGOs, particularly the Royal Society for the Protection of Birds and the Raptor Study groups, versus the Moorland Association, the National Gamekeepers Organisation and other representatives of game and landowning interests. The government statutory agencies, Natural England and Scottish Natural Heritage, have been in the centre of this long-running conflict.

Most research has been ecological, focused on impact, but increasingly the social dimensions of the conflict have been explored (e.g. Marshall *et al.*, 2007). Research has quantified illegal killing of raptors, the impact of raptors on grouse shooting and some technical solutions (Thirgood *et al.*, 2000; Thirgood and Redpath, 2008), so knowledge is plentiful.

All parties have been searching for solutions. These attempts have included advocating increased enforcement to prevent further illegal killing, testing technical solutions to reduce impact, creating demonstration sites for technical solutions, and creating stakeholder meetings to search for solutions. Enforcement is failing as shown by continued declines in raptor numbers on grouse moors. In Scotland the new law of vicarious liability has recently been introduced with the aim of making landowners, and not just their gamekeepers, responsible for wildlife crime committed on their land. Technical solutions have had some success in research trials, but have not been adopted for various reasons (Thirgood and Redpath, 2008). Demonstration sites have so far proved ineffective; possibly because the level of illegal killing is so great that raptor densities remain very low even when locally protected. A 7-year stakeholder dialogue process initiated by Natural England and facilitated by an independent third party increased levels of dialogue but ultimately failed to resolve the conflict. Conservation NGOs withdrew, deeming progress too slow and not preventing further raptor declines. The Department of Environment, Food and Rural Affairs have taken on this process with only the key stakeholders and are developing a broad management plan for harriers for consideration by the government minister.

The main impediment to success has been the unwillingness to compromise. Both sides have focused on winning, rather than seeking shared solutions. Conservationists push for enforcement or management with minimal intervention, such as habitat management or feeding raptors. Grouse managers push for active, direct control, such as a quota. Government agencies have tended to side with conservation organisations and moved forward slowly, testing the least-interventionist ideas first. Whether top-down approaches, such as vicarious liability, or bottom-up ones like seeking shared solutions through dialogue ultimately give better social and conservation outcomes remains to be seen.

References

Amar, A., *et al.* (2012). Linking nest histories, remotely sensed land use data and wildlife crime records to explore the impact of grouse moor management on peregrine falcon populations. *Biol. Conserv.*, 145, 86–94.

Etheridge, B., Summers, R. W. and Green, R. E. (1997). The effects of illegal killing and destruction of nests by humans on the population dynamics of the hen harrier *Circus cyaneus* in Scotland. *J. Appl. Ecol.*, 34, 1081–1105.

Marshall, K., White, R. and Fischer, A. (2007). Conflicts between humans over wildlife management: on the diversity of stakeholder attitudes and implications for conflict management. *Biodivers. Conserv.*, 16, 3129–3146.

Thirgood, S. and Redpath, S. (2008). Hen harriers and red grouse: science, politics and human–wildlife conflict. *J. Appl. Ecol.*, 45, 1550–1554.

Thirgood, S. J., Redpath, S. M., Rothery, P. and Aebischer, N. J. (2000). Raptor predation and population limitation in red grouse. *J. Anim. Ecol.*, 69, 504–516.

Whitfield, D. P., Fielding, A. H., Mcleod, D. R., Morton, K., Stirling-Aird, P. and Eaton, M. A. (2007). Factors constraining the distribution of Golden Eagles *Aquila chrysaetos* in Scotland. *Bird Study*, 54, 199–211.

© Adam Vanbergen.

CHAPTER FOUR

Environmental history and conservation conflicts

ROBERT A. LAMBERT
University of Nottingham & University of Western Australia

In the mid-1990s in Scotland, two historians chaired the board of the government's statutory body for nature conservation, Scottish Natural Heritage (SNH). Television broadcaster, journalist and expert on the Icelandic sagas and Viking history, Magnus Magnusson KBE, was its first Chairman from 1992; and socio-economic historian and Historiographer Royal, T. C. Smout CBE, was Deputy-Chairman from 1992 to 1997. This choice of senior management has remained something unique, quirky perhaps and yet visionary – two 'historians of people' presiding over what might rightly be perceived by many to be a purely scientific organisation concerned only with the biological management of the natural world. However, both men were firmly of the vision that people were a part of nature and found a home pioneering the emerging discipline of environmental history in Scotland. Smout reflected on those exciting days on boards and committees where he learned, 'how deep and complex were the issues surrounding nature conservation, and how little it was appreciated that they were matters of history as well as of contemporary manoeuvring and posturing round vested interest' (Smout, 2009: 1). Here was an evolving recognition and acceptance within a nature conservation policy-making world that human attitudes, values and perceptions (the socio-cultural) shaped behaviour and responses to the natural world, and that those same attitudes were both complex and had evolved over time in response to different stimuli.

Smout (2000: 2) outlined in *Nature Contested* how all conservation conflicts had a history, often ignored or forgotten in the heat of the moment: 'I was struck by the passions unleashed by the difficult cases that came before us – here was hotly contested ground – and by the essentially historical nature of many of the problems. If a wood needed saving, it was because it had a history in which human beings had once played a central part, and because today other human beings (with their own histories as foresters) wished to play a different part. When a Highland sea loch was proposed as a marine nature reserve, the anger which this aroused had to do with ancient concepts of *usufruct* (the temporary right to the use and enjoyment of the property of another, without changing the character of the property) and property, opposed to more recent concepts of

Conflicts in Conservation: Navigating Towards Solutions, ed. S. M. Redpath, R. J. Gutiérrez, K. A. Wood and J. C. Young. Published by Cambridge University Press.

heritage and public interest. Yet both sides in such conflicts tended to see them as problems with only contemporary and immediate significance, nature versus the developers, jobs versus birds, right versus wrong.'

Environment and history: strange bedfellows?

Environmental history is the history of our changing interactions with nature and the natural world over time (Hughes, 2006). It is not only about the way that we have used and tamed nature, engaged and interacted with it, but also about how nature limits and restricts us. Environmental history is often about tracing, understanding and confronting our dominant attitudes and responses (Coates, 1998) and how they have shifted through time (Thomas, 1984) and shaped our conduct and policy (White, 1967). Environmental history also blends the planet's history (the scientific story) with the people's history (the socio-cultural story), therefore offering up a much more real and meaningful understanding of the past (McNeill, 2000). Academic life in the second part of the twentieth century tended to promote a clear division between research and teaching in the Arts and Sciences, as the novelist and physicist C. P. Snow bemoaned in 1959 while delivering the Rede Lecture in Cambridge on 'The Two Cultures', separated by a gulf of incomprehension (Snow, 1959). Yet, people are a part of nature, and not to appreciate that we are so has fuelled much environmental conflict (Hughes, 2000). Indeed, hotly contested contemporary landscapes (Lambert, 2001a) or beasts (Lambert, 2002) reflect deeply held and intertwined historical forces, both natural and cultural. Environmental history exists to challenge the divide between the Arts and Sciences, and to build a bridge across.

From the late 1960s, historians of the American West sought new interpretations of both settlement and invasion (Nash, 1967). At the same time, environmental history also carved a niche for itself in British Empire studies in southern Africa, Australia and New Zealand (MacKenzie, 1988; Griffiths and Robin, 1997; Dunlap, 1999; Beinart and Hughes, 2007), often linked to the idea of ecological imperialism promoted by Crosby (1986), before being embraced by British university history departments and scholarship, most especially in Scotland (Smout, 1993b; Clapp, 1994; Sheail, 2002). This is not to deny the concurrent existence of a strong tradition in Britain for landscape history, rural history, historical geography and historical ecology (Sheail, 1980; Hoskins, 1986; Rackham, 1986; Simmons, 2001).

It is the newest of all the historical subfields (coming as it does after the post-World War II emergence of new ways of looking through social, cultural, gender and racial history), and it gives a voice to a forgotten part of the historical story: the environment (which some traditional historians perceive to be just a mere non-participant backcloth against which the human story unfolds). Indeed, political historians often berate environmental historians for dethroning people from the centre of history (Coates, 1996), when actually environmental historians place nature and people together as twin actors (McNeill, 2003). Indeed, we see humans as keystone species in that interaction. While environmental historians introspectively muse on why traditional historical scholarship has been slow to welcome this new field (Sorlin and Warde, 2007), the discipline is open to all, and embraces and blends insights from the Arts and Humanities, the Social and Natural Sciences, making it a truly 'interactive performance' (Smout, 2009).

Environmental historians are concerned with non-nation state mind-set ideas (Mosley, 2010); they try to be cross-disciplinary and incorporate social and ecological theories and skills as well as new theoretical frameworks (McNeill, 2003). Environmental history also challenges us to think just what a historical document is: using archival and documentary research techniques, but also viewing natural entities (e.g. trees and species) as documents. It views landscapes as semi-natural, shaped by natural forces in partnership with the human hand and imagination; it challenges the overuse of that culturally loaded Euro-American term: wilderness. Environmental history at its best can overcome the tendency of some scientists who look no further back than their data. Environmental historians have interests in both human culture and nature (in the past) which allows them to challenge what is perceived as natural and what is a baseline. Indeed, historians, and others, can push that baseline back further, and avoid the dangers of assuming that what happened say *x* years ago was the natural state. They seek to identify the scale of human impacts in the past, and unearth key stepping stone moments, attitudinal and value shifts that changed established ways of thinking and shaped environmental policy-making. They ask who the principal stakeholders were in our environmental past, and muse on the idea and test myths of historical sustainability, when they describe relationships between nature and culture.

Within the discipline of environmental history, the sub-field of species history, founded on 'good science, good history and pragmatism', has proved to have huge academic and popular appeal and application as we seek to understand how the fortunes of certain species have been shaped over time by a mix of natural changes and human impacts (Ritchie, 1920; Lambert, 1998). Modern species histories of so-called problem animals or invasive alien species are emerging from all over the world (Sheail, 1972; Love, 1983; Lovegrove, 1990; Jones, 2002; Lazarus, 2006; Carter, 2007; Rotherham and Lambert, 2011). In 2003, Reaktion Books launched its series 'Animal', each a natural and cultural history of an individual species. They complement studies with a longer time frame using archaeological evidence (Yalden, 1999, 2003) as we seek to understand what is native. More broad histories of nature conservation in Britain (Sheail, 1976, 1981, 1998; Moore, 1987; Evans, 1992; Marren, 2002; Sands, 2012), that look back to nineteenth-century formative roots, are another vehicle through which land-use or species-based conservation conflicts are often explored using archival and documentary research in private or public collections (Lovegrove, 2007). Nature, like us, has a history. We would do well to remember that.

Birds of prey and landed sporting estates: a conflict illuminated by history

The attitudes of some gamekeepers and estate owners today in northern England and Scotland can best be understood by the fact that the sporting estates that they own or manage are themselves a nineteenth-century creation, replacing sheep farms; and by the fact that an immense destruction of predators accompanied their heyday, when takes of game (quarry) were also vast. Environmental historians have uncovered this through the study of estate papers, family archives and, above all, Game Books (annual records of harvest kept by gamekeepers). The predator control figures are stark, column after column of birds of prey and small mammals destroyed (Smout, 2000); and, although of its time, make uncomfortable reading, even 'testing credulity' (Smout, 2009). While we must approach this evidence with the historian's critical eye, as shot or trapped predators were passed around keepered estates to obtain duplicate bounty scheme payments or proprietorial grace, the dominant culture of the Victorian and Edwardian sporting estate was that all birds or prey and predatory mammals were bad (classed as 'vermin').

This historic culture, whether perceived as right or wrong over time, resulted in a substantial modification of the natural world in the late nineteenth century and first half of the twentieth century in the upland ecosystems of Britain, the consequences of which we still wrestle with today (Holloway, 1996). The traditional utilitarian attitude to predators still survives and brings those who are charged with managing wildland and game species for economic benefit, leisure and class (sometimes with associated environmental benefits), into direct conflict with those who have non-utilitarian views (Smout, 1993a). The fortunes of some species have, in large part, undoubtedly improved in the second half of the twentieth century. Following the pesticide crisis of the 1960s, some raptors, including the osprey *Pandion haliaetus*, red kite *Milvus milvus*, white-tailed eagle *Haliaeetus albicilla* and peregrine falcon *Falco peregrinus*, have gone from persecution to reintroduction and restoration to sustainable tourism icons (Cairns and Hamblin, 2007; Lambert, 2011). However, negative attitudes to these species still exist. Others like the hen harrier *Circus cyaneus* remain seemingly trapped in a historical time warp of negative attitudes to protect sporting interests, despite national conservation endeavours by powerful mass-membership environmental NGOs such as the RSPB and county Wildlife Trusts (see Box 2).

It is important to reflect here that history, while illuminating past attitudes and responses to conflict species over time, can also be a barrier or hindrance to conflict resolution. Tradition, deeply shaped by socio-cultural forces embedded in issues of class, property and status, can hugely influence trust and responses towards land ownership and management, and any mitigation techniques that may be proposed or deployed. What is considered acceptable as a management

Figure 4.1 A very visible public manifestation of deeply held anti-raptor reintroduction attitudes (in part rooted in cultural history, as well as contemporary concerns) from farming interests and landed estate owners on the coastal strip of East Anglia, England. In response to historical patterns, conservationists now endeavour to bring back predators, itself a deeply contested practice. Photograph © James Bradley, A12 near Blyth Estuary, Suffolk, 4 January 2010.

strategy, as a way forward to ease conflict situations and promote resolution, depends on a diverse suite of factors shaped by traditional, scientific and emotional values. The dire nature of a situation, coupled with issues of historical mistrust, can lead to entrenched and immovable stances. While we are all victims of the past, we can also be prisoners of that same past. Negotiating a way out of that mental jail can be fraught with challenges for all stakeholders.

Coping with grey seals: conservation success and a contested beast

In Britain, the Atlantic grey seal *Halichoerus grypus* represents perhaps the most obvious and extreme example of how nature conservation success can lead to conflict (see Box 4). While we often lament the failings of nature conservation, rarely if ever in Britain do we address the impacts of the successes, which can generate people–people conflicts. The story of the grey seal is as much an examination of socio-cultural history as it is a history of a biological creature.

Shot for sport, food and pelts in the nineteenth century, the grey seal became the first mammal protected by the British Parliament under legislation in 1914, with more legislative protection offered in 1932 (Sheail, 1976). That came about not from massed ranks of concerned nature-lovers (they would emerge powerfully post-1960s), but from lobbying by concerned sportsmen who had noticed the decline of their chosen quarry to a perceived low of just 500 animals (Lambert, 2002; National Archives of Scotland AF56/1443). The 1914 Grey Seal (Protection) Act ended centuries of direct subsistence hunting, commercial and sporting exploitation of the grey seal. By the mid-1960s the grey seal population of 34,000 was expanding at a rate of 6% per annum, rising to 124,300 in 2000. By the mid-1930s, fishermen and fishing organisations were concerned over the impacts of rising seal numbers on fisheries, most especially along the salmon fishing rivers and estuaries of the north.

The battle lines were drawn around the Farne Islands off Northumberland in northeast England and close to the River Tweed; initially between fledgling conservation/wildlife groups, local naturalists (Hickling, 1962), and beleaguered fishing communities and organisations who talked only of an advancing national 'seal menace' that needed to be halted to protect livelihoods and recreational salmon fishing. Fishermen urged government in the 1950s to make the animal a subject of scientific enquiry, hoping that a cull would be proposed (Lambert, 2001b), but this set in motion decades of claim and counter-claim, posturing, petitions, direct action, media manipulation, propaganda and political lobbying. Thousands of seal pups died as well, from the Farnes up to the Orkney Islands off the north coast of Scotland. The government-sponsored seal culling on the Farnes in the mid-1960s was initially justified to satisfy fishing interests, but later the islands' owners themselves, the National Trust, started culling their grey seals in the 1970s to reduce the breeding female population to 1000 animals, selling it to their angry membership as a way to reduce perceived seal overcrowding on the nature reserve and prevent the crushing of vegetation and Atlantic puffin *Fratercula arctica* burrows (Lambert, 2002; National Trust archives NT 208/2/PF). On the wave of a post-war environmental, social and cultural revolution, from the 1960s onwards, and especially in the years of the largest seal culls in Scotland, the public took up the plight of the grey seal as an environmental cause in a far more popular and egalitarian crusade than the few who had sought the initial protection back in 1914. By the 1970s, this domestic campaign linked up with well-funded international seal protection crusades led by Greenpeace and the International Fund for Animal Welfare (Watson, 1982; Davies, 1990), with greatest success standing against the annual harp seal *Pagophilus groenlandicus* hunt on the pack-ice off Newfoundland (Busch, 1985; Mowat, 1997). In 1978, the proposed Scottish culls were abandoned due to an alliance of Greenpeace, local people, the media, many other wildlife enthusiasts within the British public, and a consequent loss of political will in government. The political reality of men with clubs killing baby seals was too much to stomach for all but the most resolute supporters of the cull. Naturalist and nature-writer John Lister-Kaye (1979) described it as 'a towering controversy'.

The crux of the matter now is this. Grey seals are of interest to a whole range of competing viewpoints and sectors of the population. Government sees the seal issue as a political question; to nature conservationists they are a wildlife management question; to scientists they are a biological or ecological question; to fishing communities they are an economic question; to humanitarians they are an animal welfare issue; to the wider general public they are a social/cultural question informed by things as varied as recreational seal-watching trips and wildlife television; to international agencies and NGOs they are an international marine issue (Redpath *et al.*, 2013). All of these attitudes and perceptions are rooted in history. Some are new, post-1960s environmental revolution values (Grove-White, 2001), while others are more deeply rooted in ideas of use and harvest, and myths of marine superabundance. The grey seal is an exceptionally challenging animal to manage. What might be perceived as a cherished marine icon for many thousands of recreational seal watchers around Britain from the Isles of Scilly to Shetland might just as easily be perceived as a vulgar activity, merely 'city folk gawping at vermin', by hard-pressed fishing communities.

Emerging consensus over the way forward for the grey seal in twenty-first century Britain can only come from stakeholder dialogue and local management plans (Young *et al.*, 2012), founded upon a deep understanding of past controversies, debates and bitterness, and a historical understanding of our diverse and shifting relationships with the animal over time and place (Bonner, 1982). Those charged with managing grey seal populations in the modern era must not only take into consideration all scientific and fisheries viewpoints (Matthews, 1979) but also be keenly aware of, and take into consideration, the weight of popular and contemporary support for this beast whose population has grown dramatically since the 1960s. The trial of wildlife management by public opinion is here to stay, especially with species that we have culturally constructed as most like ourselves and afforded them almost totemic status in our minds. Much action in the past has derived from whichever side pushed government the hardest, fishing organisations or conservationists, with scientists in the middle stating that the evidence is unclear as to whether killing seals would make a difference as to how much fish humans can catch, and thus calling for more research. While the BBC television *Springwatch* millions (Rollins, 2006) take great delight in watching seals (often pumping eco-tourism money into remote rural communities), they suffer no economic loss from the animal's presence. For their part, fishing communities have been blind for much of the twentieth century to their over-harvesting of the sea (Smout and Stewart, 2012), and in the face of low fish stocks, blame seal competition for declining catches. The fact that much of this conflict is a coastal one further muddies the water, as we have rallied far better to protect and understand our terrestrial ecosystems for all species to inhabit than we have our marine ecosystems.

One uncomfortable option that the nature-loving British have refused to debate was suggested by ecologist Frank Fraser Darling, as early as 1951. Darling believed that to diffuse potential conflict, absolute protection was not always the way forward for some species, and that 'an overall carefully controlled annual toll of Atlantic grey seals', would both cut out criticism from established fishery interests and yield a natural resource in meat, oil and skin that would benefit all, including the health of the seal population long term. 'The seals are extremely efficient gatherers of energy which the nation should not neglect at present' he advised (Lambert, 2002; National Trust archives NT/208). Grey and common seals *Phoca vitulina* were legally shot (by licence or police endorsement of firearms certificates, initially under the Conservation of Seals Act 1970) at netting stations or around fish farms in northern Britain, as one part of ongoing broad and inclusive management strategies since 2005 that seek to balance seal and fish conservation with the protection of inshore fishing and wildlife tourism values (Butler *et al.*, 2008). The Marine (Scotland) Act 2010 promulgated at a time when the Scottish grey seal population was estimated to be 186,000 (alongside 19,800 common seals) replaced the outdated 1970 Act, and offered improved overall protection for seals year round while delivering a new system of regulated seal management licences, and stakeholder forums, to allow the killing of animals to protect fish, fisheries and nets under strict welfare guidelines.

Environmental history, as with all histories, is replete with wise and compromise-laden 'voices from the wilderness' warning of impending conflict in years to come if bold decisions are not taken and case histories examined with openness and clarity (Darling, 1970). Frank Fraser Darling wrote an open letter to members of the Scottish Committee of the government agency, the Nature Conservancy, on 28 December 1950 in which he declared: 'My ultimate aim is to see the British stocks of the Atlantic grey seal valued as a natural resource, conserved as such and regularly used' (Lambert, 2002; National Archives of Scotland NAS AF62/929/14).

Looking back to look forward

Although a substantial part of this chapter has focused on a specific British case study of the Atlantic grey seal conflict, the themes mapped out here are common to many international conservation conflicts where nature conservation priorities and animal welfare impulses so often clash over the optimal way forward in conflict species management. Historical insights are contextually important to understanding and securing that management balancing act, and without them, or if we ignore them, we would be left floundering in an ahistorical void with no real sense or understanding of change over time. Without that important context we render the contemporary, in part, meaningless. However, we do not have to be a slave to history. People live in the present and their concerns and plans focus almost exclusively on the future. While we must never be a prisoner of the past (and at times competing parties in a conflict can dwell too much on the historical roots) we can be empowered and informed by it. It is important that we understand that fact, because ultimately it is not that history explicitly tells us what to do, rather that it will always indispensably shed new light and help us to understand and reflect upon our current position (or predicament), if we know what has happened in the past. History helps us to understand people and societies, as well as to understand environmental change and how the society that we live in now came into being; and how past societies responded to environmental pressures through changed behaviour because of shifts in public opinion or political impetus. Environmental change happens over time, and to be ignorant to the temporal (historical) dimension ultimately leaves our understanding floundering without broad context. To that extent, accessible environmental history, widely disseminated, and based on the cumulative and cooperative efforts of historians working with all types of natural and social scientists (Knight, 2000) and their data sets, can help us mould and reinvigorate future environmental policy-making decisions, by informing us of 'our' environmental past (Watson, 2002); in this instance, they provide the why and the how of environmental change and conflicts.

If we listen well to that past, and reflect on the stories it offers up, we will learn something of real value as we plan for a future of reduced and effectively managed conflicts. We will also learn a good deal, not just about nature, but also about human nature. As Smout (2000: 2) observed about conservation conflicts (or 'quarrels' over the modern British countryside), 'once we recognise our character, we can see choices more clearly'. To do that, we must first look back, to then look forward.

Acknowledgements

I am grateful to some of my fellow British environmental historians, Chris Smout, Paul Warde, Ian Rotherham, Peter Coates, John Sheail and Lucy McRobert, for discussions and reflections which helped to shape this chapter. I thank the National Trust and Department of Agriculture and Fisheries for Scotland for access to archives.

Archival sources

National Trust: Central Office, Heelis, Kemble Drive, Swindon SN2 2NA at ref: NT.

Department of Agriculture and Fisheries for Scotland: National Archives of Scotland (NAS), HM General Register House, 2 Princes Street, Edinburgh EH1 3YY at ref: AF.

References

Beinart, W. and Hughes, L. (2007). *Environment and Empire*. Oxford: Oxford University Press.

Bonner, W. N. (1982). *Seals and Man: A Study of Interactions*. Seattle: Washington Sea Grant Publication.

Busch, B. C. (1985). *The War Against the Seals: A History of the North American Seal Fishery*. Kingston and Montreal: McGill-Queen's University Press.

Butler, J. R. A., *et al.* (2008). The Moray Firth Seal Management Plan: an adaptive framework for balancing the conservation of seals, salmon, fisheries and wildlife tourism in the UK. *Aquat. Conserv.*, 18, 1025–1038.

Cairns, P. and Hamblin, M. (2007). *Tooth and Claw: Living Alongside Britain's Predators*. Dunbeath, Scotland: Whittles Publishing.

Carter, I. (2007). *The Red Kite*. Chelmsford: Arlequin Press.

Clapp, B. W. (1994). *An Environmental History of Britain since the Industrial Revolution*. Harlow: Longman.

Coates, P. (1996). On second thoughts . . . Clio's new greenhouse. *Hist. Today*, 46, 15–22.

Coates, P. (1998). *Nature: Western Attitudes since Ancient Times*. Cambridge: Polity Press.

Crosby, A. (1986). *Ecological Imperialism: The Biological Expansion of Europe, 900–1900*. Cambridge: Cambridge University Press.

Darling, F. F. (1970). *Wilderness and Plenty: The Reith Lectures 1969*. London: Ballantine Books.

Davies, B. (1990). *Red Ice: My Fight to Save the Seals*. London: Methuen.

Dunlap, T. R. (1999). *Nature and the English Diaspora: Environment and History in the USA, Canada, Australia and New Zealand*. Cambridge: Cambridge University Press.

Evans, D. (1992). *A History of Nature Conservation in Britain*. London: Routledge.

Griffiths, T. and Robin, L. (1997). *Ecology and Empire: Environmental Histories of Settler Societies*. Edinburgh: Keele University Press.

Grove-White, R. (2001). The rise of the environmental movement, 1970–1990. In *Nature, Landscape and People since the Second World War*, ed. T. C. Smout, pp. 44–51. East Linton: Tuckwell Press.

Hickling, G. (1962). *Grey Seals and the Farne Islands*. London: Routledge and Kegan Paul.

Holloway, S. (1996). *The Historical Atlas of Breeding Birds in Britain and Ireland, 1875–1900*. London: T & AD Poyser.

Hoskins, W. G. (1986). *The Making of the English Landscape*. Harmondsworth: Penguin Books.

Hughes, J. D. (2000). *The Face of the Earth: Environment and World History*. Armonk, NY: M. E. Sharpe.

Hughes, J. D. (2006). *What is Environmental History?* Cambridge: Polity Press.

Jones, K. R. (2002). *Wolf Mountains: A History of Wolves along the Great Divide*. Calgary: University of Calgary Press.

Knight, J. (2000). *Natural Enemies: People–Wildlife Conflicts in Anthropological Perspective*. London: Routledge.

Lambert, R. A. (1998). *Species History in Scotland: Introductions and Extinctions since the Ice Age*. Edinburgh: Scottish Cultural Press.

Lambert, R. A. (2001a). *Contested Mountains: Nature, Development and Environment in the Cairngorms Region of Scotland, 1880–1980*. Cambridge: The White Horse Press.

Lambert, R. A. (2001b). Grey Seals: to cull or not to cull? *Hist. Today*, 51, 30–32.

Lambert, R. A. (2002). The grey seal in Britain: a twentieth century history of a nature conservation success. *Environ. and Hist.*, 8, 449–474.

Lambert, R. A. (2011). Strangers in a familiar land: the return of the native 'aliens' and the (re)wilding of Britain's skies, 1850–2010. In *Invasive and Introduced Plants and Animals: Human Perceptions, Attitudes and Approaches to Management*, eds. I. D. Rotherham and R. A. Lambert, pp. 169–183. London: Earthscan Publishing.

Lazarus, S. (2006). *Troubled Waters: The Changing Fortunes of Whales and Dolphins*. London: Natural History Museum.

Lister-Kaye, J. (1979). *Seal Cull: The Grey Seal Controversy*. Harmondsworth: Penguin Books.

Love, J. A. (1983). *The Return of the Sea Eagle*. Cambridge: Cambridge University Press.

Lovegrove, R. (1990). *The Kite's Tale: The Story of the Red Kite in Wales*. Sandy, Bedfordshire: RSPB.

Lovegrove, R. (2007). *Silent Fields: The Long Decline of a Nation's Wildlife*. Oxford: Oxford University Press.

MacKenzie, J. M. (1988). *The Empire of Nature: Hunting, Conservation and British Imperialism*. Manchester: Manchester University Press.

Marren, P. (2002). *Nature Conservation*. London: The New Naturalist, Collins.

Matthews, L. H. (1979). *The Seals and the Scientists*. London: Peter Owen.

McNeill, J. (2000). *Something New Under the Sun: An Environmental History of the Twentieth Century*. London: Allen Lane.

McNeill, J. (2003). Observations on the nature and culture of environmental history. *Hist. Theory*, 42, 5–43.

Moore, N. W. (1987). *The Bird of Time: The Science and Politics of Nature Conservation*. Cambridge: Cambridge University Press.

Mosley, S. (2010). *The Environment in World History*. London: Routledge.

Mowat, F. (1997). *Sea of Slaughter*. Toronto: Seal Books.

Nash, R. (1967). *Wilderness and the American Mind*. New Haven: Yale University Press.

Rackham, O. (1986). *The History of the Countryside*. London: J. M. Dent.

Redpath, S. M., *et al.* (2013). Understanding and managing conservation conflicts. *Trends Ecol. Evol.*, 28, 100–109.

Ritchie, J. (1920). *The Influence of Man on Animal Life in Scotland: A Study in Faunal Evolution*. Cambridge: Cambridge University Press.

Rollins, J. (2006). It's official: half the UK loves wildlife. *Natural World*, Spring, 31.

Rotherham, I. and Lambert, R. A. (2011). *Invasive and Introduced Plants and Animals: Human Perceptions, Attitudes and Approaches to Management*. London: Earthscan Publishing.

Sands, T. (2012). *Wildlife in Trust: A Hundred Years of Nature Conservation*. London: Elliott and Thompson.

Sheail, J. (1972). *Rabbits and their History*. Newton Abbot: Country Book Club.

Sheail, J. (1976). *Nature in Trust: The History of Nature Conservation in Britain*. London: Blackie.

Sheail, J. (1980). *Historical Ecology: The Documentary Evidence*. Huntingdon: ITE Monks Wood.

Sheail, J. (1981). *Rural Conservation in Inter-War Britain*. Oxford: Clarendon Press.

Sheail, J. (1998). *Nature Conservation in Britain: The Formative Years*. London: HMSO.

Sheail, J. (2002). *An Environmental History of Twentieth-Century Britain*. Basingstoke: Palgrave.

Simmons, I. G. (2001). *An Environmental History of Great Britain from 10000 Years Ago to the Present*. Edinburgh: Edinburgh University Press.

Smout, T. C. (1993a). *The Highlands and the Roots of Green Consciousness, 1750–1990*. Occasional Paper No.1. Battleby: Scottish Natural Heritage.

Smout, T. C. (1993b). *Scotland since Prehistory: Natural Change and Human Impact*. Aberdeen: Scottish Cultural Press.

Smout, T. C. (2000). *Nature Contested: Environmental History in Scotland and Northern England since 1600*. Edinburgh: Edinburgh University Press.

Smout, T. C. (2009). *Exploring Environmental History: Selected Essays*. Edinburgh: Edinburgh University Press.

Smout, T. C. and Stewart, M. (2012). *The Firth of Forth: An Environmental History*. Edinburgh: Birlinn.

Snow, C. P. (1959). *The Two Cultures and the Scientific Revolution*. New York: Cambridge University Press.

Sorlin, S. and Warde, P. (2007). The problem of the problem of environmental history: a re-reading of the field. *Environ. Hist.*, 12, 107–130.

Thomas, K. (1984). *Man and the Natural World: Changing Attitudes in England, 1500–1800*. London: Penguin Books.

Watson, F. (2002). Seeing the wood and the trees: why environmental history matters. *Hist. Scotland*, 2, 44–48.

Watson, P. (1982). *Sea Shepherd: My Fight For Whales and Seals*. New York: W. W. Norton and Company.

White, L. (1967). The historical roots of our ecologic crisis. *Science*, 155, 1203–1207.

Yalden, D. (1999). *The History of British Mammals*. London: T. & A. D. Poyser.

Yalden, D. (2003). Mammals in Britain: a historical perspective. *British Wildlife*, 14, 243–251.

Young, J. C., Butler, J. R. A., Jordan, A. and Watt, A. D. (2012). Less government intervention in biodiversity management: risks and opportunities. *Biodivers. Conserv.*, 21, 1095–1100.

Box 4

Seal conservation and salmon fisheries in northeast Scotland

Isla M. Graham
Lighthouse Field Station, Institute of Biological & Environmental Sciences, University of Aberdeen, Cromarty, IV11 8YL, UK

In Scotland, salmon fisheries include both recreational (freshwater rod) and commercial (coastal net) fisheries. Seals can be shot legally to protect fisheries and, under the UK's Conservation of Seals Act 1970, salmon fisheries traditionally attempted to reduce the impacts of seals on fisheries by shooting seals, often at coastal haul-out sites. Harbour seals *Phoca vitulina* and grey seals *Halichoerus grypus* prey on Atlantic salmon *Salmo salar* and sea trout *Salmo trutta* in rivers and estuaries. However, salmonids are not common in the diet of seals in Scottish estuaries (Carter *et al.*, 2001; Middlemas *et al.*, 2006; Matejusová *et al.*, 2008). Nevertheless, fishery stakeholders perceive that seal predation has a significant impact on salmon stocks and catches, even though recreational fisheries, in particular, estimate the direct cost of seal interference to be low (Butler *et al.*, 2011).

The conflict between salmon fishery stakeholders and seal conservationists in Scotland has been ongoing for decades, fuelled on one side by the expansion of the grey seal population since the 1960s and recent declines in salmon catches, and on the other by declining harbour seal populations. In the Moray Firth in the early 2000s, both harbour seals and Atlantic salmon became protected under EU legislation (both species are listed under Annex II of the Habitats Directive). At the same time, because salmon stocks were declining, fisheries managers had increased their shooting effort, encouraged by a bounty scheme offered by some of the local organisations that oversee salmon fishery management (Butler *et al.*, 2008). Between 1998 and 2001 the number of seals shot annually in the Moray Firth increased by almost 40% compared with the preceding four years (Thompson *et al.*, 2007). Concern that the decline in the local harbour seal population was partly due to shooting (Thompson *et al.*, 2007) prompted the Scottish Executive to introduce a Conservation Order for harbour and grey seals in the Moray Firth in 2004, prohibiting shooting of both species except under licence.

From 2002 to 2005, key stakeholders led by a fisheries scientist negotiated a management plan to balance the competing interests of seal conservation and fisheries management in the Moray Firth (Butler *et al.*, 2008). The process involved the Scottish Government, government agencies, the statutory organisations responsible for salmon fishery management, local salmon fishermen, scientists and local wildlife tourism operators. The plan focused seal management on the lethal removal of perceived problem individuals from rivers and nets. An integrated research project found evidence for the existence of river-specialist seals (problem individuals) and their impact on salmon fisheries, lending support to this management strategy (Graham *et al.*, 2011). The plan has been very successful thus far for resolving this conflict in the northeast of Scotland, and was broadly endorsed by all stakeholders including the statutory organisations responsible for seal conservation and fishery stakeholders that had been initially resistant to seal conservation (Butler *et al.*, 2011; Young *et al.*, 2012). However, the future stability of this resolution will depend heavily upon continued investment of resources into the management process, including the provision of both financial and institutional support (Young *et al.*, 2012).

In 2011, the Marine (Scotland) Act 2010 replaced the Conservation of Seals Act 1970, and under the new legislation seals can only be shot under a specific licence. The new seal licence system is based on the model developed in the Moray Firth Seal Management Plan, thus extending this management framework throughout Scotland.

References

Butler, J. R. A., Middlemas, S. J., Graham, I. M. and Harris, R. N. (2011). Perceptions and costs of seal impacts on Atlantic salmon fisheries in the Moray Firth, Scotland: implications for the adaptive co-management of seal-fishery conflict. *Mar. Policy*, 35, 317–323.

Butler, J. R. A., *et al.* (2008). The Moray Firth Seal Management Plan: an adaptive framework for balancing the conservation of seals, salmon, fisheries and wildlife tourism in the UK. *Aquat. Conserv.*, 18, 1025–1038.

Carter, T. J., Pierce, G. J., Hislop, J. R. G., Houseman, J. A. and Boyle, P. R. (2001). Predation by seals on salmonids in two Scottish estuaries. *Fish. Manag. Ecol.*, 8, 207–225.

Graham, I. M., Harris, R. N., Matejusova, I. and Middlemas, S. J. (2011). Do 'rogue' seals exist? Implications for seal conservation in the UK. *Anim. Conserv.*, 14, 587–598.

Matejusová, I., *et al.* (2008). Using quantitative real-time PCR to detect salmonid prey in scats of grey *Halichoerus grypus* and harbour *Phoca vitulina* seals in Scotland – an experimental and field study. *J. Appl. Ecol.*, 45, 632–640.

Middlemas, S. J., Barton, T. R., Armstrong, J. D. and Thompson, P. M. (2006). Functional and aggregative responses of harbour seals to changes in salmonid abundance. *Proc. R. Soc. B Biol. Sci.*, 273, 193–198.

Thompson, P. M., Mackey, B., Barton, T. R., Duck, C. and Butler, J. R. A. (2007). Assessing the potential impact of salmon fisheries management on the conservation status of harbour seals (*Phoca vitulina*) in north-east Scotland. *Anim. Conserv.*, 10, 48–56.

Young, J. C., Butler, J. R. A., Jordan, A. and Watt, A. D. (2012). Less government intervention in biodiversity management: risks and opportunities. *Biodivers. Conserv.*, 21, 1095–1100.

CHAPTER FIVE

The political ecology of conservation conflicts

WILLIAM M. ADAMS
University of Cambridge

When I speak to conservation science audiences about the social dimensions of conservation, I am often asked why, as a social scientist, I am so intent on 'making conservation political'. The implication is that politics has no place in conservation research, is unhelpful to conservationists, and even reflects a deliberate attempt to attack or weaken conservation. Conservation scientists (and most conservationists are trained as scientists) seem to feel that conservation of itself is inherently apolitical.

I think that such questions reflect three things. First, my questioners see the conservation of living diversity as a moral necessity, something that is self-evidently right and just has to be done. In the language of conservation biology, conservation is a 'mission' (Meine *et al.*, 2006). Anything that detracts from that mission, or contextualises it as just one among other competing ideas or interests, is therefore inherently suspicious. Second, they feel that conservation is scientific, and therefore its actions, if properly based on scientific evidence (Pullin and Knight, 2001), are not the outcome of mere political choices, but the result of scientific conclusions from impartial data, which should be above dispute (except when new data become available). Third, they believe conservation can be done in a way that is essentially neutral, so that while there may be problems with what might be called 'actually existing conservation' in particular places (especially where the science is weak), these can be ironed out, and win–win solutions identified.

Some conservationists certainly feel that those speaking about conservation conflicts, or the negative impacts of projects like protected areas on local people, are troublemakers. An example of this is Spinage's hostile review of the book *Social Change and Conservation* (Ghimire and Pimbert, 1997), a volume he dismissed as 'cloaked in Marxist and neo-populist dogma' (Spinage, 1998: 265). If change is to come in conservation, Spinage argued, 'it should be based on ecological criteria and not political ideology' (Spinage, 1998: 274).

I believe he was wrong. All nature is influenced (and much is directly shaped) by human choices and the patterns and structures in economy and society.

Conflicts in Conservation: Navigating Towards Solutions, ed. S. M. Redpath, R. J. Gutiérrez, K. A. Wood and J. C. Young. Published by Cambridge University Press.

Human decisions, and therefore all conservation decisions, are inherently political. Neumann (2005: 44) noted that the environment is 'fundamentally and thoroughly social and political'. The decision to conserve species and ecosystems (or to exploit or degrade them) is a fundamentally political one. This is the basic premise of the discipline of political ecology, to explore the politics of human engagements with nature. It is, as Robbins (2004: 5) said, 'an explicit alternative to "apolitical" ecology'.

Conflict is intrinsic to conservation. While the obvious questions involved in conservation might seem to relate to the species and ecosystems being conserved, conservation involves making choices about the relations between people and nature. Conservation choices determine how nature is managed or used, and by whom. If the freedom of landowners to drain wetlands is restricted by law, nature may gain, but landowners lose. If a forest is protected, it is not available for farmers or loggers, although it may be available for scientists or tourists. If hunting is banned or restricted in particular seasons to protect a species, people who once hunted may lose out, while naturalists, tourists or tour operators gain.

Conservation therefore might focus on nature, but it involves making choices between people. Key political questions in conservation include: whose uses of nature should be allowed and whose should be prevented by rules, laws and economic relations? Whose freedom of action does conservation action protect, and whose does it constrain? Who wins from any given conservation decision, and who loses?

Political ecology

Political ecology is a valuable frame for understanding choices in conservation and arguments about them. Political ecology is not a discipline in the way that economics or political science is. It grew at 'a confluence between ecologically rooted social science and the principles of political economy' (Peet and Watts, 1996: 6). The term was used by Wolf (1972), but was established by Blaikie (1985) and Blaikie and Brookfield (1987), writing about soil erosion and land degradation. At that time, soil erosion associated with agriculture on steep land was mostly understood as a technical problem, best tackled using engineering solutions such as terracing or 'conservation' cropping patterns. Blaikie (1985) argued that loss of vegetation cover and soil were linked to social and economic processes occurring far away. Land-use practices reflected the resources, skills, assets, time horizons and technologies of land users, and their wealth or poverty. Soil erosion was therefore linked to the nature of agrarian society, issues of justice, wealth distribution, access to education, national policy, economics and wider patterns of trade, wealth and poverty linked ultimately to international political economy. Blaikie's 'political' ecology therefore linked ecologists' concerns about erosion to radical political economists' concerns about unequal development (Blaikie and Brookfield, 1987).

Political ecology is diverse. It has no single central body of theory, although much of its DNA comes from 'radical' political economy. Political ecology explicitly considers relations of power (Robbins, 2004) and the political processes and structures that determine who uses nature, how they use it, and with what outcomes (Stott and Sullivan, 2000; Robbins, 2004). It draws on a wide range of other disciplines, including environmental history, ecology and political economy, although the blending of these disciplines is not straightforward. Thus, Paulsen *et al.* (2003) explore how to 'locate the political' in political ecology, while Walker (2005) asks 'where is the ecology?'

Political ecologists seek explanations of environmental change that transcend the disciplinary boundaries of ecology and political economy. Environmental change is understood as the product of political and social processes. Of course, some environmental changes, such as variations in planetary oscillations, are driven by geophysical processes outside the scope of human action, but the impacts of these phenomena are distributed through social and economic processes: poor people and rich people often face different levels of exposure to risk. However, human action is the primary driver of much environmental change, and a factor in almost all, if only through global impacts such as anthropogenic climate change or ocean acidification. Robbins (2004: 5) puts this succinctly: 'the systems that govern use, overuse, degradation and recovery of nature are structured into a large social engine, which revolves around the control of nature and labor [*sic*]'.

Capitalism is central to those impacts (Harvey, 2005; Castree, 2008). Exploitation of natural resources for food or biofuels is the result of the decisions of a wide range of actors, from global corporations to households. These interactions reflect the working of markets, mediated by the intervention and regulation of the state and the lobbying of non-governmental actors (e.g. Box 5). The politics among actors determines not only how the environment changes, but also how the benefits and costs of these developments are shared. Resulting patterns of wealth and poverty in turn drive patterns of investment, industrialisation, urbanisation, resource consumption and pollution that in turn affect nature.

Political ecologists attempt to trace the links between ecological change and the way markets work, the extraction of profit, the incidence and persistence of poverty, the exercise of power and the nature of governance. The consequences of environmental change are therefore understood not simply in terms of biodiversity loss or transformation, but in terms of the distribution of economic benefits and costs, and issues of social vulnerability and marginalisation.

Political ecologists also consider the way these processes work across space. Scale tends to be taken for granted, and environmental management is often thought of in terms of a hierarchy from global through national to local scales. Such 'scales' do not exist as discrete and hierarchical spatial levels (Zimmerer and Bassett, 2003), because 'global processes' (e.g. the 2008 economic crash, climate change) can have significant and differentiated local impacts, and local processes (e.g. industrial pollution) can agglomerate to have impacts far away or over vast regions. Scale is in fact relational and socially constructed and the product of particular social and political processes, which reflect dominant patterns of power (Rangan and Kull, 2009). The appropriate scale for environmental management depends on social and political as much as biophysical processes (Lebel *et al.*, 2005). Choice of management scale (country, river basin, village) is highly political, because it affects the power of different groups to frame debate and influence outcomes.

A key development in political ecology has been interest in the political implications of the way nature itself is understood: the politics of knowledge (Watts and Peet, 2004). The ways ideas about nature are formed, shared and applied reflect the power of different actors. The environment and environmental change are perceived differently by stakeholders, each with their own values and interests (Stott and Sullivan, 2000). Thus, political ecologists have come to challenge the assumption that it is possible to talk of an 'environmental problem', unmediated by power/knowledge (Robbins and Monroe-Bishop, 2008: 748).

Knowledge of nature is not neutral. Knowledge underpins power and the regulation of both people and nature (Foucault, 1975). Thus strategic conservation planning, typically led by international conservation NGOs, uses expert scientific knowledge of the distribution of species and ecosystems to categorise and map both nature and people (Bryant, 2002; Fairhead and Leach, 2003). Local knowledge often has little or no place in this analysis, and local people play little or no part in the planning process. This politics of knowledge has material implications for both what nature is conserved and where, and the lives and livelihoods of people. Political ecologists have increasingly challenged the metaphors used by natural scientists (Neumann, 2008). Rarity, ecosystem type and even species are powerful and useful concepts, but the political ecologist recognises them as constituted by social interactions between scientists in research and science-based conservation bureaucracies (Hinchliffe, 2005).

The political ecology of conservation conflict

If politics is central to conservation, then so too is conflict. Some people gain from conservation and some lose. Conservation actions provide new arenas within which people compete – new arenas for conflict.

There is no single approach of political ecology to conservation conflicts. There is an interest in both ultimate drivers such as capitalism, and proximate drivers such as individual choices. The common thread is the determination to understand the links between the political economic drivers of human action (wealth, poverty, consumption, the operation of markets, government regulation) and environmental change. In between lies the everyday reality of people struggling against poverty, risk, or ill health to sustain livelihoods. The key insight of political ecology for the study of conservation conflict is its challenge to move analysis and understanding beyond the technical to explore the conflict's political dimensions. The political ecologist would say the politics is always there, you just have to look.

Political ecologists engage with conservation conflict in several ways. Below I consider three: (1) protected areas; (2) conflicts around the extension of market-based approaches to conservation; and (3) the politics of ideas.

Conflicts over protected areas

One of the most common forms of conservation conflict is between conservation organisations (governmental or non-governmental) and local communities over the creation of protected areas (e.g. Box 6). Protected Areas (PAs) have been the mainstay of international conservation strategies since the end of the nineteenth century (Adams, 2004), although their history is much longer. From the 1960s, the number and size of protected areas grew rapidly, reaching 100,000 PAs covering more than 2 million km^2 by 2005 (Chape *et al.*, 2005). Many different kinds of PAs are recognised, from strictly protected to economic reserves (Ravenel and Redford, 2005), and their socio-economic impacts vary as well as their effectiveness. However, the exclusivity of PAs, setting aside space for nature in reserves and protected zones, has often brought negative social impacts (Brechin *et al.*, 2003).

The social impacts of PAs can be understood through economic benefits and costs (Emerton, 2001). Economic benefits from activities like tourism are often slow to reach local people; costs, however, tend to be felt fast. Direct costs include crop and livestock raiding by wild animals (Woodroffe *et al.*, 2005; Box 2) and loss of land or resources in the park. Park neighbours can be exposed to corrupt rent-seeking behaviour by PA staff, particularly linked to minor infringements of park boundaries.

There is also a politics to these impacts, as conservation authorities and key individuals such as game guards interact with people from the community (Pretty, 2002). A political ecology approach has the potential to combine understanding of the political interaction of conservation staff and community members in the light of the ecological impact of different activities and the sustainability of livelihood systems. It allows understanding of the ideas that different actors hold about environmental change and its consequences, including the adequacy of scientific knowledge.

The political ecology approach is particularly useful in understanding the problem of involuntary displacement from protected areas (Adams and Hutton,

2007), which directly impacts on livelihoods of local people (Colchester, 2002). Despite the scale of the problem, and the attention paid to it, displacement still occurs (Brockington, 2002), sometimes involving the use of force (Peluso, 1993).

A well-researched example of population displacement from a PA is the Mkomazi Game Reserve in Tanzania (Brockington, 2002). This PA was created in 1951, at which time a small number of pastoralists, owning about 5000 cattle, were allowed to remain in the area. By the mid-1980s, there were almost 100,000 cattle, and the reserve was tightly integrated into the local and regional economy as a source of seasonal grazing. Conservationists believed the pastures were overstocked and threatened with permanent degradation. After several unsuccessful attempts to remove stock and their owners, the Department of Wildlife finally cleared the reserve in 1988, with significant impacts on those evicted and excluded (Brockington, 2002). The evictions were challenged in court, but the High Court found in favour of the government in 1999. A sanctuary was established in the reserve for introduced black rhinoceros *Diceros bicornis* and captive breeding of wild dogs *Lycaon pictus* began. In 2005, Mkomazi was declared a National Park. The political ecology of conflict over Mkomazi therefore has to examine not only the eviction and subsequent legal battle, but also the ideas about degradation that gave rise to it, and those that accompanied tourism development of the national park as a 'wilderness' for foreign tourists (Tanzania National Parks, 2012).

Market-based conservation and conflict

Market-based conservation strategies strive to 'make nature "pay its way"' (Eltringham, 1994). These take a variety of forms, including wildlife-based tourism and various forms of payments for ecosystem services (PES). The place of economics in conservation conflicts is discussed in Chapter 6. The political ecological approach seeks to extend economic analysis to understand how markets are developed and controlled, and how the allocation of rights and the operation of incentives lead to different outcomes for people. Market interactions are understood as politically mediated rather than being simply the outcome of a rational process of managing competing wants and scarce resources. Political ecologists understand that this process is inherently unfair.

To an economist, PES involves a voluntary and conditional transaction where a well-defined environmental service is purchased by a willing buyer, whereby payment is supposed to be conditional upon performance (Wunder, 2008). Interest in such economics-based approaches as a mechanism for promoting the conservation of biodiversity reflects the importance of the concept of ecosystem services in environmental policy following the Millennium Ecosystem Assessment (2003).

The adoption of market-based approaches to conservation, and into the services and functions traditionally led by government (Peck and Tickell, 2002), is a fundamental feature of the process of neoliberalisation: the 'increasingly all-pervasive trend to conform social and political affairs to market dynamics' (Brockington *et al.*, 2008; Büscher, 2008: 231; Castree, 2008). Conservation is involved in the creation of new commodities for capital to exploit, for example adding value in supply chains and marketing (e.g. certification and 'ethical trade'), creating ecosystem service 'products', derivatives and commodities. Conservation also provides the bureaucratic mechanisms to make schemes such as PES work; for example, securing the protection and certification of forest land, and new markets for them to interact, for example in carbon (Igoe and Brockington, 2007; Brockington *et al.*, 2008).

REDD (Reducing Emissions from Deforestation and forest Degradation) has become an important element in global policy for climate change and is based on the idea of carbon as a commodity. Forest loss and degradation accounts for up to a fifth of global greenhouse gas emissions. Advocates suggest that REDD schemes based on payments for carbon sequestration can increase the economic value, and incentivise the conservation, of forests in developing countries. However, complex technical and institutional challenges remain (Angelsen, 2008; Brown *et al.*, 2008). These are dependent on the politics of forest governance (Sandbrook *et al.*, 2010). To a political ecologist, issues of rights, governance and the distribution of benefits are fundamental to debates over REDD and over deforestation.

The economic benefits that can be derived from the exploitation of tropical timber resources have long created strong incentives for governments and individuals from within governing elites to control these resources (Ribot *et al.*, 2006). Poor governance and the capture of potential public revenue by private interests are common. The addition of new revenues from PES payments for carbon simply adds a new revenue stream from forests. There is limited potential for forest communities to enforce tenure rights over forest land, and hence access REDD revenue. Struggles for benefits reflect power relations within local communities, between local people and government, within government organisations, and within the private sector. Outcomes depend on clear regimes of rights and enforceable legal frameworks. Where there is communal control of forest resources, revenues can reach the poor, and there is evidence that forest cover is thereby maintained (Chhatre and Agrawal, 2009). Where the rule of law is weak and there is limited public accountability, REDD payments are likely to increase corruption and be annexed by those in power (Sandbrook *et al.*, 2010).

Whatever the policy appeal of the economic logic of PES, the political ecology of REDD suggests the potential for significant, intractable conflict. Thus in Cambodia, Milne and Adams (2012) examine a series of village-level PES schemes. The PES model that was adopted drew on simplistic concepts of 'community' and the possibility of 'community choice'. However, forest communities were fractured, and complex intracommunity dynamics were disguised and ignored in project development. The idea of organic communities with shared economic interests allowed planners to eliminate the need for wider participatory or deliberative processes in conservation. The market-inspired idea of 'community choice' denied local agency and silenced local voices (Milne and Adams, 2012).

Conflict over conservation ideas

Many conservation conflicts are underpinned by differences in ideas about nature, so that an apparent conflict over something material (access to a forest) can result from conflict over an idea about how the forest *should* be. Conservation problems commonly involve trade-offs, between biodiversity and development, and between the interest of one group of people and another (Hirsch *et al.*, 2011). Making decisions in such situations is highly problematic because different actors may make different assumptions and have different kinds of knowledge and understanding of the problem (Adams *et al.*, 2003). A conservation planner may be concerned about the loss of a rare mammal, but a hunter about food.

Ideas about the state of nature, or its importance, can be understood as policy narratives or discourses. Policy narratives become entrenched in the minds of researchers, scientists and recognised policy 'experts', and are supported by aid funding, and disseminated locally as community leaders learn what to say to get assistance (Leach and Mearns, 1996).

An example of the power of ideas underpinning conservation is the influence of ideas about nature in colonial Africa in the first half of the twentieth century. Neumann (2002) argues that a powerful 'Anglo-American nature aesthetic' underpinned the development of conservation in colonial Africa. This was built around two concepts: 'wilderness', as a place without people, and 'the picturesque', a pastoral landscape where people lived but in ways that did not damage the harmonies of nature. Africa was understood as 'a lost Eden in need of protection and preservation' (Neumann, 1998: 80).

Neumann (2004) traces how, far from preserving 'wilderness', conservation actually created it in colonial Tanzania. He analyses the establishment of the Selous Game Reserve under British colonial rule in the 1930s (Neumann, 2004). Problems of sleeping sickness and crop raiding by elephants *Loxodonta africana* led the colonial government to order a mass relocation of 40,000 people from the area. Elephant control was stopped inland, and intensified in the East, to make people move nearer the coast where denser settlement would reduce numbers of tsetse fly (Neumann, 2004: 207). This attempt to separate people and nature in the interests of agricultural production and disease control created the empty landscapes of the Selous, which became one of the largest protected areas in the world.

Ideas like wilderness remain powerful and exercise a continuing appeal to conservationists. In identifying key emerging issues in political ecology, Peet *et al.* (2011: 37) note the 'importance and stubborn pervasiveness of environmental representations'. These are fundamental to many conservation conflicts.

Conclusion: conservation as anti-politics

What can political ecology contribute to the management of conservation conflicts? Like most academic disciplines, it is much better fitted for understanding problems than solving them, and any discipline that tells ecologists (and political economists) to think much more broadly to understand this problem is especially vulnerable to that failing. Nonetheless, its breadth offers something that is potentially useful. Leopold (1949) observed 'the first rule of intelligent meddling is to keep all the pieces'; he was thinking about ecosystems and species. However, the same is true of conflict: if you do not have all the pieces, how can you hope to manage it? Political ecology tries to identify the pieces and work out how they fit together.

Above all, political ecology brings the notion of politics centre stage. I started this chapter by arguing that conservation is always and everywhere political because choices have to be made. I will finish by noting that one of the most subtle dimensions of its politicisation is what Ferguson (1994) calls 'anti-politics'. He analysed development policy in Lesotho, arguing that technical analyses of development projects and programmes depoliticised these interventions. Development was seen to be a technical, rather than political, problem. Yet behind this curtain of anti-politics, Ferguson (1994) argued that policies extended the power of government against the interest of the poor. Büscher (2010) describes the same tendency in conservation planning, particularly trans-boundary conservation in southern Africa. Political questions are presented as merely technical, and therefore to be answered by experts, including ecologists and economists promoting 'market-based' policies. Where there is a need for open political debate over these policies, and the winners and losers they create, there is none.

Büscher (2010) worries about conservation's 'anti-political tendencies'. Looking for and understanding the politics of conservation conflicts is likely to be the best chance to resolve them. Hiding the politics, or hiding from the politics, is a poor strategy for conflict resolution.

References

Adams, W. M. (2004). *Against Extinction, The Story of Conservation*. London: Earthscan.

Adams, W. M. and Hutton, J. (2007). 'People, parks and poverty, political ecology and biodiversity conservation. *Conserv. Soc.*, 5, 147–183.

Adams, W. M., Brockington, D., Dyson, J. and Vira, B. (2003). Managing tragedies, understanding conflict over common pool resources. *Science*, 302, 1915–1916.

Angelsen, A. (2008). *Moving Ahead with REDD, Issues, Options and Implications*. Bogor, Indonesia: Center for International Forestry Research.

Blaikie, P. (1985). *The Political Economy of Soil Erosion in Developing Countries*. London: Longman.

Blaikie, P. and Brookfield, H. (1987). *Land Degradation and Society*. London/New York: Methuen.

Brechin S. R., Wilhusen, P. R., Fortwangler, C. L. and West, P. C. (2003). *Contested Nature, Promoting International Biodiversity with Social Justice in the Twenty-first Century*. New York: State University of New York Press.

Brockington, D. (2002). *Fortress Conservation, The Preservation of the Mkomazi Game Reserve, Tanzania*. Oxford: James Currey.

Brockington, D., Duffy, R. and Igoe, J. (2008). *Nature Unbound, Conservation, Capitalism and the Future of Protected Areas*. London: Earthscan.

Brown, D., Seymour, F. and Peskett, L. (2008). How do we achieve REDD co-benefits and avoid doing harm? In *Moving Ahead with REDD, Issues, Options and Implications*, ed. A. Angelsen, pp. 107–118. Bogor, Indonesia: Center for International Forestry Research.

Bryant, R. L. (2002). Non-governmental organizations and governmentality, 'consuming' biodiversity and indigenous people in the Philippines. *Polit. Stud.*, 50, 268–292.

Büscher, B. (2008). Conservation, neoliberalism, and social science, a critical reflection on the SCB 2007 Annual Meeting. *Conserv. Biol.*, 22, 229–231.

Büscher, B. (2010). Anti-politics as political strategy. *Devel. Change*, 41, 29–51.

Castree, N. (2008). Neoliberalising nature, the logics of deregulation and reregulation. *Envir. Plann. A*, 40, 131–152.

Chape, S., Harrison, J., Spalding, M. and Lysenko, I. (2005). Measuring the extent and effectiveness of protected areas as an indicator for meeting global biodiversity target. *Phil. Trans. R. Soc. B*, 360, 443–455.

Chhatre, A. and Agrawal, A. (2009). Trade-offs and synergies between carbon storage and livelihood benefits from forest commons. *Proc. Natl Acad. Sci. USA*, 106, 17667–17670.

Colchester, M. (2002). *Salvaging Nature, Indigenous Peoples, Protected Areas and Biodiversity Conservation*. Montevideo: World Rainforest Movement.

Eltringham, S. K. (1994). Can wildlife pay its way? *Oryx*, 28, 163–168.

Emerton, L. (2001). The nature of benefits and the benefits of nature, why wildlife conservation has not economically benefited communities in Africa. In *African Wildlife and Livelihoods, The Promise and Performance of Community Conservation*, eds. D. Hulme and M. Murphree, pp. 208–226. London: James Currey.

Fairhead, J. and Leach, M. (2003). *Science, Society and Power, Environmental Knowledge and Policy in West Africa and the Caribbean*. Cambridge: Cambridge University Press.

Fazey, I., Pettorelli, N., Kenter, J., Wagatora, D. and Schuett, D. (2011). Maladaptive trajectories of change in Makira, Solomon Islands. *Global Environ. Chang.*, 21, 1275–1289.

Ferguson, J. (1994). *The Anti-Politics Machine, Development, Depoliticization and Bureaucratic Power in Lesotho*. Minneapolis: University of Minnesota Press.

Foucault, M. (1975). *Discipline and Punish, The Birth of the Prison*. Paris: Gallimard.

Ghimire, K. and Pimbert, M. (1997). *Social Change and Conservation, Environmental Politics and Impacts of National Parks and Protected Areas*. London: Earthscan.

Harvey, D. (2005). *A Brief History of Neoliberalism*. Oxford: Oxford University Press.

Hinchliffe, S. (2005). Reconstituting nature conservation, towards a careful political ecology. *Geoforum*, 39, 88–97.

Hirsch, P. D., Adams, W. M., Brosius, J. P., Zia, A. M., Bariola, N. and Dammert, J. L. (2011). Acknowledging conservation trade-offs and embracing complexity. *Conserv. Biol.*, 25, 259–264.

Igoe, J. and Brockington, D. (2007). Neoliberal conservation, a brief introduction. *Conserv. Soc.*, 45, 432–449.

Leach, M. and Mearns, R. (1996). *The Lie of the Land, Challenging Received Wisdoms on the African Environment*. Oxford: James Currey.

Lebel, L., Garden, P. and Imamura, M. (2005). The politics of scale, position, and place in the governance of water resources in the Mekong region. *Ecol. Soc.*, 10, 18.

Leopold, A. (1949). *A Sand County Almanac: And Sketches Here and There*. New York: Oxford University Press.

Meine, C., Soulé, M. and Noss, R. E. (2006). 'A mission-driven discipline': the growth of conservation biology. *Conserv. Biol.*, 20, 631–651.

Millennium Ecosystem Assessment (2003). *Ecosystems and Human Well-being, A Framework for Assessment*. Washington: Island Press.

Milne, S. A. and Adams, W. M. (2012). Market masquerades, uncovering the politics of community-level payments for environmental services in Cambodia. *Dev. Change*, 43, 133–158.

Neumann, R. P. (1998). *Imposing Wilderness: Struggles over Livelihood and Nature Preservation in Africa*. Berkeley, CA: University of California Press.

Neumann, R. P. (2002). The postwar conservation boom in British colonial Africa. *Environ. Hist.*, 7, 22–47.

Neumann, R. P. (2004). Nature-state-territory, towards a critical theorization of conservation enclosures. In *Liberation Ecologies, Environment, Development, Social Movements*, eds. R. Peet and M. Watts, pp. 195–217. London: Routledge.

Neumann, R. P. (2005). *Making Political Ecology*. London: Hodder Arnold.

Neumann, R. P. (2008). Probing the (in)compatibilities of social theory and policy relevance in Piers Blaikie's political ecology. *Geoforum*, 39, 728–735.

Peck, J. and Tickell, A. (2002). Neoliberalizing space. *Antipode*, 34, 380–404.

Peet, R. and Watts, M. (eds.). (1996). *Liberation Ecologies: Environment, Development, Social Movements*. London: Routledge.

Peet, R., Robbins, P. and Watts, M. J. (2011). Global nature. In *Global Political Ecology*, eds. R. Peet, P. Robbins and M. J. Watts, pp. 1–47. London: Routledge.

Peluso, N. (1993). Coercing conservation, the politics of state resource control. *Global Environ. Chang.*, 3, 199–217.

Pretty, J. (2002). People, livelihoods and collective action in biodiversity management. In *Biodiversity, Sustainability and Human Communities, Protecting Beyond the Protected*, eds. T. O'Riordan and S. Stoll-Kleeman, pp. 61–86. Cambridge: Cambridge University Press.

Pullin, A. S. and Knight, T. M. (2001). Effectiveness in conservation practice. Pointers from medicine and public health. *Conserv. Biol.*, 15, 50–54.

Rangan, H. and Kull, C. A. (2009). What makes ecology 'political'? Rethinking scale in political ecology. *Prog. Hum. Geog.*, 33, 28–45.

Ravenel, R. M. and Redford, K. H. (2005). Understanding IUCN Protected Area categories. *Nat. Area. J.*, 25, 381–389.

Ribot, J., Agrawal, A. and Larson, A. (2006). Recentralizing while decentralizing: how national governments reappropriate forest resources. *World Dev.*, 34, 1864–1886.

Robbins, P. (2004). *Political Ecology: A Critical Introduction*. Malden: Blackwell Publishing.

Robbins, P. and Monroe-Bishop, K. (2008). There and back again, epiphany, disillusionment and rediscovery in political ecology. *Geoforum*, 39, 747–755.

Sandbrook, C., Nelson, F., Adams, W. M. and Agrawal, A. (2010). Carbon, forests and the REDD paradox. *Oryx*, 44, 330–334.

Spinage, C. (1998). Social change and conservation misrepresentation in Africa. *Oryx*, 32, 265–276.

Stott, P. and Sullivan, S. (2000). *Political Ecology, Science, Myth and Power*. London: Arnold.

Walker, P. A. (2005). Political ecology, where is the ecology? *Prog. Hum. Geog.*, 29, 79–82.

Watts, M. and Peet, R. (2004). Liberating political ecology. In *Liberation Ecologies, Environment, Development, Social Movements*, second edition, eds. R. Peet and M. Watts, pp. 3–43. London: Routledge.

Wolf, E. (1972). Ownership and political ecology. *Anthropol. Quart.*, 45, 201–205.

Woodroffe, R., Thirgood, S. and Rabinowitz, A. (2005). *People and Wildlife, Conflict or Coexistence?* Cambridge: Cambridge University Press.

Wunder, S. (2008). Payments for environmental services and the poor, concepts and preliminary evidence. *Environ. Dev. Econ.*, 13, 279–297.

Zimmerer, K. S. and Bassett, T. J. (2003). *Political Ecology, An Integrative Approach to Geography and Environment-Development Studies*. New York: Guilford.

Box 5

Conservation, culture, kids and cash crops in the Solomon Islands

Jasper O. Kenter and Ioan Fazey
School of Environment/CECHR University of Dundee, UK

The Solomon Islands (SI) are globally significant biologically and culturally (Lamoreux *et al.*, 2006), yet remain poorly studied. This case study explains the conflicts and complex dynamics between nature conservation, local culture, population growth and economic development in Kahua, a remote region of the SI with around 5000 inhabitants.

In most of SI, commercial logging has led to devastating environmental impacts and severe social disputes over land rights. Kahua communities have resisted commercial logging thus some intact coastal rainforest remains (Bayliss-Smith *et al.*, 2003). However, Kahua forests are under increasing threat from population growth, cash-cropping and mining companies. Local people are dependent on forests for ecosystems services including clean water, shifting agriculture and wild foods and for cultural services such as materials for traditional dwellings and locations of ancestral remains. Cash-cropping through copra (dried coconut) and cocoa provide the main sources of income. This has expanded in recent years through support by SI government and aid agencies which has increased money flowing into communities.

In 2005 community leaders formed the Kahua Association (KA) with the aim of addressing emerging tensions and conflicts within and between communities about resource use and to find more sustainable forms of development. The KA worked with external researchers, including large scale interdisciplinary, participatory action research (PAR) co-managed by the KE and researchers. Community workshops and focus groups led by local facilitators/research assistants (RAs) were used to bridge literacy, language and culture issues and encourage learning at all levels (Fazey *et al.*, 2010). These participatory methods resulted in conceptual models of the complex social-ecological dynamics (Fazey *et al.*, 2010). Novel group-based deliberative approaches to valuing ecosystem services focused on understanding the impacts of cash crops (Kenter *et al.*, 2011) and to engage participants with scenarios involving mining and conservation. Researchers and local assistants also surveyed biodiversity and produced a local language bird guide, which was only the third book to appear in the Kahua language. Overall, the focus of these projects has been to: (1) understand complex dynamics of change in Kahua; (2) build capacity; and (3) build awareness around conservation and sustainability.

Key dynamics relate to desire for prosperity, money to communities, the sociological differences between traditional subsistence and cash cropping, land rights, and population growth. Many motives drive desire for prosperity (e.g. fees, travel, domestic housing). Cash crops are important for income, but displace subsistence activities away from more fertile soils. Combined with population growth and demand for subsistence foods, the amount of food per capita produced is declining, with increasing demand for imported, less nutritious foods. At certain times of the year people experience hunger, which is the result of uncertainty in global markets, reliance on only two cash crops, high cost of imported food, lack of financial education, and seasonality in cash crop harvests and its revenues.

'Sharing' is a principle of Kahua culture, along with asking for permission, care and dialogue. Socially, traditional cropping was a collective affair where people shared the fruits of their labour. However, cash-cropping and monetary economies have increased individualism, leading to disputes over money and land and reducing trust (Fazey *et al.*, 2010). Environmental degradation has also resulted from village extension and cash-cropping (Garonna *et al.*, 2009), which is impacting ecosystem services. For example, in some places people now spend considerable time finding building materials and potable water. Accumulating money, therefore, comes at a cost to natural, social and cultural capital (Fazey *et al.*, 2010).

Research co-managed by the KA has increased local capacity to understand these dynamics. Over 30 RAs have been trained, some of whom have become active in the KA, and community leaders actively seek sustainable modes of development that fit better with local culture than increased cash-cropping. Also, the research projects, which have included around 150 discussion groups of 5–20 people, have substantially raised awareness of the value of conservation. Enthusiasm by community leaders initially led to an EU-sponsored micro project that included work on soil fertility, water source protection, and a Ngali-nut harvesting cooperative, but this proved too ambitious. Although there were some positive outcomes, distrust around money management led to most activities being abandoned (Schuett and Fazey, 2010). Now, KA leaders have decided to focus on small-scale conservation demonstration projects. While these improve forest management practices, they do not resolve conflicts between forest conservation and economic and social interests. Research has helped communities analyse problems and consider options; however, more practical support that is culturally sensitive is needed than can be given by the irregular visits of researchers. Possible options include community conservation agreements, where communities accept conservation obligations but are supported by community development projects, such as improved medical or educational facilities, or communally organised forms of conservation-sensitive exploitation (e.g. production of rattan furniture).

References

Bayliss-Smith, T., Hviding, E. and Whitmore, T. (2003). Rainforest composition and histories of human disturbance in Solomon Islands. *Ambio*, 32, 346–352.

Fazey, I., *et al.* (2010). A three-tiered approach to participatory vulnerability assessment in the Solomon Islands. *Global Environ. Chang.*, 20, 713–728.

Garonna, I., Fazey, I., Brown, M.E. and Pettorelli, N. (2009). Rapid primary productivity changes in one of the last coastal rainforests: the case of Kahua, Solomon Islands. *Environ. Conserv.*, 36, 1–8.

Kenter, J. O., Hyde, T., Christie, M. and Fazey, I. (2011). The importance of deliberation in valuing ecosystem services in developing countries – evidence from the Solomon Islands. *Global Environ. Chang.*, 21, 505–521.

Lamoreux, J. F., *et al.* (2006). Global tests of biodiversity concordance and the importance of endemism. *Nature*, 440, 212–214.

Schuett, D. and Fazey, I. (2010). *Evaluation of the Kahua and European Union Forest Resource Management and Conservation Project*. University of St Andrews.

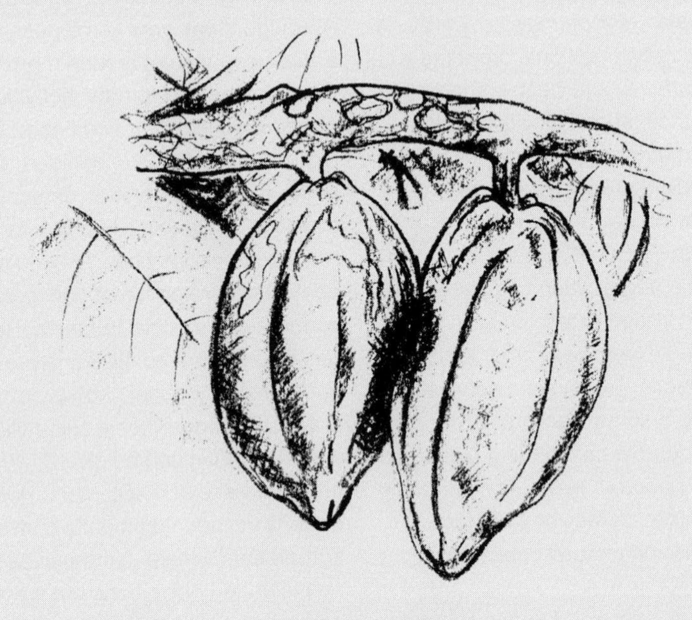

© Adam Vanbergen.

CHAPTER SIX

Understanding conservation conflicts: an economic perspective

NICK HANLEY
University of Stirling

Economics is concerned with understanding human behaviour and with analysing the consequences of the interactions between economic 'agents', be they firms, consumers, bird watchers or trade unions. As a field of study, it is interested in modelling how agents respond to changes in incentives, and how institutions like markets and firms help us manage competing wants given our limited resources. Economics has developed a body of theory and empirical methods which can be applied to analyse the consequences of policy-making on both individual and social well-being. As such, it offers an interesting insight into conflicts. Economics can help us understand why conservation conflicts occur, through measuring the benefits and costs to different parties, and showing how identification of these values can help mitigate conflicts by determining appropriate compensation for losses. Here, I describe the potential contributions that economics can make to conflict management before presenting two case studies to provide some illustration.

Understanding why conservation conflicts occur

Economists would point to three interrelated issues that help explain why conservation conflicts occur. These are incentives, property rights and market failures.

Incentives

Sporting estate managers in Scotland, farmers in African villages and forest owners in Malaysia all respond to incentives of various types. Profit incentives can drive land uses which are detrimental to conservation interests. For example, changes in the price of different outputs from land, such as timber and corn, alter the production plans of farmers, increasing deforestation in Central America (Barbier and Burgess, 1996). Increases in ivory prices heighten incentives for elephant poaching in Kenya, while rising bushmeat prices can encourage local people to allocate more time to illegal hunting in the Serengeti (Campbell *et al.*, 2001). Profits from palm oil production increase forest clearance for replanting

Conflicts in Conservation: Navigating Towards Solutions, ed. S. M. Redpath, R. J. Gutiérrez, K. A. Wood and J. C. Young. Published by Cambridge University Press.

with palm oil trees, with consequent negative effects on biodiversity in peat swamps in SE Asia (Jaenicke *et al.*, 2008). High prices for agricultural crops under the Common Agricultural Policy in the 1970s and 1980s drove intensification of land management in Europe, with resultant losses in farmland species and landscape quality (Bowers and Cheshire, 1983).

Changes to incentives can also produce beneficial actions to conservation. For example, Payment for Ecosystem Service (PES) schemes which reward less-intensive moorland management to benefit water supply companies in England can also benefit moorland birds and plants (Bain *et al.*, 2011). Paying for carbon sequestration in tropical forests through REDD+ payments can reduce deforestation pressures in Indonesia (Busch, 2013; but see Chapter 5).

While many other factors help explain people's actions with a bearing on conservation (including social conventions and institutions), incentives certainly matter. Such incentives may be deliberately established, in that affecting ecosystem quality or biodiversity is their principal aim – such as payments within a PES scheme – or can be incidental, whereby incentives designed for a different purpose, such as the support of farm incomes, turn out to have indirect effects on environmental outcomes. When these indirect effects are undesirable, it is common to refer to such incentives as 'perverse' (TEEB, 2009).

Property rights

Property right arrangements can lead to unsustainable environmental pressures. The most famous example is Hardin's (1968) *Tragedy of the Commons*, where he pictures an unregulated grazing area where many farmers have access to this land for grazing their cattle. As cattle numbers increase, the quality of the grazing land declines. While each individual farmer faces the full costs of actions to reduce grazing pressures (because one fewer cow grazed by Farmer Jones means he loses all the income associated with that cow), the benefits of reduced grazing by Farmer Jones are available to all. Everyone else has the incentive to increase their own stocking rates as Farmer Jones cuts his. As Farmer Jones knows this, he chooses not to reduce his own stocking rate. All farmers thus behave in this strategic manner, and land is degraded to the point of zero or negative profits.

This pattern also applies to open-access resources, access to and use of which is not regulated or managed by formal or informal institutions. The over-use of deep sea fisheries beyond the reach of national laws is an example of problems related to open access. This open access encourages too much 'effort' (too many boats, too much gear, too many hours at sea fishing) from the viewpoint of the management of the fishery to maximise net social benefits because a decision to reduce effort on the part of one fisherman leaves more fish for others to catch. As each individual cannot guarantee that they will benefit from reducing their individual effort, none does. Too much effort is applied, profits fall to zero, and in many cases the fishery is pushed to a point of population collapse (Smith, 2012).

The distinction between open-access resources and common property resources is important. In the latter, while many potentially have access to a resource, rules and customs develop to regulate this access for the collective good (Ostrom, 1990). A good example would be common grazings in England. Here, individuals in a collective have well-defined access rights to an area of land to use it in a particular fashion, even though they do not own it. However, if the goal of the collective is to maximise joint income, there is no guarantee this goal will not result in ecological damage. Indeed, there is evidence of such damage by over-grazing on common land in the UK (Hanley *et al.*, 1998).

Market failures

Economists define market failure as a situation where the operation of market forces leads to an outcome that does not serve society's best interest (Hanley *et al.*, 2007). Three types of market failure are relevant to conservation conflicts. The first relates to public goods. A public good is a good – such as clean air or a beautiful view – whose consumption is *non-excludable* (if it is provided for one, then it is provided for all, irrespective of whether they have paid for it), and *non-rival* (the benefits obtained from it do not depend on the number of users or people who benefit). Biodiversity is an example of a public good, and shows how the existence of public goods creates problems for markets. A forest owner who decides to leave old trees untouched for the benefit of insects and birds is reducing biodiversity loss associated with logging (see Box 5). Yet the market will typically not reward the owner for this. Although many conservationists enjoy the birds which such an action allows to thrive, the market offers no economic incentive for this pro-conservation action. Because no one can be excluded from the knowledge that biodiversity has been enhanced, the forest owner finds it impossible to charge for the extra biodiversity his actions have resulted in. However, the market will reward the owner for more timber production. Incentives are thus skewed towards the production of market-valued goods (timber) and away from public goods (biodiversity) (Fig. 6.1).

Landowners and land managers might try to develop markets for some environmental benefits flowing from their land (e.g. by charging hill walkers in Scotland to park their cars or by charging eco-tourists in Kenya for wildlife safaris), but the size of benefits which can be captured in such ways will always be less than the total value to society of the public good. The market thus provides incentives which are skewed from the viewpoint of social benefit.

Externalities are another relevant market failure. An *externality* arises whenever one agent's actions impact on the well-being of another, without a price being

Figure 6.1 Managing land for biodiversity or livestock grazing can result in a dramatic effect on land cover and ecological quality in the UK uplands. The bird reserve to the left of the fence maintains a diverse heather moorland plant community. The farmer who manages the land to the right of the fence maintains a species-poor grassland community, but the grass cover allows increased revenues from livestock farming.

paid or received for this impact. Externalities are the unintended side effects of production and consumption. If a farmer applies fertiliser to increase crop growth, but this then contributes to eutrophication in a nearby lake, then the farmer's actions impose an external cost on those who use the lake for recreation because water quality is diminished. However, the lack of private property rights over water quality means that the farmer pays no environmental price for this effect (for the cost they impose on recreationalists), and thus has an insufficient incentive to reduce fertiliser use for environmental reasons. Similarly, when a sporting estate allows an increase in its red deer population, it may result in increased forest damage on neighbouring estates, which go uncompensated (Hanley and Sumner, 1995). All of these are example of externalities leading to conservation conflict.

Finally, economists see *information failures* as a type of market failure that is important for explaining the emergence of conflicts. Such failures can be of two types. In the case of *moral hazard*, governments and agencies find it very costly (or impossible) to observe the actions of agents whose behaviour they are trying to influence. This means that agents no longer face the full costs of their actions. For instance, conservation groups have complained for decades about illegal persecution of raptors on grouse moors in the UK (Thirgood and Redpath, 2008; Box 3). Because it is hard for authorities to observe the actions of estate staff over large areas of moorland, this reduces the expected costs of poisoning, nest destruction and egg removal, and so increases the extent of illegal persecution (see the first case study below). A second type of information failure is *adverse selection*. Here, the government finds it hard or very costly to discover the 'type' of each agent whose behaviour they seek to influence. For instance, in EU agri-environmental schemes, farmers are offered payments for contracts to undertake a series of management actions. The costs of these actions, such as reducing stocking rates or fertiliser applications, vary among farmers, but this is private information not available to the regulator (Hanley *et al.*, 2012a). This means that many farmers get over-compensated for their actions, which raises the costs of achieving conservation policy objectives, and gives rise to the idea of land managers being paid 'money for nothing' if they do not have to change their current actions to meet the terms of the contract.

Measuring the gains and losses from conservation actions

Conservation conflicts often involve instances where those who benefit from a pro-conservation action, such as the designation of marine protected areas (see Box 6), are different from those who incur costs as a result of such actions. The costs imposed by conservation typically include *opportunity costs*. Opportunity costs are the highest-valued benefits forgone from using a scarce resource in a particular way. For instance, if designating an area of coastal waters as a marine protected area involves banning commercial fishing, then fishermen will be adversely affected because of lost income from fishing. This is the opportunity cost of designating the reserve from the viewpoint of the fishermen. Those who like to observe marine biodiversity while scuba diving may benefit from the creation of marine protected areas, but because those who benefit are different to those who lose out, a conflict can occur over proposals for such a designation unless fishermen are compensated for their losses.

The costs of conservation actions can include other kinds of direct and indirect impacts on people (Barua *et al.*, 2013). For example, establishing a protected area for elephant or lion conservation in Africa could result in increased crop and livestock losses to people living around the protected area (see Box 2). Yet the beneficiaries are likely to be a different set of agents (e.g. safari company employees and owners, park visitors). Revenue-sharing from park use could be an important device to compensate losers in such cases (Bush *et al.*, 2009).

The framework of cost–benefit analysis provides a systematic way of setting out not only who gains and who loses from a particular action, such as designating marine protected areas or creating a new tiger reserve in India, but also how much they gain or lose. The method involves expressing gains and losses in comparable, commensurable terms in monetary equivalents (Krutilla, 2005). Through measurement of the size of losses to different parties, and at different points in time, conflicts can be eased through the design of compensation schemes, funded either by beneficiaries or by the state. The 'Kaldor–Hicks Potential Compensation Test' which underlies the cost–benefit analysis criterion only approves projects or policies where the sum of gains is large enough to compensate the sum of all losses. However, it is important to note that its application only asks about the potential for compensation of losers, rather than requiring such compensation actually to take place.

Better communication of the costs of conservation actions to some parties can improve our appreciation of the likely reaction of these parties. For example, the Endangered Species Act in the US creates an expected cost for private landowners if designated species are found on their land when it is designated as 'critical habitat', because the Act has the potential to impose constraints on land use in such cases. This creates an incentive for landowners to oppose designation and to remove or destroy species on their land before designation occurs (Brown and Shogren, 1998). A similar situation occurred with the Wildlife and Countryside Act in the UK in the 1980 and 1990s. Landowners whose land included a designated 'Site of Special Scientific Interest' (SSSI) were issued a notification of 'potentially damaging operations' which might damage the ecological quality of the site. However, if a landowner wanted to carry out such an operation (such as draining a wetland, or re-seeding an ancient grassland with more modern grass species), then the Nature Conservancy Council (NCC) could only stop such an action in most cases by offering monetary compensation to the landowner to not carry out the operation. Due to budget constraints binding on the NCC, this system led to a loss of Sites of Special Scientific Interest because the NCC could not afford the payments needed to protect all of them. It also produced an incentive for landowners to threaten a potentially damaging operation in order to obtain payments because the NCC had no way of knowing if these threats were real (Spash and Simpson, 1993), a problem of hidden information.

In many cases, market values may not reveal the full costs of conservation to affected parties. For example, establishing protected areas in Africa to conserve endangered species might mean that local people cannot access such areas for fuelwood, building materials and bushmeat. Market prices could be used to measure aspects of these losses, such as the market value of fuelwood needed to substitute for what had been extracted from the protected area, but market prices would under-value the benefits of accessing such areas during times of temporary food shortages. Moreover, they would fail to reflect cultural values associated with bushmeat hunting and consumption. Economic theory tells us that minimum *willingness to accept compensation* (WTAC) is the correct measure of loss. This WTAC may well exceed the market value of losses due to conservation actions (Bush *et al.*, 2013). See the protected areas in Uganda case study below for more detail on this point.

How can economics help in the management of conservation conflicts?

Economics suggests at least two ways of helping to manage conservation conflicts: realigning incentives, and creating markets for ecosystem services.

Realigning incentives

As noted above, conservation often imposes costs on private landowners, and markets do not reward them for the biodiversity and ecosystem services such as water quality, carbon sequestration, recreation opportunities or landscape quality which their land 'produces'. Realigning incentives involves the government, or some other institution, providing a positive financial reward for environmental benefits and imposing a price (tax) on negative environmental impacts such as pollution. PES schemes work this way by rewarding land managers for environmental outputs such as water quality and biodiversity (Jack *et al.*, 2008; OECD, 2010). Such payments are offered by governments (e.g. in European and Australian agri-environmental programmes) and by the private sector (e.g. private water companies in England paying farmers to change land management in order to reduce total organic carbon run-off). Conservation organisations such as Ducks Unlimited offer positive payments to farmers in Canada to change land management in a way that improves habitat for waterfowl (Brown *et al.*, 2011). However, problems can exist in terms of 'end of contract' issues, whereby conservation gains created during a PES contract are lost at the end of the contract period; and where PES schemes operate more as income support mechanisms with little additional environmental benefit being produced by participants (Kleijn *et al.*, 2006; Acs *et al.*, 2010).

A good example of a PES scheme is provided by REDD (Reducing Emissions from Deforestation and Forest Degradation (Busch, 2013)). This scheme started in an attempt to create a financial incentive to reward developing countries for the carbon stored in forests. The scheme offers incentives to reduce deforestation and increase reforestation, and to help offset the opportunity costs of forest conversion to agriculture. The scheme was further developed to include forest conservation, sustainable management of forests and enhancement of forest carbon stocks (REDD+). The establishment of REDD+ trading in carbon credits will provide a financial incentive for forest managers to increase biodiversity on their land and reduce species loss (Busch, 2013).

Creating markets for ecosystem services

Markets for ecosystem services can be created through government action which sets a cap on biodiversity or wetland losses, and then allows landowners to trade entitlements within this cap. For example, if the government creates a rule for urban areas within a country of no net loss of wetlands, then farmers who can create new wetlands could offer credits based on their wetland creation to those who want to drain existing wetlands for housing (such arrangements operate in some states of the USA). A market in wetland credits is then established, with those creating new wetlands being the sellers, and those wishing to convert wetland to housing being the buyers. So long as ecological equivalence is somehow established between gains and losses, then a 'no net loss in wetland hectares' rule which allows this kind of market would be a cheaper way of achieving a wetlands conservation objective than an approach which did not allow trading or any loss of wetlands to development. Such markets provide an economic incentive to create new wetlands, and offset the costs of protecting existing wetlands, although there are clearly challenges in creating new habitats with ecological equivalence. Biodiversity offsetting is another example of such a market (Maron *et al.*, 2012; Womble and Doyle, 2012). Here, farmers or developers whose actions reduce the habitat of some species could buy 'credits' to offset these losses; credits are earned by other agents who create or restore enough 'equivalent hectares' of this habitat.

Ecosystem service markets can also arise in the absence of a government setting a cap, if those who are potential suppliers of such services can contract with potential buyers in exchange for some 'unit' of ecosystem service. An example is the emergence of voluntary markets for carbon credits from new forest planting in the UK. Sellers of credits are forestry companies agreeing to plant a certain area with trees and to manage those trees in a manner which 'delivers' some quantity of estimated additional CO_2 sequestration. Buyers are companies or individuals who wish to offset their CO_2 emissions from activities such as running a transport business or flying to Australia on holiday. Governments can help such markets grow by facilitating the establishment of codes of conduct, monitoring and verification schemes, as in the Woodland Carbon Code (FC, 2011).

Example case studies

Hen harriers on sporting estates in Scotland

In the case of the conflict around hen harriers (see Box 3), the interests of grouse moor owners, who would prefer fewer hen harriers, are at odds with those of conservation organisations, who would prefer more hen harriers. This is because an economic incentive exists for owners to reduce harrier numbers (more grouse can be shot, so people will pay more for a day's shooting); while moorland owners receive no economic benefits from higher hen harrier numbers, even though conservationists might be willing to pay for this (Hanley *et al.*, 2010).

There are several features of this conflict which relate to the earlier discussion in this chapter. Market failure is present because grouse moor managers cannot earn a financial return on much of the biodiversity that their grouse moors 'produce'. Indeed, maintaining hen harrier populations comes at a financial cost to them. Information problems linked to the costs of observing illegal persecution actions tip the financial incentives further towards persecution. Markets do not exist which charge those who enjoy seeing hen harriers for this benefit because of the public good nature of biodiversity.

Economists might suggest various ways of dealing with this conflict. One solution from an economic perspective would be to pay moorland owners an eco-bonus for each pair of harriers counted on their land, funded by the government who acts to collect funds from those who benefit – but also from all other taxpayers. Or, those who benefit from harriers could come together in clubs to offer such payments or to buy land and manage it for biodiversity rather than for grouse production. Economic techniques such as choice experiments could also help moorland owners discover how they could price alternative hunting experiences which offer lower numbers of grouse to be shot in return for more varied moorland landscapes in which shooting takes place (Bullock *et al.*, 1998; Hanley *et al.*, 2012b). Finally, cost–benefit analysis could be used to establish which parties benefit and which lose under current management actions and whether those who currently lose would be able to compensate moorland owners for income lost due to less-intensive management techniques.

Protected areas in Uganda

The establishment of protected areas (PA) such as national parks and game reserves is a common means of protecting biodiversity from habitat loss and hunting pressures in many developing countries. PAs enhance conservation by applying land-use restrictions such as banning bushmeat hunting, collecting fuelwood and restricting the conversion of land for agriculture. Unfortunately, properly enforced land-use restrictions impose potentially high (opportunity) costs on local resource users (Norton-Griffiths and Southey, 1995), who may be among the poorest. Communities adjacent to PAs in Uganda normally consume, exchange or sell timber and non-timber forest products sourced locally as part of their livelihood strategies. In national parks, however, legislation precludes the hunting of wild animals and the extraction of timber and other forest products. Without access to PA resources or an alternative source of revenue, many rural households face increased levels of impoverishment.

Exclusionary management practices in PAs tend to create conflicts between local people and authorities (Hulme and Murphee, 2001). Information on the costs of loss of access is a key indicator when determining how much effort must be deployed to mitigate the effect of loss of access to local people. Such information can be used to devise mechanisms that can help maintain the well-being of local residents whilst also securing conservation objectives. Bush *et al.* (2013) use *contingent valuation* to estimate the costs of PA designation to communities living around four PAs in Uganda. Contingent valuation is a technique for estimating individuals' minimum WTAC for changes in environmental entitlements. The authors found that the costs of conservation estimated in this way were considerably larger than costs estimated using market prices of forgone access benefits. The contingent valuation technique can also be used to explore how variations in socio-economic characteristics relate to variations in the costs of conservation, as measured by WTAC. Bush *et al.* (2013) found that household education level had a positive and significant correlation with WTAC, while households with a larger area of agricultural land to cultivate stated significantly higher WTAC amounts. Net annual total household income and asset variables did not have a statistically significant relationship with WTAC. However, income from the protected area had a positive and statistically significant impact on WTAC.

Efforts either to share the benefits of PAs, for example through revenue-sharing agreements between national park authorities and local communities (Lindsey *et al.*, 2006), or to provide community resources such as micro credit schemes or wells funded by national park revenue are ways of reducing conflicts. Revenue-sharing schemes could be used as aspects of the marketing strategy of national parks for foreign visitors, including big game hunters. Cost–benefit analysis could highlight the multiple impacts of PA designation on local households and show how these vary by distance or socio-economic group. Choice experiments, a technique closely related to contingent valuation, could be used to quantify the improvements in alternative livelihood options such as livestock, waged employment or cropping, which would be needed to offset specific losses in the benefits of access to PAs by local people (Moro *et al.*, 2013).

Conclusions

Economics provides ways of understanding conservation conflicts in terms of incentives, property rights allocations and market failures. Economics also offers tools for measuring the gains and losses of conservation actions to different groups or individuals and suggests approaches to help manage conflicts by realigning incentives and creating markets for ecosystem services.

While these insights are useful, they by no means constitute an adequate stand-alone explanation for the causes of conflict, for which we need to understand the behavioural motivations and institutions which underlie the current situation. Neither are they an adequate basis for managing all conflicts. However, as an economist, I believe that the ideas discussed here are too important for the conservation community and policy makers to ignore. For example, placing monetary values on changes in species abundance, distribution and endangered status, and valuing the landscapes within which species exist, has proved useful for analysing the impacts of future scenarios (for example, of changing agricultural product prices, or changes in agri-environmental schemes) on social benefits and costs, and how these are distributed across stakeholders, as well as through making an economic case for conservation (TEEB, 2009; United Kingdom National Ecosystems Assessment, 2011). Moreover, economics helps us understand why conservation conflicts emerge, and how to resolve them.

Acknowledgements

I thank the Crawford School of Public Policy, Australian National University, for hosting me during the writing of this chapter.

References

Acs, S., *et al.* (2010). The effect of decoupling on marginal agricultural systems: implications for farm incomes, land use and upland ecology. *Land Use Policy*, 27, 550–563.

Bain, C. G., *et al.* (2011). *IUCN UK Commission of Inquiry on Peatlands*. Edinburgh: IUCN UK Peatland Programme.

Barbier, E. B. and Burgess, J. (1996). Economic analysis of deforestation in Mexico. *Environ. Dev. Econ.*, 1, 203–239.

Barua, M., Bhagwat, S. A. and Jadhav, S. (2013). The hidden dimensions of human–wildlife conflict: health impacts, opportunity and transaction costs. *Biol. Conserv.*, 157, 309–316.

Bowers, J. and Cheshire, P. (1983). *Agriculture, the Countryside and Land Use*. London: Methuen and Co.

Brown, G. M. and Shogren, J. (1998). Economics of the Endangered Species Act. *J. Econ. Perspect.*, 12, 3–20.

Brown, L., Troutt, E., Edwards, C., Gray, B. and Hu, W. (2011). A uniform price auction for conservation easements in the Canadian Prairies. *Environ. Resour. Econ.*, 50, 49–60.

Bullock, C. H., Elston, D. A. and Chalmers, N. A. (1998). An application of economic choice experiments to a traditional land use – deer hunting and landscape change in the Scottish Highlands. *J. Environ. Manage.*, 52, 335–351.

Busch, J. (2013). Supplementing REDD+ with biodiversity payments: the paradox of paying for multiple ecosystem services. *Land Econ.*, 89, 655–675.

Bush, G., Colombo, S. and Hanley, N. (2009). Should all choices count? Using the cut-offs approach to edit responses in a choice experiment. *Environ. Resour. Econ.*, 44, 397–414.

Bush, G., Hanley, N., Moro, M. and Rondeau, D. (2013). Measuring the local opportunity costs of conservation: a provision point mechanism for willingness-to-accept estimates of hypothetical household loss of access to protected areas in Uganda. *Land Econ.*, 89, 490–513.

Campbell, K., Nelson, V. and Loibooki, M. (2001). *Sustainable Use of Wildland Resources: Economic, Ecological and Social Interactions*. Chatham, Kent: Natural Resources Institute.

FC (2011). *Woodland Carbon Code: Requirements for Voluntary Carbon Sequestration Projects*. Edinburgh: Forestry Commission.

Hanley, N. and Sumner, C. (1995). Bargaining over common property resources: applying the Coarse Theorem to red deer in the Scottish Highlands. *J. Environ. Manage.*, 43, 87–95.

Hanley, N., Kirkpatrick, H., Oglethorpe, D. and Simpson, I. (1998). Paying for public goods from agriculture: an application of the Provider Gets Principle to moorland

conservation in Shetland. *Land Econ.*, 74, 102–113.
Hanley, N., Shogren, J. and White, B. (2007). *Environmental Economics in Theory and Practice.* Second edition. Basingstoke: Palgrave MacMillan.
Hanley, N., Czajkowski, M., Hanley-Nickolls, R. and Redpath, S. (2010). Economic values of species management options in human–wildlife conflicts: hen harriers in Scotland. *Ecol. Econ.*, 70, 107–113.
Hanley, N., Banerjee, S., Lennox, G. and Armsworth, P. (2012a). How should we incentivise private landowners to produce more biodiversity? *Oxford Rev. Econ. Pol.*, 28, 93–113.
Hanley, N., Moro, M., Brennan, D., Redpath, S. and Irvine, J. (2012b). A choice experiment analysis of preferences for red grouse shooting in the Scottish uplands. *Discussion Papers in Economics.* Stirling: University of Stirling.
Hardin, G. (1968). The tragedy of the commons. *Science*, 162, 1243–1248.
Hulme, D. and Murphee, M. (2001). Community conservation in Africa. In *African Wildlife and Livelihoods; The Promise and Performance of Community Conservation*, ed. D. Hulme and M. Murphee, pp. 1–8. Portsmouth, NH: Heinemann.
Jack, B., Kousky, C. and Sims, K. (2008). Designing payments for ecosystem services: lessons from previous experience with incentive-based mechanisms. *Proc. Natl Acad. Sci. USA*, 105, 9465–9470.
Jaenicke, J., Rieley, J. O., Mott, C., Kimman, P. and Siegert, F. (2008).Determination of the amount of carbon stored in Indonesian peatlands. *Geoderma*, 147, 151–158.
Kleijn, D., *et al.* (2006). Mixed biodiversity benefits of agri-environment schemes in five European countries. *Ecol. Lett.*, 9, 243–254.
Krutilla, K. (2005). Using the Kaldor–Hicks Tableau format for cost–benefit analysis and policy evaluation. *J. Pol. Anal. Manag.*, 24, 864–875.
Lindsey, P. A., Frank, L. G., Alexander, R., Mathieson, A. and Romanach, S. S. (2006). Trophy hunting and conservation in Africa: problems and one potential solution. *Conserv. Biol.*, 21, 880–883.
Maron, M., *et al.* (2012). Faustian bargains? Restoration realities in the context of biodiversity offset policies. *Biol. Conserv.*, 155, 141–148.
Moro, M., *et al.* (2013). An investigation using the choice experiment method into options for reducing illegal bushmeat hunting in western Serengeti. *Conserv. Lett.*, 6, 37–45.
Norton-Griffiths, M. and Southey, C. (1995). The opportunity cost of biodiversity conservation in Kenya. *Ecol. Econ.*, 12, 125–139.
OECD (2010). *Paying for Biodiversity.* Paris: Organisation for Economic Cooperation and Development.
Ostrom, E. (1990). *Governing the Commons: The Evolution of Institutions for Collective Action.* Cambridge: Cambridge University Press.
Smith, M. D. (2012). The new fisheries economics: incentives across many margins. *Ann. Rev. Resour. Econ.*, 4, 379–402.
Spash C. and Simpson I. (1993). Protecting Sites of Special Scientific Interest: intrinsic and utilitarian values. *J. Environ. Manage.*, 39, 213–227.
TEEB (2009). *The Economics of Ecosystems and Biodiversity.* Nairobi: United Nations Environment Programme.
Thirgood, S. and Redpath, S. (2008). Hen harriers and red grouse: science, politics and human wildlife conflict. *J. Appl. Ecol.*, 45, 1550–1554.
United Kingdom National Ecosystems Assessment (2011). *Technical Report.* Cambridge: Cambridge University Press.
Womble, P. and Doyle, M. (2012). The geography of trading ecosystem services: a case study of wetlands and stream compensatory mitigation markets. *Harvard Law Rev.*, 36, 229–296.

Box 6

Conflicts in marine protected areas: the case of leatherback turtles

Cristina Pita[1] and Isidora Katara[2,3]
[1] *Centre for Environmental and Marine Studies, University of Aveiro, Portugal*
[2] *Dalhousie University, Department of Biology, Life Science Centre, Canada*
[3] *Instituto Português do Mar e da Atmosfera, Portugal*

Marine protected areas (MPAs) describe a wide range of marine areas (e.g. marine parks, reserves, sanctuaries) where human activity has been limited for conservation purposes. The establishment of MPAs has increased greatly in recent years, with MPAs being widely implemented for fisheries management and conservation, and some 5000 statutory MPAs existing globally (Wood, 2007).

Controversy and conflicts are associated with most MPAs due to the resource-use restrictions that such management tools impose. MPAs can generate or exacerbate conflicts between different resource users, as well as between users of the same resource, competing for a limited space (e.g. marine tourism versus fishing industry, between different fisheries operating in the same area) and between resource users and conservation interests (e.g. biodiversity conservation versus fisheries interests). Here we focus solely on the latter and provide examples of successful conflict resolution for the conservation of marine turtles.

Marine turtles are among the most iconic creatures inhabiting the oceans and their conservation can be described as a 'wicked' problem: a problem for which different stakeholders disagree on both the problem and its solution (Salwasser, 2004). Leatherback turtles (*Dermochelys coriacea*) have often been the focus of conflict due to their fragile conservation status. Egg poaching, coastal development, habitat degradation and climate change threaten leatherback nesting activities and hatchling survival, whereas juvenile and adult mortality can be partially attributed to incidental catch in fishing gear (e.g. Witherington and Frazer, 2003; James *et al.*, 2005).

Worldwide leatherback populations are collapsing, and it is estimated that half of the global population of leatherbacks has been lost since the 1980s (Spotila *et al.*, 2000). However, there are examples of successful initiatives to mitigate and/or resolve conflicts related to nesting sites and to interactions between turtles and fisheries. Conservation efforts began in the North Atlantic in the early 1990s, in the form of nesting site protection. Since then, the decrease of nesting females was halted with remarkable recoveries at many sites around the Caribbean. Along the beaches of Sandy Point National Wildlife Refuge, an MPA in the US Virgin Islands, we encounter an example of successful conservation. Egg and adult turtle harvesting had almost depleted the nesting populations, but since a recovery project began, the average nesting population increased by > 500% (Alexander *et al.*, 2004). Recovery results from the length of the project (> 30 years) and the profitable ecotourism activity that developed in the area, which served as an incentive to abandon the consumption of leatherbacks and enhanced public engagement in conservation (Tisdell and Wilson, 2002; Dutton *et al.*, 2005).

A good example of fishers' involvement in the conservation of leatherbacks comes from the USA. The US National Marine Fisheries Services (NMFS NOAA) in collaboration with the Blue Water Fishermen's Association[1] developed a framework, in 2001, to reduce sea turtle mortality associated with pelagic longline fishing. This effort was triggered by the closure of one of the most important fishing grounds for swordfish, the Northeast Distant Fishing Area, due to increased catch (over the allowable limit) of sea turtles (Witzell, 1999). In response, fishers, scientists and engineers collaborated to develop gear and tactics to minimise interactions between sea turtles and the fishery. This resulted in the implementation of several mitigation measures with the potential to reduce turtle captures by 90% (Watson *et al.*, 2005). This successful endeavor compelled policy makers to designate the proposed by-catch mitigation measures as mandatory for the entire US longline fishing fleet. Plus, NMFS launched a fishery-oriented education programme to disseminate the results of this research at a global scale.

Conflicts contribute to the high rate of MPA failure. These examples show that MPAs can be successful in resolving conservation conflicts affecting threatened species. However, an important element for the successful implementation of MPAs is the involvement of stakeholders from the early stages of planning and implementation. Stakeholder participation needs to be meaningful and decision-makers need to work closely with resource users, in an open and transparent way, in order for MPAs to be viewed as legitimate.

References

Alexander, J., Garrett, K., Garner, S., Conrad, J. and Coles, W. (2004). Tagging and nesting research of leatherback sea turtles (*Dermochelys coriacea*) on Sandy Point, St. Croix, U.S. Virgin Islands. WIMARCS Publication. 41 pp.

Dutton, D., Dutton, P., Chaloupka, M. and Boulon, R. (2005). Increase of Caribbean leatherback turtle (*Dermochelys coriacea*) nesting population linked to long-term nest protection. *Biol. Conserv.*, 126, 186–194.

James, M., Ottensmeyer, A. and Myers, R. (2005). Identification of high use habitat and threats to leatherback sea turtles in northern waters: new directions for conservation. *Ecol. Lett.*, 8, 195–201.

Salwasser, H. (2004). Confronting the implications of wicked problems: changes need in Sierra Nevada National Forest planning and problem solving. In *USDA Technical Report PSW-GTR-193*, eds. D. Murphy and P. Stine, pp. 7–22.

[1] A group of swordfish longline fishery stakeholders created with the aim of achieving a sustainable swordfish fishery and developing efficient management schemes.

Spotila, J., Reina, R., Steyermark, A., Plotkin, P. and Paladino, F. (2000). Pacific leatherback turtles face extinction. *Nature*, 405, 529–530.

Tisdell, C. and Wilson, C. (2002). Ecotourism for the survival of sea turtles and other wildlife. *Biodivers. Conserv.*, 11, 1521–1538.

Watson, J., Epperly, S., Shah, A. and Foster, D. (2005). Fishing methods to reduce sea turtle mortality associated with pelagic longlines. *Can. J. Fish. Aquat. Sci.*, 62, 965–981.

Witherington, B. and Frazer, N. (2003). Social and economic aspects of sea turtle conservation. In *The Biology of Sea Turtles*, eds. P. Lutz, J. Musick and J. Wyneken, pp. 356–378. Boca Raton: CRC Press.

Witzell, W. (1999). Distribution and relative abundance of sea turtles caught incidentally by the US pelagic longline fleet in the western North Atlantic Ocean, 1992–1995. *Fish. B NOAA*, 97, 200–211.

Wood, L. (2007). *MPA Global: A Database of the World's Marine Protected Areas*. Sea Around Us Project. UNEP-WCMC & WWF.

© Catherine Young.

CHAPTER SEVEN

Anthropological approaches to conservation conflicts

ANDREW WHITEHOUSE
University of Aberdeen

Conservation is a relational practice. Thus, it is commonly supposed that *nature* conservation is primarily concerned with nature, particularly nature in the sense of that which is external to human society. Following this model, which is somewhat redolent of the '*Platonic cave*' (i.e. the conceptualisation of society as existing in the political world of 'the cave' away from the external world of nature, which can only come to be known via the specialist techniques of science or philosophy; see Latour, 2004), conservation has been dominated by the natural sciences, particularly ecology, in assessing the work that needs doing and how best it might be achieved. The potential contribution of the social sciences has arguably been perceived as limited to providing insights into the 'human dimensions'; for example, by helping to incorporate the interests of 'local people' and other 'stakeholders' into the management plans that have already been developed by conservationists. They have dealt with the politics of 'the cave' and the natural scientists have ventured into the world of nature to find out what is going on there before returning to society to explain what needs to be done.

While this is an illustrative simplification, social science has largely been subservient to natural science in the development of conservation. The reasons for this, I argue, arise from conservation being conceptualised as the management of a detached nature that is understood by means of natural science. When one analyses the *practice* of conservation, it soon becomes clear that it is an activity that is less about people mastering a detached nature than about assessing how best to regulate human activities in relation to their environment. This shift in emphasis, from conservation as an understanding of nature and its management to conservation as primarily about human activity in relation to the environment, still demands the skills of natural scientists, but places the social sciences much more at the heart of conservationist endeavours. It is with this in mind that I make a case for the role that anthropology can play in understanding conservation and its conflicts, and in both critiquing and improving its practices. I also consider what anthropology can offer in rethinking what conservation might be and what it can achieve.

Conflicts in Conservation: Navigating Towards Solutions, ed. S. M. Redpath, R. J. Gutiérrez, K. A. Wood and J. C. Young. Published by Cambridge University Press.

What is anthropology and how can it contribute to understanding conservation?

Anthropology in its broadest sense is the study of humanity in all its diversity, including social and biological elements. As such, it encompasses both the natural and social sciences. It has long been conventional, although not necessarily ideal, to separate the discipline into biological and social divisions and also to detach these from archaeology. Here, I primarily discuss the role of *social* anthropology in addressing conservation conflicts, but one must also emphasise the growing affinity that social anthropology has with ecology. Indeed, now is an apposite time for anthropology to play a more prominent role in conservation because, over recent years, anthropology, in common with other social sciences, has become far more concerned with the environment and with human relations with non-humans. As such, anthropology has become an ecological discipline (see e.g. Milton, 1993, 1996; Descola and Palsson, 1996; Ellen and Fukui 1996; Ingold, 2000, 2011; Campbell, 2005).

A characteristic of traditional anthropology is its methodological approach, which is centred on long-term fieldwork and participant observation. While the discipline is actually rather diverse in its methods, a guiding principle has always been that to properly understand the complexity of human life, one needs to spend extended periods of time living alongside people. Anthropology is thus also an inherently collaborative discipline that has emphasised working *with* people rather than *on* people. This emphasis on fieldwork is something that anthropologists share with field scientists in conservation biology and ecology. The 'field science approach' is critical to understanding ways of life (whether human or non-human) because it is based on a wholeness of experience and a proximity to the world that provides an empirical basis for understanding and knowing, instead of being detached, reductive and under controlled conditions. As such, it is integral to understanding how different ways of life, both human and non-human, come into conflict, the effects of these conflicts and how they might be resolved.

There has been a long-standing anthropological interest in environmental and ecological questions, initially using approaches such as cultural ecology in which culture was explained as an adaptation to the environment (e.g. Steward, 1955; Rappaport, 1967; Harris, 1977). Recently, there has been a proliferation of alternative approaches to studying human–environmental relations, ranging from political ecology (e.g. Escobar, 1999) to ethnographic studies of environmentalists and the global dimensions of environmental issues (Milton, 1996, 2000; Berglund, 1998; Tsing, 2005; Gatt, 2011, 2013), to phenomenology (Tilley, 1994; Ingold, 2000, 2011), and to studies of biologists (Helmreich, 2009; Candea, 2010). I turn to some of the theoretical questions that have concerned ecological and environmental anthropology later, but will focus first on the ways in which a growing interest in environmental issues and conflicts has emerged in anthropological work.

In an early article on anthropology and conservation, Orlove and Brush (1996) outlined the ways they considered the discipline could contribute to conservation, particularly in relation to plants. They focused on how anthropologists could assist conservation, particularly by documenting local knowledge and practices in relation to plants and animals. Here there are parallels with applied work in development anthropology, in which anthropologists take up the practical role of bringing different forms of knowledge into dialogue. Indeed, anthropological studies of conservation have closely paralleled studies of economic development in terms of offering both assistance (e.g. Sillitoe, 1998) and critique (e.g. Hobart, 1993). This is not surprising given that conservation has emerged as a counterbalance to development, but, despite this apparent opposition, both are often perceived similarly by local communities, namely, as a set of values and knowledge practices imposed from outside (e.g. Arce and Long, 1993; Einarsson, 1993; Satterfield, 2002; Whitehouse, 2004, 2009, 2012). Because anthropology has traditionally been concerned with studies in small communities and with differences in knowledge and practice, it is perhaps not surprising that, just as with studies of development, an emphasis has often been placed on how particular 'indigenous' communities were affected by the actions of institutional conservation (e.g. Nygren, 2003; Theodossopoulos, 2003). Such approaches have stressed the incommensurability between knowledge characterised as local or traditional on the one hand, and knowledge considered to be scientific on the other, or between the understandings of local people and those of conservationists (Richards, 1992; Ellis, 2003; Novellino, 2003). Anthropologists have thus tended to be more concerned with how conservation impacts upon 'their people' rather than with working with conservationists themselves. Such approaches have been valuable in critiquing conservation, its logic and methods (Argyrou, 2005), particularly the 'fortress conservation' approach that has often led to the exclusion of local people from land that conservationists mistakenly imagined as 'pristine wilderness' or considered best conserved by removing people (Brockington, 2002). It also perhaps reflects a tendency in anthropology to explain how non-Western practices and ideas 'make sense' in a way that Western practices and ideas do not. Often there is an assumption that when it comes to relating to the environment in sustainable ways, non-Westerners have 'got it right' whereas the West has 'got it wrong'. For example, Berglund and Anderson (2003) focus on the myth-making of conservation, particularly the way it tends to naturalise or 'de-humanise' space in order for it to be classified as parks or reserves. While this criticism is important, it rests on rather vague and perhaps unduly homogeneous notions of what conservation involves and what conservationists themselves think and do. We have long needed more sophisticated accounts of what goes on when conservationists deal with other people.

Anthropological studies of conservation conflicts

A pioneering anthropological study of a conservation conflict was Satterfield's *Anatomy of a Conflict* (2002), which explored the contestations surrounding old growth forest and timber production in Oregon (see Chapter 11; Box 19). It was notable in that its focus was squarely on the conflict, the relations involved and the ideas and assumptions of the groups and individuals involved rather than on one side or the other. Satterfield was most concerned with understanding why the conflict was happening and how it unfolded, and as such it was a distinctly symmetrical account of a conflict and not of a group of people and how certain external factors, such as policies or designations, impacted upon them. She contextualises the conflict within a longer history of both forest management and environmentalist thinking in the region and examines how different interests and arguments are informed by symbols and ideals. She portrays what forests mean for different people and how they believe the land should, ideally, be. These symbols and ideals are often central to many conflicts, which are as much about meaning as about resources or gaps in understanding, and the sort of symmetrical ethnography developed by Satterfield is a powerful means to understand their emergence and deployment.

Whitehouse's (2004, 2009, 2012) work in Islay in the west of Scotland follows Satterfield's (2002) lead by focussing on the relations *between groups* rather than on one particular group and the external factors that impact upon them. This approach emphasises what different groups and individuals had in common rather than how the interests of 'outside' agencies and their representatives diverge from those of the core groups. Contestations are often underpinned by shared ideas and assumptions that enable different positions to be taken and interpreted by all parties. In the example of Islay, where conflicts emerged over the designation of sites for conservation by government (see Box 13) and the protection of geese that farmers regarded as agricultural pests (Box 7), it was clear that farmers and conservationists shared ideas about the relations between the island and the outside world. Both associated the outside world with change. Conservation was regarded as a recent arrival in Islay that brought with it new regulations and procedures as well as new people. The contestations that emerged between farmers and conservationists tended to relate to the difficulties in negotiating the changes that conservation was perceived to bring and that 'outsider' conservationists used to achieve their aims. Conflict was thus resolved in part through conservationists learning to appear more local and thus more negotiable through, for example, employing local people and developing co-management strategies. While different people and organisations positioned themselves differently in relation to one another, these *shared* assumptions about change and the outside world underpinned the conflict and its resolution.

An approach that emphasises gaps in understanding but that does so through an attention to the knowledge and practice of both indigenous peoples and conservationists is Wishart's (2004) account of a 'Story about a muskox' in northern Canada. The story centred on a Gwich'in elder who killed a protected muskox *Ovibos moschatus*, only to be apprehended by a wildlife officer who confiscated the meat. The elder protested that he was hunting in Gwich'in country and that meat was vital at a time of year when supplies were running low. The wildlife officer was adamant that muskoxen were protected and were part of an important experimental reintroduction. Although the dispute was resolved when the meat was returned to the Gwich'in, Wishart was interested in what the dispute revealed about the different ways that wildlife managers and Gwich'in conceived the land, its animals and how humans should behave in relation to them. For the wildlife managers, the muskoxen were an important component of an archetypal Arctic fauna, and so it was appropriate to reintroduce them to suitable areas. Their presence was also intended to provide good wildlife viewing opportunities for eco-tourists who visit the region in the hope of seeing and photographing charismatic animals. The Gwich'in, by contrast, considered the land and its animals to be sentient and so their relations with the animals they hunted were freighted with ideas of appropriate ways of behaving. For them, the 'experimenting' the wildlife managers were doing with the muskoxen was deeply troubling, as it was a form of interference that the animals were likely to find unacceptable. The idea that muskoxen should inhabit the area was also problematic because Gwich'in considered that they interfered with caribou *Rangifer taramdus* by eating their food and smelling bad. The caribou were also very sensitive to human intrusion, including visiting eco-tourists. The presence of muskoxen in the area thus represented an inappropriate breakdown in relations between humans and caribou. Here, Wishart considered what both Gwich'in and conservationists attend to and what they consider to be appropriate ways of behaving in relation to land and animals in order to bring about convivial ecological relations, an idea I return to later.

While anthropologists generally have not studied conservationists themselves, there are some notable exceptions (Vivanco, 2006). For example, Gatt (2011, 2013) describes the ways in which Friends of the Earth International is constituted through the actions of its members in different situations, from international meetings to online communities to local groups. This work is important because it provides us with insights into how environmental organisations work at a variety of levels, what motivates their members and activists and how their values and ideas intersect with organisational rhetoric. Anthropologists studying environmental issues often leave out or fail to properly contextualise organisations and activists, and Gatt (2011, 2013) provides a coherent approach to doing just this for global environmentalism.

Accounting for relationships that underlie conflict is consistent with a concern for place (Campbell, 2005; Ingold, 2005). Recent studies in environmental anthropology have focused more on the social and ecological relations that prevail within a particular place than on any particular bounded group of people. Through this has come an understanding of how people interact with places in diverse ways and how these places come to be contested. For example, Krause (2010) examined the varied relations people had with the Kemi River in northern Finland. When numerous dams were constructed to provide electricity, this displaced people, modified landscapes, affected the behaviour of wildlife and disrupted the rhythms of the river. The hydropower companies were criticised by many river dwellers, who described them as 'river robbers'. Krause's study was sensitive, however, not only to these grievances but also to the ways in which hydropower workers attended to the river and were themselves river dwellers. These conflicts were not simply about outside impositions on local communities, but emerged through practices that were always emplaced in one way or another.

Perhaps the most influential anthropological account of conservation conflict in recent times has been Tsing's (2005) *Friction: An Ethnography of Global Connection*. Indeed, this has been one of the most important texts in the whole of the discipline because of the innovative way that it describes and examines the threads that draw together the local and the global, as they emerge in conflicts over deforestation in Indonesia (see also Lowe, 2006). Contemporary environmental disputes are rarely entirely local, so Tsing asks how the global should be studied. Her approach draws on the metaphor of 'friction', which she defines as 'the grip of worldly encounter' (Tsing, 2005: 1). This conceptualisation counters prevalent assumptions that, in an era of globalisation, worldly movements, connections and encounters are the product of *flows*. Instead, different ideas, interests, species and ways of life rub up against one another in Tsing's (2005) account, sometimes 'combusting' and sometimes merging or entangling. By following these lines of connection as they emerge in Indonesia, she is able to show how a 'local issue' is in fact a manifestation of transnational knowledge, ideas and concerns that interact in complex and sometimes unexpected ways.

Rethinking conservation

The above discussion portrays some of the ways in which anthropologists have attempted to understand and critique conservation and its conflicts. In this section, I speculate on some of the ways that anthropology might be helpful for conservationists in order to rethink their aims and practices in ways that could temper conflicts. In particular, I explore how certain theoretical developments in ecological anthropology might be applied to the conceptualisation and practical work of conservation.

Over the past 20 years, anthropological theory has moved from approaches grounded in a dualism of nature and culture, or nature and society towards more relational ways of thinking (e.g. Ingold, 2000, 2011; Latour, 2004). This is not to say that the former have entirely disappeared, but it has become far harder for the division of the world into nature and culture to pass uncontested at least within anthropology and other social sciences. The question is: with what do we replace these concepts, which are at the heart of so much academic work, and which indeed underwrite the division of academic labour between the natural sciences on the one hand, and the arts, humanities and social sciences on the other?

Recent work has attempted to focus on understanding life, whether human or non-human, as emerging from an unfolding field of relations rather than as the enactment of a pre-ordained natural or cultural plan (Ingold, 2000). This has significance for conservation, in part because the breaking down of the nature–culture dualism makes it harder to conceive of the practice of conservation as involving the management of a detached, objectified and classified nature. Rather, it moves the focus back to the sorts of relations that emerge between humans and their environment and to how these are conducted. As mentioned earlier, conservation can be viewed as a means of regulating these relations, even though it is more often seen as being concerned with the management of a detached nature. The difference in emphasis is critical because it encourages us to think about conservation as primarily ethical and political (Latour, 2004). Such a shift in focus could help conservation to develop the kind of 'land ethic' that Aldo Leopold (1949) advocated over 60 years ago, but which has yet to be properly realised. Many peoples throughout the world place ethics at the centre of their interactions with their environment. For example, many hunter-gatherers understand the world to be sentient and as such great store is set by acting to maintain, or conserve, cordial and appropriate relations with non-humans (Bird-David, 1990; Howell, 1996; Ingold, 2000). Conservation might thus be re-conceived as conserving not 'biodiversity' in an objectified and neatly categorised 'nature', but convivial relations with the beings that co-inhabit our environment.

An approach to conservation focused on maintaining ethical relations would necessarily be centred on place and on the direct interactions that people have with their surroundings. In one sense this lies at the heart of what has come to be called 'community conservation' (see Campbell, 2005), in which the sort of anthropological concern with 'local knowledge and practices' that Orlove and Brush (1996) emphasised is added to scientific and technocratic practices, and in which local stakeholders are consulted about their interests. However, this still takes place within a discourse of conservation that is essentially global rather than place-centred. Local concerns are measured against the value of biodiversity that is understood in terms of a limited global space, rather than in terms of the unfolding and unbounded relations of place (Ingold, 2005). While place-based conservation could still heed the wider implications of local practices, the situational requirements of particular places with their own assemblages of humans and non-humans would be paramount. While global scarcity may remain a driving force in conservation for the foreseeable future, it has resulted in wildlife management systems that focus on protecting relatively small areas of habitat for scarce species. This approach can lead to conflict by prioritising global interests over those of local people and has also done little to mitigate species loss, much of which occurs outside of protected areas (Hoffman *et al.*, 2010). More widespread place-based conservation that incorporates conservation into local practices and concerns could potentially reduce conflict, or at least render the power relations of conflict more local and negotiable. It could also reduce the loss of species in 'unprotected' areas, as well as addressing the loss of wider ecosystem services.

A further point that follows from this is that conservation should not simply involve questions of what should be conserved and how, but for whom. Conservation has sometimes been driven, at least since the publication of Rachel Carson's *Silent Spring* (1962), by the everyday relations with non-humans that *some* people have as much as by the rare and endangered. However, far more attention has been paid to relatively privileged people in the developed world who are able to provide the financial and political support for conservation institutions and initiatives. A place-based, ethical conservation would need to be concerned with the everyday lives and relations of all people.

The breakdown of the nature–society dualism also means that, on the political level, the concerns and interests of non-humans would assume much greater prominence. Non-humans would be not so much 'objects of concern' as 'stakeholders' themselves, to borrow the parlance of policy (Latour, 2004). Illuminating the concerns and interests of non-humans is a complex and potentially varied process, but the sciences are clearly essential to the task. As such, the processes of exploration that potentially bring non-humans into society already take place but, according to Latour, this incorporation is hidden behind ideologies that place other species in the external realm of nature. What is required is a wider acknowledgement that non-humans have political effects and interests that the sciences and other forms of exploration bring into society and that require political representation within democratic processes. While this representation would not be without controversy, it would need to attend to the sciences and to the place-based concerns of both humans and non-humans to be credible.

Finally, consideration should be given to how conservation practice could learn from the field methods of anthropology. Often, conservation conflicts have at their root differences in knowledge. On such occasions conservationists, like those I worked with in Islay, are inclined to say that education is the key to resolving conflicts because it would enable local people to come to a better understanding of what conservationists were doing, and to a greater appreciation of the significance of the species they are trying to conserve and the work required to achieve this. Anthropologists take an entirely different approach.

When embarking on fieldwork, they assume *they* are the ones who will be educated by the people they encounter. Sometimes an anthropologist entering the field is likened to a child gradually learning how to behave. This is not to say that learning does not happen on both sides, but it is the anthropologist – at least initially – who lacks competence in that particular place. Anthropology is, in some respects, a discipline founded on a *lack* of expertise. The sort of ethical, relational and place-based conservation that I have advocated would necessarily require conservationists to assume their own ignorance of how people live in a place, what is important to them and how to maintain or restore their environmental relations. Such an approach would not put an end to conflict (cf. Latour, 2004). Indeed, conflict is both inevitable and perhaps even desirable if it helps to draw non-humans into human social life and to make it possible for different interests to be expressed. An important question, then, is not how to avoid conflicts but how to understand the role they play and to consider the ways in which they might be channelled towards conserving, rather than destroying, the relational conditions of human and non-human ecology.

References

Arce, A. and Long, N. (1993). Bridging two worlds: an ethnography of bureaucrat–peasant relations in western Mexico. In *An Anthropological Critique of Development*, ed. M. Hobart, pp. 179–208. London: Routledge.

Argyrou, V. (2005). *The Logic of Environmentalism: Anthropology, Ecology and Postcoloniality.* Oxford: Berghahn.

Berglund, E. (1998). *Knowing Nature, Knowing Science: An Ethnography of Environmental Activism*. Cambridge: White Horse Press.

Berglund, E. and Anderson, D. (2003). Introduction: towards an ethnography of ecological underprivilege. In *Ethnographies of Conservation: Environmentalism and the Distribution of Privilege*, eds. E. Berglund and D. Anderson, pp. 1–15. Oxford: Berghahn.

Bird-David, N. (1990). The giving environment: another perspective on the economic system of gatherer-hunters. *Curr. Anthropol.*, 31, 189–196.

Brockington, D. (2002). *Fortress Conservation: The Preservation of the Mkomazi Game Reserve, Tanzania*. Oxford: James Currey.

Campbell, B. (2005). Introduction: changing protection policies and ethnographies of environmental engagement. *Conserv. Soc.*, 3, 280–322.

Candea, M. (2010). 'I fell in love with Carlos the meerkat': engagement and detachment in human–animal relations. *Am. Ethnol.*, 37, 241–258.

Carson, R. (1962). *Silent Spring*. London: Penguin.

Descola, P. and Palsson, G. (1996). *Nature and Society: Anthropological Perspectives*. London: Routledge.

Einarsson, N. (1993). All animals are equal but some are cetaceans: conservation and culture conflict. In *Environmentalism: The View from Anthropology*, ed. K. Milton, pp. 73–84. London: Routledge.

Ellen, R. and Fukui, K. (1996). *Redefining Nature: Ecology, Culture and Domestication*. Oxford: Berg.

Ellis, D. M. (2003). Promoting consumption in the rainforest. In *Ethnographies of Conservation: Environmentalism and the Distribution of Privilege*, eds. E. Berglund and D. Anderson, pp. 119–137. Oxford: Berghahn.

Escobar, A. (1999). After nature: steps to an anti-essentialist political ecology. *Curr. Anthropol.*, 40, 1–30.

Gatt, C. (2011). Becoming Friends of the Earth: an anthropology of global environmentalism. Unpublished thesis, University of Aberdeen.

Gatt, C. (2013). Enlivening the supra-personal actor: vectors at work in a transnational environmentalist federation. *Anthropol. Action*, 20, 17–27.

Harris, M. (1977). *Cannibals and Kings*. New York: Vintage.

Helmreich, S. (2009). *Alien Ocean: Anthropological Voyages in Microbial Seas*. Berkeley: University of California Press.

Hobart, M. (1993). Introduction: the growth of ignorance? In *An Anthropological Critique of Development*, ed. M. Hobart, pp. 1–30. London: Routledge.

Hoffman, M., *et al.* (2010). The impact of conservation on the status of the world's vertebrates. *Science*, 330, 1503–1509.

Howell, S. (1996). Nature in culture or culture in nature? Chewing ideas of 'humans' and other species. In *Nature and Society: Anthropological Perspectives*, eds. P. Descola and G. Palsson, pp. 127–144. London: Routledge.

Ingold, T. (2000). *The Perception of the Environment: Essays in Livelihood, Dwelling and Skill*. London: Routledge.

Ingold, T. (2005). Epilogue: towards a politics of dwelling. *Conserv. Soc.*, 3, 501–508.

Ingold, T. (2011). *Being Alive: Essays on Movement, Knowledge and Description*. London: Routledge.

Krause, F. (2010). Thinking like a river: an anthropology of water and its uses along the Kemi River, Northern Finland. Unpublished thesis, University of Aberdeen.

Latour, B. (2004). *Politics of Nature: How to Bring the Sciences into Democracy*. Cambridge, MA: Harvard University Press.

Leopold, A. (1949). *A Sand County Almanac*. Oxford: Oxford University Press.

Lowe, C. (2006). *Wild Profusion: Biodiversity Conservation in an Indonesian Archipelago*. Oxford: Princeton University Press.

Milton, K. (1993). *Environmentalism: The View from Anthropology*. London: Routledge.

Milton, K. (1996). *Environmentalism and Cultural Theory*. London: Routledge.

Milton, K. (2000). Ducks out of water: nature conservation as boundary maintenance. In *Natural Enemies: People–Wildlife Conflicts in Anthropological Perspective*, ed. J. Knight, pp. 229–246. London: Routledge.

Novellino, D. (2003). Contrasting landscapes, conflicting ontologies: assessing environmental conservation on Palawan Island (The Philippines). In *Ethnographies of Conservation: Environmentalism and the Distribution of Privilege*, eds. E. Berglund and D. Anderson, pp. 171–188. Oxford: Berghahn.

Nygren, A. (2003). Nature as contested terrain: conflicts over wilderness protection and local livelihoods in Rio San Juan, Nicaragua. In *Ethnographies of Conservation: Environmentalism and the Distribution of Privilege*, eds. E. Berglund and D. Anderson, pp. 33–49. Oxford: Berghahn.

Orlove, B. and Brush, S. (1996). Anthropology and the conservation of biodiversity. *Annu. Rev. Anthropol.*, 25, 329–352.

Rappaport, R. (1967). *Pigs for the Ancestors: Ritual in the Ecology of a New Guinea People*. New Haven: Yale University Press.

Richards, P. (1992). Saving the rain forest? Contested futures in conservation. In *Contemporary Futures: Perspectives from Social Anthropology*, ed. S. Wallman, pp. 138–153. London: Routledge.

Satterfield, T. (2002). *Anatomy of a Conflict: Identity, Knowledge, and Emotion in Old-Growth Forests*. Vancouver: University of British Columbia Press.

Sillitoe, P. (1998). The development of indigenous knowledge: a new applied anthropology. *Curr. Anthropol.*, 39, 223–252.

Steward, J. (1955). *Theory of Culture Change: The Methodology of Multilinear Evolution*. Urbana: University of Illinois Press.

Theodossopoulos, D. (2003). *Troubles with Turtles: Cultural Understandings of the Environment on a Greek Island*. Oxford: Berghahn.

Tilley, C. (1994). *A Phenomenology of Landscape*. Oxford: Berg.

Tsing, A. (2005). *Friction: An Ethnography of Global Connection*. Princeteon: Princeton University Press.

Vivanco, L. (2006). *Green Encounters: Shaping and Contesting Environmentalism in Rural Costa Rica*. Oxford: Berghahn.

Whitehouse, A. (2004). Negotiating small differences: conservation organisations and farming in Islay. Unpublished thesis, University of St Andrews.

Whitehouse, A. (2009). 'A disgrace to a farmer': conservation and agriculture on a nature reserve in Islay, Scotland. *Conserv. Soc.*, 7, 165–175.

Whitehouse, A. (2012). How the land should be: narrating progress on farms in Islay, Scotland. In *Landscapes Beyond Land: Routes, Aesthetics, Narratives*, eds. A. Arnason, N. Ellison, J. Vergunst and A. Whitehouse, pp. 160–177. Oxford: Berghahn.

Wishart, R. (2004). A story about a muskox: some implications of Tetlit Gwich'in human–animal relationships. In *Cultivating Arctic Landscapes: Knowing and Managing Animals in the Circumpolar North*, eds. D. Anderson and M. Nuttall, pp. 79–92. Oxford: Berghahn.

Box 7

Too many geese?

Jesper Madsen
Department of Bioscience, Aarhus University, DK-8000 Aarhus, Denmark

Having recovered from low population sizes during the last century, many populations of wild geese in western Europe and North America are burgeoning following improved protection, land-use changes that create food for geese, and climate change. Because geese often congregate to forage on farmland, which has become more intensively cultivated, conflicts between goose conservation and agricultural interests have become common. Solutions to conflicts have differed widely among countries depending on the political willingness to pay economic compensation to farmers. However, even in countries where compensation or subsidies are provided to farmers, conflicts tend to continue because goose populations are still increasing. Hence, in several countries farmers ask for more economic support, but politicians are reluctant to provide aid because of budget cuts and fear of creating entitlements. Consequently, farmers in some countries have requested that populations be controlled to stop agricultural losses. This has been suggested for breeding geese in Scotland and the Netherlands and with migratory pink-footed geese *Anser brachyrhynchus* in Norway. With regard to the breeding stocks, the political decision to cull populations lies with the national governments; however, with migratory species, the issue is international and there is little European history of transboundary coordination of wildlife management.

The strategic plan for 2009–2017 of the African–Eurasian Waterbird Agreement (AEWA) under the Convention for Migratory Species addresses conflicts caused by migratory waterbirds, recognising that international coordination and flexible instruments are needed. The Svalbard pink-footed goose has been selected as the first test case for an AEWA International Species Management Plan for the following reasons. First, conflicts have escalated because of increasing goose numbers and spring-staging geese which feed on pasture grass and newly sown crops (Tombre *et al.*, 2013). Second, these geese have degraded tundra habitats in Svalbard and this damage might be increasing. Third, Norwegian farmers and management authorities agreed to set a population target of about 40,000 without consulting other range states (Denmark, the Netherlands and Belgium). Therefore, an internationally coordinated adaptive management process was initiated that included nature agencies from the four range states, organisations (representing hunters, BirdLife, and farmers) and scientists. This international working group reached consensus on problems, objectives, and alternative actions to ‘maintain a favourable conservation status for Svalbard pink-footed geese while taking into account economic and recreational interests’ (Madsen and Williams, 2012). After much consultation among affected parties, this plan was endorsed by the Meeting of the Parties to AEWA in 2012 and implemented by the range states. One of the most controversial objectives was to stabilise the population at 60,000 individuals, using hunting as the tool. The species was legal game in Norway and Denmark, but it was protected in the Netherlands and Belgium. The Netherlands and Belgium did not want a hunting season; hence, it was between Denmark and Norway to achieve the objective. The agreed population level was a social construct (Williams and Madsen, 2013). Preliminary model evaluations suggested that a lower target could be justified from a population viability perspective, provided that hunting could be effectively regulated, so Norwegian representatives argued for a target below 60,000. Oppositely, BirdLife representatives argued that in principle, populations should be allowed to fluctuate naturally, but also recognised that continued increase of the goose population would potentially cause loss of biodiversity in Svalbard and lead to more conflict; hence, they accepted the stabilisation target. An important reason the parties agreed to the target and the tool was the implementation of an adaptive management framework to predict changes, monitor, evaluate and revise the harvest of geese (Johnson *et al.*, 2014). Hence, scientific credibility was important. Furthermore, Denmark and Norway agreed to make an emergency closure of the hunting season if the population was predicted to decline below the target. The coming years will show if hunting regulations will work. If they do not, the working group will discuss alternative actions.

Adaptive management has provided the structure to deal with these politically delicate issues. Building trust between parties and having a joint learning process have been important cornerstones in the conflict resolution. Hunters have expressed a desire to be seen as partners contributing to management of the system. Science-led projects have been established in Norway and Denmark to explore how voluntary agreements with hunting organisations can be used to regulate the harvest. Interestingly, hunters know that if adaptive management is successful, they will have to reduce their harvest in order to meet population objectives, which demonstrates commitment to joint stakeholder desires.

References

Johnson, F., Jensen, G. H., Madsen, J. and Williams, B. (2014). Uncertainty, robustness, and the value of information in managing an expanding Arctic goose population. *Ecol. Model.*, 273, 186–199.

Madsen, J. and Williams, J. H. (2012). International Species Management Plan for the Svalbard population of the pink-footed goose *Anser brachyrhynchus*. *AEWA Technical Report* 48. Bonn, Germany: African–Eurasian Waterbird Agreement.

Tombre, I. M., Eythorsson, E. and Madsen, J. (2013). Stakeholder engagement in adaptive goose management; case studies and experiences from Norway. *Ornis Norveg.*, 36, 17–24.

Williams, J. H. and Madsen, J. (2013). Stakeholder perspectives and values when setting waterbird population targets: implications for flyway management planning in a European context. *PLoS ONE*, 8, e81836.

© Catherine Young.

CHAPTER EIGHT

Law and conservation conflicts

ARIE TROUWBORST
Tilburg Law School

Conflict is a dominant notion within the legal discipline. In this chapter, I review some roles that law and legal research play with respect to conservation conflicts. I first introduce briefly law and legal research and then discuss the respective roles of law and legal research with regard to conservation conflicts. Although I focus my discussion on Europe, many if not most of the general principles are applicable elsewhere.

Law

Although most people have a general idea of what law is about, a universally accepted definition is lacking. In essence, law is a set of agreements, of various origins and serving various purposes. Those purposes include promoting and safeguarding different societal interests, for example the equality and individual freedoms of citizens, and the protection of health, safety and the natural environment – including biodiversity. One representative definition describes law as the sum of all rules and prescriptions for the promotion and protection of societal interests that have been publicly (governmentally) established and/or recognised and are, where necessary, enforced through penalties (Uylenburg and Vogelezang-Stoute, 2008).

Law serves several functions. Principally, it provides norms and safeguards. Law indicates what conduct is and is not permitted in society, and offers citizens the possibility to seek compliance by others with the rules involved. Furthermore, law provides a means for avoiding, mitigating and settling conflicts in society. Last but not least, governments use law as an instrument to achieve all manner of policy objectives – including nature conservation objectives.

A major source of law is legislation (i.e. the issuing of written, binding rules) by governments and other public entities at international, national, subnational and lower governmental levels. Other sources include judicial rulings (court decisions) and *customary law* (unwritten rules created through practice). Within domestic legal systems, a distinction can be drawn between *private law* (primarily concerned with relations between citizens or other private parties like

Conflicts in Conservation: Navigating Towards Solutions, ed. S. M. Redpath, R. J. Gutiérrez, K. A. Wood and J. C. Young. Published by Cambridge University Press.

companies) and *public law* (primarily concerned with the promotion of public interests). Within public law, in turn, a distinction can be made between *constitutional* and *administrative law* (concerning inter alia the powers of governmental bodies and the relationships between them) and *criminal law*. A final distinction to bear in mind is that between substance (meaning the actual content of legal rights and obligations) and procedure (meaning the ways in which rules come about), rules are applied and legal disputes are settled. At the international level, most law is generated through the conclusion of binding agreements (treaties) between two or more national governments. These commonly come by the names of treaty, convention, agreement or protocol. A special, and for 28 European states particularly influential, branch of international law is the law of the European Union (EU). EU law consists mainly of several treaties and a large body of secondary legislation adopted on the basis of those treaties, particularly in the form of (legally binding) regulations and directives.

The repeated use of the word *binding* above indicates the distinguishing feature of law as compared with other, non-binding instruments in the wider domain often referred to as *policy*. Policy instruments that are *not* legally binding include political declarations, statements of intent, strategies, guidelines, recommendations, memoranda of understanding and codes of conduct.

Legal research

As an academic discipline, law is somewhat odd. Explaining the nature of law itself is certainly easier than explaining the nature of legal research, or *legal scholarship* as it is often called. Readers are accordingly advised to adjust their expectations of the following discussion to the reality that '[l]egal researchers have always struggled to explain the nature of their activities to colleagues in other disciplines' (Chynoweth, 2008: 28). My main ambition here is to keep the struggle brief.

Legal research may be described as the process by which legal scholars seek to define the meaning, boundaries and impacts of existing or potential laws. However, it is not difficult to come across statements like the one by Smits (2012: 112–113) that legal research knowledge is acquired in a way that is 'not strictly prescribed by research methods: the legal discipline is primarily a practice, in which the community of academic colleagues (the forum) decides on what is to be seen as reliable knowledge'. Although the majority of legal scholars probably does not share this view, it is a curious fact that in published legal research 'methodology is often largely absent or at least not made explicit' (Van Hoecke, 2013: vi). There is an ongoing and lively debate regarding the methodologies applied within the discipline of law, and the extent to which legal research actually qualifies as *scientific* research (Ross, 1958; Chynoweth, 2008; Samuel, 2008; Fisher *et al.*, 2009; Siems and Mac Síthigh, 2012; Smits, 2012; Van Hoecke,

2013). For the present purposes, it is not necessary, however, to delve too deeply into this debate.

Many legal studies focus on some or other version of the question: 'what is the law?' The answer to such questions is chiefly pursued through the analysis of legislation, *case law* (i.e. court decisions) and other legally relevant texts. Concretely, a scholarly study of a particular set of rules could aim to do any or all of the following: (1) identify and compare different interpretations of the rules; (2) identify and clarify ambiguities within the applicable rules; (3) arrange the rules in a logical and coherent structure; (4) identify and describe their relationship with other rules; or (5) determine the rules' consequences in a particular setting. Such interpretive, qualitative exercises are often referred to as *doctrinal research*, or more informally as *black-letter law* research. However, most studies are not strictly confined to black-letter analysis, even if the latter is often a major component. As Chynoweth (2008: 30) observes:

> In practice, even doctrinal analysis usually makes at least some reference to other, external, factors as well as seeking answers that are consistent with the existing body of rules. For example, an uncertain or ambiguous legal ruling can often be more easily interpreted when viewed in its proper historical or social context, or when the interpreter has an adequate understanding of the industry or technology [or other non-legal context] to which it relates. As the researcher begins to take these extraneous matters into account, the enquiry begins to move . . . in the direction of interdisciplinary research. There comes a point . . . when the epistemological nature of the research changes from that of internal enquiry into the meaning of the law to that of external enquiry into the law as a social entity. This might involve, for example, an evaluation of the effectiveness of a particular piece of legislation in achieving particular social goals or an examination of the extent to which it is being complied with.

Hence, studies within the legal discipline often contain a mix of methodological approaches. It is, for instance, 'not uncommon that a legal researcher starts with an historical introduction, then turns to an analysis of the relevant case law and finally engages with socio-political considerations' (Siems and Mac Síthigh, 2012: 668). At any rate, it will be apparent that the process of analysis habitually employed in legal research 'owes more to the subjective, argument-based methodologies of the humanities than to the more detached data-based analysis of the natural and social sciences' (Chynoweth, 2008: 30). Indeed, the evaluation of legislation's effectiveness, mentioned above, is at the limits of what the average legal academic feels comfortable doing. When it comes to actually measuring such effectiveness with hard data, this is usually no longer done by legal scholars, who tend to lack the capability (and desire) to engage seriously with statistics (e.g. Donald *et al.*, 2007; Baruch-Mordo *et al.*, 2011).

Plate 1 Snow leopards *Panthera uncia* coexist alongside livestock in Asian mountains. Livestock populations in snow leopard habitats are increasing, while wild prey populations are declining. Livestock losses to snow leopards and other predators result in monetary losses for local communities and negative perceptions of snow leopards (Box 1). In addition, mining for minerals in snow leopard habitats has become a land use conflict between herding communities, mining companies, governments, and conservationists in countries like Mongolia and Kyrgyzstan. Finally, large hydroelectric dams have become sources of conflict between local communities, conservationists, the state, and hydropower companies in Himalayan countries. Upper left, upper centre and lower centre images: Kulbhushansingh Suryawanshi. Upper right image: Steve Redpath.

Plate 2 The spotted owl is at the center of a conflict involving logging and conservation of old growth forests in western North America (Box 19). However, like most conflicts, it is more complex than a simple dichotomy of wildlife vs. commercial interests. It involves other issues such as the risks to rural communities from wildfire and reduced economic stability when logging is constrained for reasons of biodiversity conservation; logging has been hypothesised to reduce these risks. This conflict led to national legislation to implement a trial program in northeastern California, USA to test the efficacy of certain types of logging to reduce these risks and, hopefully, ameliorate conflict among people having different goals and values for forests (see Chapter 19). The program facilitated onsite consultation with concerned citizens and others for such program elements as the general structure and functioning of the program and specific inspections for salvage logging of trees following fire. Upper right drawing of owl by Viktor Bakhtin. Other photographs by Antony Cheng.

Plate 3 The aquatic plants of chalk rivers in lowland Britain support an ecological community of high conservation value, as well as socio-economically important game fisheries. These aquatic plants are eaten by mute swans (*Cygnus olor*), large herbivorous birds that can consume up to 4kg of fresh vegetation each day. Conservationists and anglers have accused swans of damaging chalk rivers and their plant communities, and have asked the government agencies that oversee river management to act. However, swans are native birds that are popular with bird watchers and the general public and are legally protected from harm or interference. Thus, a conflict has arisen between conservationists on one side who seek to protect the swans and anglers on the other (Box 9). To address this conflict, ecologists have worked with river managers, bird watchers, conservationists and anglers to monitor aquatic plant communities, quantify the effects of grazing and evaluate potential management options. All photographs by Kevin Wood.

Plate 4 Gray wolf (*Canis lupus*) populations in the Upper Midwest and the Northern Rockies, USA recovered enough to have been removed from the US Endangered Species List. Always the subject of controversy, wolves continue to foster strong feelings, both pro and con, in their recolonised areas (Box 15). Lawsuits by wolf advocates kept the wolf on the Endangered Species List based on legal technicalities for several years after their biological recovery. This process allowed wolf populations to far exceed prescribed recovery levels and engendered strong anti-wolf resentment by many ranchers, hunters and guides. These people decry wolf depredations on livestock and big game, which are much increased because of the higher wolf populations than originally prescribed by recovery teams. Public wolf hunting seasons by each state to attempt to stabilise or reduce wolf numbers also continue to fuel the controversy, with animal rights and welfare groups pitted against wolf hunters and trappers and state departments of natural resources. There seems no reason to believe that controversy over wolves and wolf management will ever abate. Upper right and left – photos by Dave Mech. Lower right photo by Sacha Dench.

Plate 5 The harbour seal, *Phoca vitulina*, and grey seal, *Halichoerus grypus*, prey on Atlantic salmon, *Salmo salar*, in rivers and estuaries, reducing catches and damaging fishing gear of netsmen and fishermen. This situation has led to an ongoing conflict between seal conservationists and salmon fisheries in Scotland, resulting in an innovative management plan involving all key stakeholders to balance the competing interests of seal conservation and fisheries management in the Moray Firth (Box 4). Photos of seals and netsmen: Rob Harris. Photos of fisherman and river: Juliette Young.

Plate 6 Large areas of moorland in northern England and east and central Scotland are intensively managed for red grouse (*Lagopus l. scoticus*). On these estates the aim is to maximise the number of grouse available for shooting. There is an ongoing, contentious conflict between grouse managers and conservationists over illegal killing of raptors, especially hen harrier (*Circus cyaneus*), peregrine falcon (*Falco peregrinus*) and golden eagle (*Aquila chrysaetos*) on grouse moors (Box 3). On one hand, landowners fear that uncontrolled birds of prey will make shooting economically unviable, resulting in reduced income and job losses. On the other hand, raptor conservationists seek to enhance the status of raptors through protection. The conflict also reflects differences between groups about politics, land ownership and decision making. Although all parties have been searching for solutions, none has yet been found due to unwillingness to compromise. Lower centre image: Juliette Young. Top right: Pete Moore.

Plate 7 White sharks aggregate seasonally at select coastal sites, some of which are human recreational areas. Although sharks rarely injure or kill people, when they do it induces negative public perceptions and economic impacts that compel stakeholders to urge local authorities to kill sharks to reduce attacks. In Cape Town, a novel community initiative, set-up by local business owners, known as the "Shark Spotter" programme has been implemented as a non-lethal alternative that balances the desires of water users with those of conservationists (Box 16). The Shark Spotter programme provides an example of successful conflict mitigation related to shark bites, but sustainable resolution requires ongoing investment keeping people and sharks apart. Upper left and centre image: Jon-Erik Munro. Right image: Morne Hardenberg. Lower left image: Sean Greer.

Plate 8 As environmental and social changes accelerate, fishers see a global change in fish markets and often face reduced catches. These debilitating factors, across most aquatic habitats, are set against a dramatic increase in cormorant (*Phalacrocorax carbo*) populations resulting from such factors as conservation legislation and changes in fish-stocks. The birds are often considered competitors with people for scare fish resources. This conflict is not simply between fishermen and cormorants but is far more complex – one where some fisheries interests believe that "conservationists" (a term often used to include ecological scientists) want to unfairly protect birds at the expense of people's livelihoods or leisure activities (Box 16). Cormorants are a conspicuous species and have often become a symbol of growing frustrations and antagonism over wider social, economic and environmental upheavals and "top-down" approaches to biodiversity protection and management – often imposed on local people by distant "others." The complexity and deep-rooted, multi-faceted antagonisms in this conflict, coupled with the breadth of geographic scales involved, has led to the conflict being exceptionally difficult to resolve. All photos by INTERCAFE, except lower right (Erich Prowald).

As in most other disciplines, legal studies can be classified along a gradient between *fundamental* and *applied* research. The former refers to research producing pure academic knowledge concerning the content and operation of the law, primarily for a scholarly readership. By contrast, applied research is produced primarily to suit the professional needs of practitioners, policy makers and the like. It is typically conducted with a certain purpose, which is often to facilitate a future change in the law itself or in the way it is interpreted or applied. This type of research is sometimes referred to as *law in context, law reform research* or *socio-legal research*.

Within the legal discipline, *environmental law* represents a particularly challenging field, inter alia because of its particularly distinct international dimension and interdisciplinary character, and because within the legal discipline it intersects many areas. Each of these traits entails its own methodological challenges (Fisher *et al.*, 2009). Environmental law is essentially a thematically focused cross-section of all common branches of law. Horizontally, it transcends the boundaries between private, constitutional, administrative and criminal law. Vertically, it cuts across national and international legal frameworks. A chemical plant emitting waste water containing contaminants into a river can be taken as a small illustration here. These emissions may cause farmers downstream, using river water to irrigate their crops, to sue the company for damages (private law). Another issue is whether the plant holds a permit to emit contaminated waste water, and if so, whether it is complying with the permit conditions (administrative law). In case it is violating the rules, this may lead to prosecution (criminal law). Finally, if the competent governmental authorities are lax when it comes to enforcement, this may constitute a breach of the country's international obligations, for instance under a transboundary river treaty, or EU water quality standards. The strong international dimension of environmental law hardly needs explanation, given the transboundary or even global nature of many environmental problems. Indeed, many legal scholars specialise in *international environmental law* (e.g. Birnie *et al.*, 2009; Bodansky, 2010), whereas within the latter field of expertise, international biodiversity or wildlife law has emerged as a research domain in its own right (e.g. Bowman *et al.*, 2010). Regarding the interdisciplinary nature of environmental law, it suffices to note the difficulty of conducting a worthwhile study of an environmental problem's legal dimensions without at least a basic understanding of the wider ecological and socio-economic context.

Law and conservation conflicts

Law will be a factor of some significance in respect of many actual conservation conflicts, whereby the circumstances of each case will determine the law's precise influence and which of a conflict's parties will be most favoured by it. That law frequently plays a role in conservation conflicts appears to be affirmed in the scientific literature, which moreover affirms that this role can be both negative and positive from a conservation point of view (Young *et al.*, 2005, 2007, 2010; Madden, 2008; Baruch-Mordo *et al.*, 2011; Redpath *et al.*, 2013).

Law influences conservation conflicts at many levels, ranging from conflicts between individual farmers and protected area managers over what to do about particular crop-raiding elephants, to conflicts between blocks of countries over the issue of international ivory trade (Couzens, 2014). The law involved may be constitutional, administrative, criminal, private and/or international law. What law is relevant will also be influenced by the stakeholders involved, which may include individual citizens, companies, local communities, NGOs, government agencies, countries, or intergovernmental organisations. Relevant rules will certainly not always be in the field of environmental or wildlife law, but may also concern rules regarding, for instance, agriculture and fisheries, land-use planning, participation in decision-making, and access to justice (for an example of the latter, see Epstein and Darpö, 2013). Occasionally, the legal framework governing a conservation conflict appears, or is, conflicting itself. An example concerns the long-standing conflict between reindeer-herding Saami people and conservationists over the management of wolves *Canis lupus*, brown bears *Ursus arctos*, Eurasian lynx *Lynx lynx* and wolverines *Gulo gulo* in northern Fennoscandia (Box 8). International nature conservation law prescribes large carnivore conservation in the area (Trouwborst, 2010; Epstein, 2013), whereas international human rights law, specifically indigenous peoples' rights, may be interpreted as prescribing large carnivore control.

Altogether, the effect of law on conservation conflicts themselves is not straightforward. On the one hand, law can cause or exacerbate such conflicts. This is the case, for instance, when land-use planning legislation enables human activities that cause adverse impacts on biodiversity (e.g. Madden, 2008), or vice versa when conservation legislation is (perceived as) unduly restrictive of certain human activities (e.g. Heydon *et al.*, 2010). On the other hand, law can be instrumental in mitigating conservation conflicts and minimising their harmful effects on nature (i.e. it can be a tool for 'conflict management' as described above). The primary means for doing so is legislation, which can take any shape, ranging from compensation schemes for damages caused by wildlife to the prescription of technical standards to be applied in agriculture, fisheries, mining and forestry operations. Another means is the settlement of disputes in court. Courts settle concrete disputes by issuing judgments that are binding for all parties to the dispute and, if delivered in last instance, final. This is usually done by courts of administrative, criminal and private law at various levels within domestic legal systems, and less frequently by dispute settlement mechanisms at the international level, which include binding arbitration and courts such as the International Court of Justice (ICJ), the International Tribunal for the Law of the Sea (ITLOS), and the Court of Justice of the EU. One high-profile example is the whaling dispute between Australia and Japan that was recently ruled on by the ICJ.

Some care should be taken not to confuse the judicial settlement of a dispute with the resolution of a conservation conflict. When a court has issued a decision in a legal dispute and no further appeal is possible, the average legal academic would say the conflict is resolved. This, however, obviously concerns the *legal* conflict only, that is, a disagreement between parties over a legal issue, often involving opposing interpretations of particular rules of law. The settlement of such a legal dispute is, evidently, not identical with the resolution of a conservation conflict as defined previously. Court proceedings are generally an *ultimum remedium*, and thus tend to provide a fairly reliable indicator of the tenacious nature of the underlying conflict. Moreover, it is normal for at least one of the parties to a legal dispute to be profoundly unhappy with the resulting judgment. In brief, even if they resolve associated legal disputes, court decisions will usually leave the underlying conservation conflicts unresolved. This squares with the sobering observation by Redpath *et al.* (2013: 107) that '[t]o our knowledge, no conservation conflict has ever been fully resolved in the sense that conflict is eliminated'. Indeed, whereas human–wildlife 'impacts can be resolved, for example, through legislation, mitigation, or technical solutions, conflicts are more challenging to resolve' (Redpath *et al.*, 2013: 100). At any rate, by providing a degree of certainty regarding the legal position, court decisions do reduce or eliminate one point of contention.

Legal research and conservation conflicts

Legal research has an apparent contribution to make in the context of conservation conflicts. This contribution concerns the mapping and understanding of conflicts' legal dimensions, as well as addressing conflicts themselves. The latter is done chiefly by charting concrete legal pathways likely to further the prevention or management of particular conservation conflicts, for instance the adoption of certain interpretations of applicable legislation, or a legal instrument's amendment. When studying a given conservation conflict, legal scholars typically focus on questions like the following:

(1) Which current legal instruments, and which of their provisions, are applicable?
(2) What is the relevance of this law for the particular conflict?
(3) In what ways can this law be interpreted, and which of these interpretations is/are most compatible with the applicable rules governing interpretation?
(4) What new legal solutions could be devised to manage the conflict (e.g. the adoption of authoritative interpretations, the modification of existing legal instruments, and/or the adoption of new ones)?

I provide here a few examples to illustrate the diversity of published research addressing conservation conflicts within the legal discipline. Some studies focus broadly on law and conservation conflicts generally (e.g. Madden, 2008). Other legal publications address a particular theme, for instance by-catch of seabirds in fisheries (Trouwborst, 2008), or particular (groups of) species such as elephants and whales (Couzens, 2014). Yet other studies detail the legal dimensions of one particular conflict, for instance concerning wolves in Wisconsin (Sanders, 2013; Box 15), or a single killer whale *Orcinus orca* marooned in The Netherlands (Trouwborst *et al.*, 2013a). Besides, all sorts of combinations of the above angles occur, a representative instance being a recent legal analysis centering on the theme of species reintroductions under EU law, using the Eurasian beaver *Castor fiber* as a case study (Pillai and Heptinstall, 2013).

Finally, it is appropriate that I provide the reader with a brief impression of the actual content of such research. I do this with reference to several studies that focus on the application of the EU's 1992 Directive 92/43/EEC on the Conservation of Natural Habitats and of Wild Fauna and Flora ('Habitats Directive') to the conservation and management of wolves in EU member states. Wolves are listed by default in the Directive's Annex II (species for which 'Special Areas of Conservation' must be designated as part of the Natura 2000 network) and Annex IV ('strictly protected species'), and these legal regimes apply in most member states (see Trouwborst, 2010; Epstein, 2013). Article 12 of the Directive requires these states to 'take the requisite measures to establish a system of strict protection' for wolves, including the institution and enforcement of domestic prohibitions on killing and capturing. Exemptions (*derogations*) from these prohibitions may only be allowed when three conditions are met: (1) the derogation is made for one of several purposes enumerated in Article 16(1) of the Directive; (2) there is 'no satisfactory alternative'; and (3) the derogation is 'not detrimental to the maintenance of the populations of the species concerned at a favourable conservation status in their natural range'. Several interpretation questions regarding the meaning of terms such as 'satisfactory alternative' and 'favourable conservation status' are still outstanding.

Within this context, Sweden (Box 8) offers an apt example of a conservation conflict, between proponents and opponents of wolf culling, where legal issues have taken centre stage. The issue of contention is Sweden's national wolf policy, which has in recent years included a cap on the national wolf population and allowed quota hunting. Much of the legal debate has revolved around the question of what scope, if any, there is for wolf hunting under Article 16(1)(e) of the Habitats Directive, especially given the unfavourable conservation status of the wolf population involved. Provided the second and third of the aforementioned conditions are also met, Article 16(1)(e) permits derogations in order 'to allow, under strictly supervised conditions, on a selective basis and to a limited extent, the taking or keeping of certain specimens... in limited numbers specified by the competent national authorities'. The Swedish government has come into conflict over this issue with the European Commission, whose task it is to monitor compliance with EU law by the member states. In 2011, the Commission opened a so-called infringement procedure against Sweden, a procedure that can end up before the Court of Justice of the EU. In parallel, national court proceedings in Sweden have focused on the same question. Several scholarly studies within the legal discipline have been dedicated to describing, comparing and assessing the legal merits of the opposing interpretations of the Habitats Directive by the Swedish government and the European Commission, and of the position taken hitherto by the Swedish judiciary (Darpö, 2011; Epstein and Darpö, 2013; Epstein, 2013).

Other legal studies that can serve as illustrations in the same European wolf context concern The Netherlands, one of the last European countries expected to be naturally recolonised by wolves (Trouwborst and Bastmeijer, 2010; Bastmeijer and Trouwborst, 2013; Trouwborst *et al.*, 2013b; Trouwborst, 2014b). One of the steps taken by the competent Dutch authorities in the face of this expected wolf comeback has been the commissioning of a legal research report to map the bandwidth of available policy options regarding wolf management (Trouwborst *et al.*, 2013b). Clearly, initial clarification of the limits imposed by the law may well avoid unnecessary discussions and litigation later on. The questions addressed in the report include the following:

(1) What is the legal status of wolves returning to the Netherlands?
(2) What lawful options exist for dealing with wolves preying on livestock?
(3) Is a zoning policy, whereby wolves are tolerated in designated parts of the country and removed from others, a legally viable option?
(4) At what stage of recolonisation are Special Areas of Conservation to be designated for wolves?
(5) What is the legal position of wolf–dog hybrids and of measures to counter hybridisation?

Going into detail as regards the answers to the above questions is not necessary for present purposes, nor feasible within the scope of the present chapter. However, it suffices to concisely address the last question regarding hybridisation between wolves and domestic dogs. A recent study dedicated to this issue (Trouwborst, 2014a) concludes that addressing hybridisation through preventive and mitigation measures is in conformity with the obligations of EU member states under the Habitats Directive, and may indeed be essential in order to comply with these obligations. This includes dealing with feral and stray dogs and captive hybrids, and removing hybrid animals from the wild. At the same time, it appears that the national prohibitions on the killing and capturing of wolves and other strictly protected species, as prescribed by the Directive, also cover free-ranging wolf–dog hybrids. This entails that the removal of such hybrid animals from the wild is subject to the rules concerning derogations from strict protection. Like the open-ended language of Article 16(1)(e) of the Directive cited above when discussing the Swedish example, this intricate and seemingly paradoxical state of affairs concerning wolf–dog hybrids is a good example of the kind of issue that intrigues legal scholars.

Concluding remarks

It is clear that the role of law in respect of conservation conflicts should not be underestimated, and that legal research has a meaningful contribution to make. Nevertheless, the legal discipline's importance should not be overestimated either. Conservation conflicts 'arise from a wide range of interacting factors' (Young *et al.*, 2010), with law being just one of an often complex 'mix of biological, social, historical, legal, geographic, political, economic, ethical, institutional, financial, cultural, and management factors' (Madden, 2008: 191). It is appropriate to conclude the chapter on this note, the gist of which is adequately captured in the postscript of an international law textbook (Lowe, 2007: 290):

> Lawyers have a contribution to make. They offer one way of going about resolving some of the most crucial problems that face the world. But it is only one way among many. There are many times when it is much better to call upon a politician, or a priest, or a doctor, or a plumber or, in the present context, a government agency, an ecologist, an economist or a negotiator.

References

Baruch-Mordo, S., Breck, S. W., Wilson, K. R. and Broderick, J. (2011). The carrot or the stick? Evaluation of education and enforcement as management tools for human–wildlife conflicts. *PloS ONE*, 6, e15681.

Bastmeijer, C. J. and Trouwborst, A. (2013). Welkom terug? De relevantie van het Europese gebiedenbeschermingsrecht voor de wolf en andere 'terugkomers'. *Milieu en Recht*, 40, 80–91.

Birnie, P., Boyle, A. and Redgwell, C. (2009). *International Law & the Environment*. Third edition. Oxford: Oxford University Press.

Bodansky, D. (2010). *The Art and Craft of International Environmental Law*. Cambridge, MA: Harvard University Press.

Bowman, M., Davies, P. and Redgwell, C. (2010). *Lyster's International Wildlife Law*. Second edition. Cambridge: Cambridge University Press.

Chynoweth, P. (2008). Legal research. In: *Advanced Research Methods in the Built Environment*, eds. L. Ruddock and A. Knight, pp. 28–38. Oxford: Wiley-Blackwell.

Couzens, E. (2014). *Whales and Elephants in International Conservation Law and Politics: A Comparative Study*. London/New York: Routledge.

Darpö, J. (2011). Brussels advocates Swedish grey wolves: on the encounter between species protection according to Union law and the Swedish wolf policy. *SIEPS Eur. Policy Anal.*, 2011(8), 1–19.

Donald, P. F., Sanderson, F. J., Burfield, I. J., Bierman, S. M., Gregory, R. D. and Waliczky, Z. (2007). International conservation policy delivers benefits for birds in Europe. *Science*, 317, 810–813.

Epstein, Y. (2013). Population based species management across legal boundaries: the Bern Convention, Habitats Directive, and the gray wolf in Scandinavia. *Georgetown Int. Environ. Law Rev.*, 25, 549–587.

Epstein, Y. and Darpö, J. (2013). The wild has no words: environmental NGOs empowered to speak for protected species as Swedish courts apply EU and international environmental law. *J. Eur. Environ. Plan. Law*, 10, 250–261.

Fisher, E., Lange, B., Scotford, E. and Carlarne, C. (2009). Maturity and methodology: starting a debate about environmental law scholarship. *J. Environ. Law*, 21, 213–250.

Heydon, M. J., Wilson, C. J. and Tew, T. (2010). Wildlife conflict resolution: a review of problems, solutions and regulation in England. *Wildl. Res.*, 37, 731–748.

Lowe, V. (2007). *International Law*. Oxford: Oxford University Press.

Madden, F. M. (2008). The growing conflict between humans and wildlife: law and policy as contributing and mitigating factors. *J. Int. Wildl. Law Policy*, 11, 189–206.

Pillai, A. and Heptinstall, D. (2013). Twenty years of the Habitats Directive: a case study on species reintroduction, protection and management. *Environ. Law Rev.*, 15, 27–46.

Redpath, S., *et al.* (2013). Understanding and managing conservation conflicts. *Trends Ecol. Evol.*, 28, 100–109.

Ross, A. (1958). *On Law and Justice*. London: Stevens & Sons.

Samuel, G. (2008). Is law really a social science? A view from comparative law. *Cambr. Law J.*, 67, 288–321.

Sanders, J. D. (2013). Wolves, lone and pack: Ojibwe treaty rights and the Wisconsin wolf hunt. *Wisconsin Law Review*, 2013(6), 1263–1294.

Siems, M. and Mac Síthigh, D. (2012). Mapping legal research. *Cambr. Law J.*, 71, 651–676.

Smits, J. (2012). *The Mind and Method of the Legal Academic*. Cheltenham: Edward Elgar.

Trouwborst, A. (2008). Seabird bycatch – deathbed conservation or a precautionary and holistic approach? *J. Int. Wildl. Law Policy*, 11, 293–333.

Trouwborst, A. (2010). Managing the carnivore comeback: international and EU species protection law and the return of lynx, wolf and bear to Western Europe. *J. Environ. Law*, 22, 347–372.

Trouwborst, A. (2014a). Exploring the legal status of wolf–dog hybrids and other dubious animals: international and EU law and the wildlife conservation problem of hybridization with domestic and alien species. *Rev. Eur. Comp. Int. Environ. Law*, 23, 111–124.

Trouwborst, A. (2014b). Country report: The Netherlands – gearing up for wolf comeback. *IUCN Academy of Environmental Law E-Journal*, 5, 224–230.

Trouwborst, A. and Bastmeijer, C. J. (2010). Lynxen en wolven: Het natuurbeschermingsrecht en de terugkeer van grote roofdieren naar Nederland. *Milieu en Recht*, 37, 272–283.

Trouwborst, A. and Bastmeijer, C. J., with the cooperation of Backes, C. W. (2013b). *Wolvenplan voor Nederland: Naar een gedegen juridische basis*. Tilburg/Maastricht: Tilburg University and Maastricht University.

Trouwborst, A., Caddell, R. and Couzens, E. (2013a). To free or not to free? State obligations and the rescue and release of marine mammals: a case study of 'Morgan the Orca'. *Transnatl Environ. Law*, 2, 117–144.

Uylenburg, R. and Vogelezang-Stoute, E. M. (2008). *Toegang tot het Milieurecht: Een Inleiding voor Niet-Juristen*. Fifth edition. Deventer: Kluwer.

Van Hoecke, M. (2013). *Methodologies of Legal Research: What Kind of Method for What Kind of Discipline?* Oxford/Portland: Hart Publishing.

Young, J., *et al.* (2005). Towards sustainable land use: identifying and managing the conflicts between human activities and biodiversity conservation in Europe. *Biodivers. Conserv.*, 14, 1641–1661.

Young, J., *et al.* (2007). Conflicts between biodiversity conservation and human activities in the central and eastern European countries. *Ambio*, 36, 545–550.

Young, J. C., *et al.* (2010). The emergence of biodiversity conflicts from biodiversity impacts: characteristics and management strategies. *Biodivers. Conserv.*, 19, 3973–3990.

Box 8

Wolves in Sweden

Jens Frank
Department of Ecology, Swedish University of Agricultural Sciences, Sweden
Grimsö Wildlife Research Station, 730 91 Riddarhyttan, Sweden

After legal protection in 1965, wolves *Canis lupus* in Sweden increased from below 10 (Wabakken *et al.*, 2001) to approximately 400 individuals in 2013 (Svensson *et al.*, 2013). Wolf populations occur in the area of reindeer *Rangifer tarandus* husbandry in the north, but are mainly concentrated in the south central part of Sweden. The landscape is dominated by boreal forests and villages with and without livestock. Sweden's wolf conflict may ultimately stem from the different values humans place on wolves regarding protecting wolves or minimising wolf impacts on farmers and hunters. At present the conflict is focused less on these underlying dimensions and more on how wolves are managed. The conflict centers on four main themes:

(1) Attacks on livestock (mostly sheep but also cattle and horses), resulting in 500–600 livestock attacks annually (Karlsson *et al.*, 2013). In addition, about 200–300 reindeer are killed annually.
(2) Competition for game, such as moose *Alces alces* and roe deer *Capreolus capreolus* (Sand *et al.*, 2012).
(3) Attacks on pets, mainly hunting dogs that work relatively far from the hunter (Karlsson *et al.*, 2013).
(4) Fear of attacks on humans. Although there have been no wolf attacks on humans in Sweden since 1821, fear is widespread with 33% of adult humans afraid of wolves (Johansson and Karlsson, 2011). A major concern is fear of attacks on children.

In northern Sweden the conflict is exclusively connected to free-ranging reindeer husbandry. Reindeer grazing is widespread and governed by the availability of suitable land. Reindeer herders are worried about increasing wolf presence. The estimated number of reindeer killed by large carnivores in Sweden ranges from 25,000–50,000 per year. The main concern, however, is that wolves repeatedly disturb grazing reindeer, creating indirect losses and increasing costs of husbandry. The conflict between wolf conservation and reindeer herding is amplified by industrial forestry, mining, wind turbine arrays and tourism because these activities reduce available grazing grounds.

In southern Sweden the conflict consists of a mix of the four themes. Like the reindeer husbandry area, pressure is put on the rural communities by other sectors such as industrial farmers competing with local small-scale livestock operations, and industrial forestry demanding low moose densities to reduce tree damage, which reduces both wolf hunting opportunities and game hunting. Besides the Habitats Directive, many other laws and regulations are involved (e.g. food and health, animal welfare and EU regulations on grazing practices). Local economies are also affected by urbanisation and declining inhabitants, which leads to reduction of services in the countryside. For example, access to health care, schools, banks or supplies is declining. Additionally, wolves have increased. While wolves are not the sole, or major reason for the difficulties for people living in the countryside, they add to the burden of rural people. On a recent flight back to Sweden from an EU meeting on large carnivores in Brussels, I sat next to a farmer who attended the same meeting. She repeatedly stressed that farmers' main concern was wolf management, not wolves per se. This concern is often heard from hunters, farmers, conservationists and authorities in Sweden. Unfortunately, a feeling of not knowing what to expect has been enhanced by the difficulties in defining the jurisdicial term 'favourable conservation status'. Different agencies as well as lawyers and experts within the same agency have made very different interpretations. In practice, this has led to several wolf hunts authorised by the Swedish EPA being stopped by last minute court rulings. The lack of concrete management plans with politically specified goals for the wolf population and specified goals for impacts on, for example, livestock husbandry and hunting with dogs makes it hard for both stakeholders and managers to predict future conditions.

There are studies on the effects of wolf predation on game species (roe deer and moose) and non-game species, wolf genetics, human fear of wolves, wolf attacks on livestock, public attitudes towards wolves, wolf management actions and trust in authorities involved in management. There have been many initiatives to improve dialogue between stakeholders and authorities. In short, the more successful initiatives have in common that stakeholders left the process with a feeling of having been listened to and having had an influence on the outcome. In less successful ones, stakeholders had the experience that decisions were communicated to them, but that they had few or no possibilities to affect the outcome. During the last 10 years, large carnivore management in Sweden has been decentralised from the Environmental Protection Agency (EPA) in Stockholm to all 21 counties. This process is the result of a political decision and the parliament has stated that the process should continue. In each county a 'wildlife board' consisting of stakeholders and politicians has been formed to participate in the strategic but county-specific decisions. The government has not been very clear on what mandate the wildlife boards should have. The imprecise directive has led to increased conflicts and distrust between stakeholders and regional authorities.

However, there seems to be political awareness and motivation to make large carnivore policy in Sweden more predictable for national and regional authorities as well as stakeholders. This is likely to push the process forward into a discussion on practices instead of interpretation of vague policies.

References

Johansson, M. and Karlsson, J. (2011). Subjective experience of fear and the cognitive interpretation of large carnivores. *Hum. Dimens. Wildl.*, 16, 15–29.

Karlsson, J., Danell, A., Månsson, J., Svensson, L. and Hellberg, R. (2013). Wildlife damages in Sweden 2012. Report 2013–1.

Sand, H., Wikenros, C., Ahlqvist, P., Strømseth, T. H. and Wabakken, P. (2012). Comparing body condition of moose selected by wolves and human hunters: consequences for the extent of compensatory mortality. *Can. J. Zool.*, 90, 403–412.

Svensson, L., *et al.* (2013). *The Wolf in Scandinavia: Status Report of the 2012–2013 Winter*. Oulu, Norway: Høgskolen i Hedmark, Viltskadecenter, Grimsö forskningsstation, Vilt- och fiskeriforskningen (summary in English).

Wabakken, P., Sand, H., Liberg, O. and Bjärvall, A. (2001). The recovery, distribution, and population dynamics of wolves on the Scandinavian peninsula, 1978–1998. *Can. J. Zool.*, 79, 710–725.

CHAPTER NINE

The relevance of psychology to conservation conflicts

HERBERT H. BLUMBERG
University of London

Psychology is the scientific study of human thought, behaviour and emotion. Accordingly, this makes it applicable to understanding conflicts and their resolution, at all levels from intrapersonal to international.

Research on conservation conflict in the psychological literature began mainly in the 1980s and has been increasing gradually since then. Notwithstanding a substantial psychological literature on conservation, of the papers published between 2003 and 2012, only about 10% of those retrieved via the PsycINFO database using the joint keywords 'environment*' (i.e. environment, environmental, etc.) and 'conflict' were actually about and relevant to conservation conflicts. Furthermore, there has hitherto apparently been no systematic overview of this literature particularly as regards application to conflict resolution (Coleman and Deutsch, 2012; see also: Deutsch, 1998 and Deutsch *et al.*, 2006). I base much of the discussion in this chapter on approximately 270 publications on psychological aspects of conservation conflict published between 2003 and 2012 plus a number from earlier decades in order to give an overview of relevant aspects of psychology-specific fields and their interface with other disciplines and to demonstrate the diversity of approaches and intersections that psychology brings to conservation issues.

An overview of some areas of psychology and its interface with other disciplines

Psychology is a diverse field. Cognitive psychologists, for example, might point to the importance of how a particular situation is *perceived* and to common biases and heuristics (mental shortcuts) in perception. Developmental specialists might examine lifespan changes, including how people are socialised into their respective, sometimes conflicting, groups. Other specialities are concerned with human motivation, principles of learning and individual differences. Individual differences include not only demographic variables such as age and geographical location but also some potentially conflict-relevant personality

Conflicts in Conservation: Navigating Towards Solutions, ed. S. M. Redpath, R. J. Gutiérrez, K. A. Wood and J. C. Young. Published by Cambridge University Press.

dimensions, such as agreeableness, extraversion, conscientiousness and openness to others' views. A substantial part of contemporary psychology is concerned with neuro-biological underpinnings of the phenomena being studied, but this is not considered here.

Social psychology, in particular, has contributed much to understanding environmental conflict, especially by experiments relating to cooperation but also other areas, ranging from the more 'social' dynamics (group processes, leadership and conformity, friendship, etc.) to the more 'psychological' dimensions (interpersonal perception). It includes areas in between these main topics such as attitude change, bargaining, prejudice, communication, and also work concerned with various social issues.

The remainder of the present section is concerned with the interface between psychology and other disciplines. In the next section I return in more detail to contributions from the main areas of psychology.

Although it may seem hyperbolic to view psychology as the centre of the academic universe, the discipline does border on the natural sciences, the arts and humanities, and the social sciences. Because psychological phenomena have increasingly figured within other disciplines' analyses, I bring together some examples from these interfaces to help bridge the boundaries between disciplines.

Anthropology

Sivaramakrishnan and Vaccaro (2006) introduce a special issue of the journal *Social Anthropology* devoted to documenting and analysing struggles over land, water, agricultural and industrial decline, and especially issues related to the legal rights held by various entities. Understanding such struggles is enhanced by considering the psychology of social interaction and of interpersonal perception and its biases. The work also endeavours to evaluate 'newly imagined... post-industrial landscapes where mobility and place-attachment create a constant tension within and between humans' (Sivaramakrishnan and Vaccaro, 2006).

International relations

Shifts in international boundaries, for example, have implications for environmental responsibilities. The social psychology of how such shifts are effected, publicised and perceived plays an important part in whether conflict emerges and how it is resolved. International or intercultural hostilities can of course reduce cooperation on environmental issues and, worse, devastate natural environments. As one way in which such international hostilities can be addressed, Langholtz (2010) describes major changes in the post-Cold War world, particularly UN peacekeeping, which has dramatically expanded and largely decentralised, with peacekeepers from over 100 nations. Thus psychological knowledge about, among other things, intergroup processes and reactions to stress is helpful. Skilled mediation and the promotion of identities that transcend 'us–them' prejudices are often essential.

Law (and legal geography, i.e. land-use regulation)

Analysing reception by people of laws promoting biodiversity, Mouro and Castro (2010) found psycho-social variables to be important. For instance, when people had 'vested interests' relating to ecology, whether people 'identified' with a place provided a particularly good prediction of attitudes toward the relevant laws. How people view their own 'non-conservationist' behaviour may help in formulating communications and policies. Analysis of over 1500 cases of individual transgressors of environmental laws in Spain that indicated an implicit conflict between the transgressors and those holding the values that underpin the laws in question showed that the transgressors usually simply questioned the legitimacy of the law being broken (Martín *et al.*, 2008; see also Gifford, 2007; Collis, 2009; Gillespie, 2011; Valverde, 2011).

Architecture

Architecture includes, among other things, effects of density and crowding on privacy, conflict and violence (linking to a substantial psychological literature on these topics). Of increasing interest is design on a human scale, not only open spaces, corridors and lighting, but also ecological aspects related to odours, floors, desks and overall energy conservation (Walden, 2009).

Biology

Biology–psychology interfaces include exploring how reciprocal trust and cooperation can be favoured by natural selection in various species, including cultural aspects as well as technology and communication in humans (Bowles and Gintis, 2003; Kurzban, 2003; Richerson *et al.*, 2003; Smith, 2003).

Criminology/forensics

This area of research relates especially to environmental harm and its prevention (White, 2011). Forensic psychologists may, for instance, need to evaluate criminal responsibility when environmental harm occurs.

Ethics and philosophy

This research is concerned, for example, with the 'cultural construction of nature' (Atran and Medin, 2008), which can relate to cultural differences and resultant intergroup conflict and stereotyping. Also, early career experience is thought to influence decisions about ethical conflicts. Environmental dimensions related to leadership, rewards, career direction, etc., are associated with ethical decisions such as perceptions of climate change (Mumford *et al.*, 2007).

Education

Education related to conservation and peace forms a large literature, much of which relates to fostering conflict resolution and cooperation. Johnson and Johnson (2006) discuss the importance of consistency between means and ends, exemplified by establishing a classroom management system with conflicts managed constructively by means of having a cooperative learning environment, teaching students and staff how to manage conflicts positively, and inculcating civic values. Conflict resolution is thus not imposed hierarchically, but is reached with democratic structures that remain in place to help prevent future conflicts from occurring. The authors also provided general advice about quality decision-making concurrence among opposing parties in conflict (Johnson and Johnson, 2000). If, for example, a fair ecological balance represents a goal in a particular instance, then the means of achieving this goal are best also carried out with fair regard for the needs and wishes of all parties concerned with it. (For an overview of psychology-based conflict resolution that is rooted in decision theory, see Edwards *et al.*, 2007.)

A subset of education is 'eco-education', which can concurrently promote sustainability, diversity and peace (Timpson *et al.*, 2012). Jurin (2012) covers all aspects of sustainability including those bearing on conflicts (Blumberg, 2013b). For additional material on conflict resolution related to sustainability see Sugiura (2005), Sandy (2006), Haste (2010), Nowak *et al.* (2010) and Tzou *et al.* (2010).

Applications of psychology for resolving conservation conflicts

The psychological research areas most relevant to conservation conflicts are those concerned with social psychological phenomena – 'commons dilemmas', trust and communication and attitude change – and with the psychology of motivation, personality, psychopathology and human development. I explore each in turn in this section.

Commons dilemmas

Perhaps the psychology research most relevant to resource conservation is that related to 'commons dilemmas' (see Pratarelli and Johnson, 2012). This research typically deals with communal or publicly owned resources that get over-used, ranging from naturally occurring ones such as vegetation used for grazing to ethereal resources such as replenishable points in a multi-party game-like procedure. The conflict is manifest in mixed-motive systems (i.e. situations eliciting a choice between cooperation and competition) where parties need to cooperate so as not to deplete the resource, but also may profit from individual defection (Deutsch, 1977). The principles are plausibly applicable to a wide variety of situations where resources require conservation (Hardin, 1968).

Such situations are often simulated as laboratory experiments where the resource may be a growing pool of points that participants can 'harvest'. In a commons dilemma, such as the Prisoners Dilemma Game (PDG), everyone will benefit if all cooperate to avoid over-harvesting a resource in common (see Chapter 14). However, if everyone else cooperates, any given individual will be better off by taking a large harvest. Unlike prototypical PDG as typically studied in laboratory-based studies, however, resource dilemmas tend to include more than two parties, to be iterative and to have a resource pool that is updated with each round (Blumberg *et al.*, 2012). Blumberg *et al.* (2012) note that merely appealing for cooperation through persuasive messages may be effective in social dilemmas (Rosen and Haaga, 1998). Indeed, just learning the metaphor of the commons dilemma may be more effective in influencing behaviour than reading a seemingly more informative environmental passage (Mio *et al.*, 1993).

Let us turn to research related both to cooperation and to effective communication when there is a potential conflict between the public good and individual behaviour. For example, what effect do you think it might have if you encountered the following sign on entering Petrified Forest National Park in Arizona:

> Your heritage is being vandalized every day by theft losses of petrified wood of 14 tons a year, mostly a small piece at a time.

Cialdini (2003) established that a sign similar to the one just quoted, emphasising that a behaviour was common (albeit socially undesirable), actually markedly *increased* the likelihood of that behaviour compared with a (no sign) control group. By contrast, a sign that simply stated the prohibition and the reasons for it was most effective of all. Thus, care about implicit norms can help resolve conflicts of interest such as that between conservation goals of park managers and the goals of many tourists.

Blumberg *et al.* (2006: 204–205) concluded that 'ongoing careful analysis is required as to preferred behaviours and the means of achieving them – by individual actions, by social norms, by legislation, or [by] other means. Descriptive norms should be consistent with prescriptive ones'. In another experiment by Cialdini (2003), people were more likely to leave litter (breaking an implicit prescriptive norm) in an area that was already littered (descriptive norm). This finding, moreover, interacted with the salience of a littering norm: seeing a model drop litter in a clean area elicited particularly low levels of littering from participants and seeing a model litter in an already-littered area elicited especially high levels of littering (Blumberg *et al.*, 2006; Oceja and Berenguer, 2009). Hence, in potential conflicts between those promoting conservation norms and those who may, possibly unwittingly, violate them, the promoters need to take into account both the potential behaviour of individuals and the overall distributions of how others behave and interact among themselves.

Trust

Turning to other social psychological work involving trust, Canning and Hanmer-Lloyd's (2007) research on bargaining identified, in a qualitative analysis of some real-world examples, that efforts towards environmental (green) adaptation can enhance or undermine buyer–seller mutual trust depending on circumstance. For instance, they cite business partners who tried to reduce costs by (a) reducing packaging waste (conservation of source materials) and (b) improving the environmental design of a product. Apparently trust was better preserved in the former instance because customers happened to find the partners to be more forthright (and to have more personal commitment to the conserving adaptations) in explaining what they were doing and taking responsibility for doing so.

Linked to trust is the study of prejudice. For instance, intergroup bias or a negative (or positive) inclination with regard to particular environmental policies is in principle crucial to understanding environmental conflict, yet little research has focused explicitly on this application. Nevertheless, wherever conservation-related conflict occurs between different groups or categories of people – environmentalists versus farmers (e.g. Box 7), indigenous versus corporate bodies, or even hostilities having the destruction of ecosystems as a 'byproduct' – the substantial general literature on resolving intergroup conflict seems likely to be relevant (see Chapter 12). Indeed, even conflict within what is initially the same group, such as conservation groups interested in birds and rivers (Box 9) may lead to the sub-groups perceiving each other as 'out-groups'. Fisher (2006) suggests that intergroup conflict invariably can benefit from skilled third-party mediators who help clarify, confront and resolve (or at least manage) such conflicts by encouraging trust-building, seeking mutually acceptable solutions, and transforming situations and relationships to facilitate sustainable solutions (see also Chapter 16).

Communication and attitude change

In addition to the above work on trust, some relevant psychology work has concerned communication (social interaction) and attitudes. For example, a welcome, if unsurprising, finding is that environmental awareness can be facilitated by network information technology (Scharl, 2006). 'With regard to environmental communication ... [interactive electronic communication] facilitates transitions from ... conflict positions to shared meaning' (Scharl, 2004: vi); such shifts away from conflict can be associated with integrating parties' positions and explicitly recognising and treating complexity, uncertainty and risk.

General principles of communication and attitude change, useful for almost anyone concerned with the social interaction aspects of conservation conflicts, have formed a large literature dating at least to the 1950s (see Perloff, 2010). One focus has been on the elements of communication and on a variety of processes (such as consistency within communications and accommodating behaviour rather than threat). The matters analysed as especially relevant, however, vary widely across conservation examples as follows.

Boholm (2008) analysed a variety of aspects of public meetings concerned with land use and environmental planning of a Swedish railway tunnel. The work confirmed, for instance, the need for differing parties to distinguish between risks and actual hazards. Heinen and Shrivastava (2009) analysed attitudes and awareness towards endangered species around Kaziranga National Park in India. While different parties had different concerns and no one wished to lose crops to wildlife, the parties (and also the researchers) concluded that a local approach to resolution and education helped resolve conflicts. Similar conclusions emerged among people near Kibale National Park and wetlands in Uganda (Hartter, 2009), and in the relationship between Canadian trappers and industrial development companies (Webb *et al.*, 2008). Environmental attitudes also relate to the time perspective (Milfont and Gouveia, 2006). For instance, environmental preservationist views were, as expected, associated with future – and 'biospheric' – orientations rather than with short-term profit (see Chapter 10).

Obviously, one must keep the context of competition, conflict and attitude change in mind. Field experiments studying how best to provide helpful hints for Californians to use less water on their lawns so fish can have more are one thing (cf. McKenzie-Mohr and Oskamp, 1995), but starker realities may be another. With field experiments there can be sufficient time, leeway and potential goodwill to derive solutions without immediate threat either to parties or to mediators. By contrast, one might cite a seemingly intractable conflict in Cambodia between communities dependent on sustainable use of high-biodiversity rainforest and illegal loggers who operate with legal impunity and impose lethally armed security guards (Plokhii, 2012).

Motivation

Another key area of psychology which can inform conflicts relates to motivation. Both primary motives – those necessary for living, such as needs linked to maintaining a sustainable environment – and secondary ones, such as self-esteem built on altruism, could in principle be linked to conservation awareness and attempted resolution of related problems. One contribution has been to separate the measurement of altruistic and specifically 'biospheric' motivation (the latter emphasising the inherent value of 'nature'). The conclusion is that both are needed for 'pro-environment' behaviour (Jurin and Hutchinson, 2005; de Groot and Steg, 2008). Conflicts may arise between environmental and utilitarian values, such as are manifest in the effort to allow more off-highway vehicle recreational use in American Forests and National Parks (Wilson, 2008).

Personality

The 'Big Five' factors of personality – agreeableness, neuroticism (or anxiety), conscientiousness, extraversion and openness – account for much of the individual personality variability in a variety of contexts. One could hypothesise that a readiness to resolve environmental conflict constructively (a) might relate to conscientiousness, agreeableness and openness, and (b) perhaps might not be substantially correlated in either direction with anxiety/neuroticism or extraversion, and (c) might well be negatively associated with classic authoritarianism, a personality dimension largely independent of the Big Five but relevant not only to uncompromising prejudice and to anti-progressive politics but also to 'conserving' and hence arguably of interest to study further in the present context.

Mindfulness would seem to be particularly important for environmental harmony, including e.g. the minimisation of conservation conflicts (Jacob *et al.*, 2009). This would be expected to be the case for all of the following three strands of the psychological literature on the topic: experimental work linked, for example, to the reduction of prejudice and illness as in Langer's research, almost-meditative social action as in Nhat Hanh's work, and therapeutic applications (Hochman, 2007; Blumberg, 2012a).

Psychopathology

Climate change, and therefore the resolution of different views about related policy, may have adverse effects on mental health (not to mention various species' physical health and indeed existence) (Wilmoth, 2012). Similarly, almost any conservation-linked conflict may impact on parties' mental health apart from its other impacts.

Developmental psychology

Education to promote awareness of the importance of conservation and of dealing with related conflicts may be regarded as within the remit of either developmental psychology or on the interface between the disciplines of psychology and education. Some work focuses on action to provide an environmentally secure setting for children, beyond avoiding pollutants (Nelson, 2005), and into developing a sustainable daily routine (Weisner, 2010). Landmark work on children in war zones and child soldiers (related to conflicts that can indirectly but profoundly affect stable ecology) has been done by Wessels and by others, as reviewed in part by van der Veer (2009) considering an ecological framework for psycho-social assistance.

Global psychology

Global psychology as described by Stevens (2007) represents a new specialty of relevance to conservation conflicts. This is distinct from peace psychology (Christie *et al.*, 2001; Blumberg *et al.*, 2006) and political psychology. Here the concern is to deal globally with three issues in particular: intergroup conflict, threats to the natural environment (Winter, 2007) and risks to physical and mental health especially among women, children, migrants and others.

Concluding comment

There is no single unifying paradigm for work on psychological aspects of conservation conflicts. There is, however, a broad contemporary literature covering both a wide variety of psychological sub-disciplines and the interfaces between psychology and other disciplines that are linked to environmental topics. There are some common themes, such as how to facilitate constructive communication and, more generally, to foster cooperation and achieve environmental goals while maintaining peace and justice for the parties concerned. Finally, the field of psychology can contribute to the theory and methods needed to address conservation conflicts.

Acknowledgements

I thank Alison Britton, Joanna Britton, Erik Spindler and several reviewers for their helpful comments on a draft of this chapter. This chapter is based in part on papers by Blumberg (2012b, 2013a).

References

Atran, S. and Medin, D. (2008). *The Native Mind and the Cultural Construction of Nature*. Cambridge, MA: MIT Press.

Blumberg, H. H. (2012a). Mindfulness. In *Encyclopedia of Peace Psychology* (Vol. 2), ed. D. Christie, pp. 644–648. Hoboken, NJ: Wiley-Blackwell.

Blumberg, H. H. (2012b). *Psychological Aspects of Environmental Conflicts and Their Resolution*. Coventry: Conflict Research Society.

Blumberg, H. H. (2013a). Conservation conflicts and their resolution. In *Thirteenth International Symposium on Contributions of Psychology to Peace*. Kuala Lumpur: International Islamic University Malaysia.

Blumberg, H. H. (2013b). Understanding and moving towards a sustainable world: a comprehensive, psychologically informed text. *PsycCRITIQUES*, 58, 23.

Blumberg, H. H., Hare, A. P. and Costin, A. (2006). *Peace Psychology: A Comprehensive Introduction*. Cambridge: Cambridge University Press.

Blumberg, H. H., Kent, M. V., Hare, A. P. and Davies, M. F. (2012). *Small Group Research: Implications for Peace Psychology and Conflict Resolution*. New York and London: Springer.

Boholm, Å. (2008). The public meeting as a theatre of dissent: risk and hazard in land use and environmental planning. *J. Risk Res.*, 11, 119–140.

Bowles, S. and Gintis, H. (2003). Origins of human cooperation. In *Genetic and Cultural Evolution of Cooperation*, ed. P. Hammerstein, pp. 429–444. Cambridge, MA: MIT Press.

Canning, L. and Hanmer-Lloyd, S. (2007). Trust in buyer-seller relationships: the challenge of environmental (green) adaptation. *Eur. J. Marketing*, 41, 1073–1095.

Christie, D. J., Wagner, R. V. and Winter, D. D. (2001). *Peace, Conflict, and Violence: Peace Psychology for the 21st Century*. Upper Saddle River, NJ: Prentice Hall.

Cialdini, R. B. (2003). Crafting normative messages to protect the environment. *Curr. Dir. Psychol. Sci.*, 12, 105–109.

Coleman, P. T. and Deutsch, M. (2012). *The Psychological Components of a Sustainable Peace*. New York and London: Springer.

Collis, C. (2009). The geostationary orbit: a critical legal geography of space's most valuable real estate. *Sociol. Rev.*, 57, 47–65.

de Groot, J. I. M. and Steg, L. (2008). Value orientations to explain beliefs related to environmental significant behaviour: how to measure egoistic, altruistic, and biospheric value orientations. *Environ. Behav.*, 40, 330–354.

Deutsch, M. (1977). *The Resolution of Conflict: Constructive and Destructive Processes*. New Haven, CT: Yale University Press.

Deutsch, M. (1998). Constructive conflict resolution: principles, training, and research. In *The Handbook of Interethnic Coexistence*, ed. E. Weiner, pp. 199–216. New York: Continuum Publishing.

Deutsch, M., Coleman, P. T. and Marcus, E. C. (eds.) (2006). *The Handbook of Conflict Resolution: Theory and Practice*. Second edition. San Francisco: Jossey-Bass.

Edwards, W., Miles, R. F. and von Winterfeldt, D. (2007). *Advances in Decision Analysis: From Foundations to Applications*. Cambridge: Cambridge University Press.

Fisher, R. J. (2006). Intergroup conflict. In *The Handbook of Conflict Resolution: Theory and Practice*. Second edition, eds. M. Deutsch, P. T. Coleman and E. C. Marcus, pp. 176–196. San Francisco: Jossey-Bass.

Gifford, R. (2007). Dedicated to government-level solutions: can peace be made through resource treaties? *Peace and Conflict: J. Peace Psychol.*, 13, 127–128.

Gillespie, J. (2011). Exploring the limits of the judicialization of urban land disputes in Vietnam. *Law Soc. Rev.*, 45, 241–276.

Hardin, G. (1968). The Tragedy of the Commons. *Science*, 162, 1243–1248.

Hartter, J. (2009). Attitudes of rural communities toward wetlands and forest fragments around Kibale National Park, Uganda. *Hum. Dimens. Wildl.*, 14, 433–447.

Haste, H. (2010). Citizenship education: a critical look at a contested field. In *Handbook of Research on Civic Engagement in Youth*, eds. L. R. Sherrod, J. Torney-Purta and C. A. Flanagan, pp. 161–188. Hoboken, NJ: John Wiley and Sons Inc.

Heinen, J. T. and Shrivastava, R. J. (2009). An analysis of conservation attitudes and awareness around Kaziranga National Park, Assam, India: implications for conservation and development. *Popul. Environ.*, 30, 261–274.

Hochman, J. (2007). Return to nature. *J. Mental Imagery*, 31, 62–66.

Jacob, J., Jovic, E. and Brinkerhoff, M. B. (2009). Personal and planetary well-being:

mindfulness meditation, pro-environmental behaviour and personal quality of life in a survey from the social justice and ecological sustainability movement. *Soc. Indic. Res.*, 93, 275–294.

Johnson, D. W. and Johnson, R. T. (2000). Civil political discourse in a democracy: the contribution of psychology. *J. Peace Psychol.*, 6, 291–317.

Johnson, D. W. and Johnson, R. T. (2006). Conflict resolution, peer mediation, and peacemaking. In *Handbook of Classroom Management: Research, Practice, and Contemporary Issues*, eds. C. M. Evertson and C. S. Weinstein, pp. 803–832. Mahwah, NJ: Lawrence Erlbaum Associates Publishers.

Jurin, R. R. (2012). *Principles of Sustainable Living: A New Vision for Health, Happiness, and Prosperity.* Champaign, IL: Human Kinetics.

Jurin, R. R. and Hutchinson, S. (2005). Worldviews in transition: using ecological autobiographies to explore students' worldviews. *Environ. Educ. Res.*, 11, 485–501.

Kurzban, R. (2003). Biological foundations of reciprocity. In *Trust and Reciprocity: Interdisciplinary Lessons from Experimental Research*, eds. E. Ostrom and J. Walker, pp. 105–127. New York: Russell Sage Foundation.

Langholtz, H. J. (2010). International relations, technology, conflict, and a changing world, can create unseen opportunities for service. In *Change (Transformation) in Government Organizations*, ed. R. R. Sims, pp. 233–254. Charlotte: IAP Information Age Publishing.

Martín, A. M., *et al.* (2008). Individual breaches of environmental laws in cases from public administration files. *Deviant Behav.*, 29, 611–639.

McKenzie-Mohr, D. and Oskamp, S. (1995). Psychology and sustainability: an introduction. *J. Soc. Issues*, 51, 1–14.

Milfont, T. L. and Gouveia, V. V. (2006). Time perspective and values: an exploratory study of their relations to environmental attitudes. *J. Environ. Psych.*, 26, 72–82.

Mio, J. S., Thompson, S. C. and Givens, G. H. (1993). The commons dilemma as metaphor: memory, influence, and implications for environmental conservation. *Metaphor Symb. Act.*, 8, 23–42.

Mouro, C. and Castro, P. (2010). Local communities responding to ecological challenges – a psycho-social approach to the Natura 2000 Network. *J. Commun. Appl. Soc.*, 20, 139–155.

Mumford, M. D., *et al.* (2007). Environmental influences on ethical decision making: climate and environmental predictors of research integrity. *Ethics Behav.*, 17, 337–366

Nelson, R. M. (2005). Justice, lead, and environmental research involving children. In *Ethics and Research with Children: A Case-based Approach*, ed. E. Kodish, pp. 161–178. New York: Oxford University Press.

Nowak, A., Bui-Wrzosinska, L., Coleman, P. T., Vallacher, R., Jochemczyk, L. and Bartkowski, W. (2010). Seeking sustainable solutions: using an attractor simulation platform for teaching multistakeholder negotiation in complex cases. *Negotiation J.*, 26, 49–68.

Oceja, L. and Berenguer, J. (2009). Putting text in context: the conflict between pro-ecological messages and anti-ecological descriptive norms. *Span. J. Psych.*, 12, 657–666

Perloff, R. M. (2010). *The Dynamics of Persuasion: Communication and Attitudes in the 21st Century.* Fourth edition. New York: Routledge.

Plokhii, O. (2012). Death of a forester: Chut Wutty sought to save one of Indochina's last great ecological sanctuaries. It cost him his life. *Newsweek [International Edition]*, 160, 44–49.

Pratarelli, M. E. and Johnson, S. M. (2012). Exploring the paradox in Hardin's 'Tragedy of the Commons': a conservation psychology classroom exercise. *Ecopsychology*, 4, 158–165.

Richerson, P. J., Boyd, R. T. and Henrich, J. (2003). Cultural evolution of human cooperation. In *Genetic and Cultural Evolution of Cooperation*, ed. P. Hammerstein, pp. 357–388. Cambridge, MA: MIT Press.

Rosen, J. and Haaga, D. A. F. (1998). Facilitating cooperation in a social dilemma: a persuasion approach. *J. Psych.*, 132, 143–153.

Sandy, S. V. (2006). The development of conflict resolution skills: preschool to adulthood. In *The Handbook of Conflict Resolution: Theory and Practice*. Second edition, eds. M. Deutsch, P. T. Coleman and E. C. Marcus, pp. 356–388. Hoboken, NJ: Wiley Publishing.

Scharl, A. (2004). *Environmental Online Communication*. London: Springer.

Scharl, A. (2006). Catalyzing environmental communication through evolving internet technology. In *The Environmental Communication Yearbook* (Vol. 3), ed. S. P. Depoe, pp. 235–242. Mahwah, NJ: Lawrence Erlbaum Associates.

Sivaramakrishnan, K. and Vaccaro, I. (2006). Postindustrial natures: hyper-mobility and place-attachments. *Soc. Anthropol.*, 14, 301–317.

Smith, E. A. (2003). Human cooperation: perspectives from behavioral ecology. In *Genetic and Cultural Evolution of Cooperation*, ed. P. Hammerstein, pp. 401–427. Cambridge, MA: MIT Press.

Stevens, M. J. (2007). Orientation to a global psychology. In *Toward a Global Psychology: Theory, Research, Intervention, and Pedagogy*, eds. M. J. Stevens and U. P. Gielen, pp. 3–33. Mahwah, NJ: Lawrence Erlbaum Associates.

Sugiura, J. (2005). Environmental education using Persuasion Game and its transferability. *Jpn. Psych. Rev.*, 48, 134–139.

Timpson, W. M., AlAndejani, J., Gabriel, M. and Schafer, S. (2012). Stories on the edge: transforming education with case studies of sustainability, diversity, and peace. In *Transformative Eco-education for Human and Planetary Survival*, eds. R. L. Oxford and J. Lin, pp. 25–40. Charlotte, NC: IAP Information Age Publishing.

Tzou, C., Scalone, G. and Bell, P. (2010). The role of environmental narratives and social positioning in how place gets constructed for and by youth. *Equity Excell. Educ.*, 43, 105–119.

Valverde, M. (2011). Seeing like a city: the dialectic of modern and premodern ways of seeing in urban governance. *Law Soc. Rev.*, 45, 277–312.

van der Veer, G. (2009). Review of 'A world turned upside down. Social ecological approaches to children in war zones'. *Intervention*, 7, 69–70.

Walden, R. (2009). The school of the future: conditions and processes – contributions of architectural psychology. In *Schools for the Future: Design Proposals from Architectural Psychology*, ed. R. Walden, pp. 75–122. Ashland, OH: Hogrefe and Huber Publishers.

Webb, S. M., Davidson, D. J. and Boyce, M. S. (2008). Trapper attitudes and industrial development on registered traplines in West-Central Alberta. *Hum. Dimens. Wildl.*, 13, 115–126.

Weisner, T. S. (2010). Well-being, chaos, and culture: sustaining a meaningful daily routine. In *Chaos and its Influence on Children's Development: An Ecological Perspective*, eds. G. W. Evans and T. D. Wachs, pp. 211–224. Washington, DC: American Psychological Association.

White, R. (2011). Ecological crime. In *The Routledge Handbook of Deviant Behaviour*, ed. C. D. Bryant, pp. 456–462. New York: Routledge/Taylor and Francis Group.

Wilmoth, G. H. (2012). Addressing the mental health effects of climate change. *PsycCRITIQUES*, 57, 15.

Wilson, P. I. (2008). Preservation versus motorized recreation: institutions, history, and public lands management. *Soc. Sci. J.*, 45, 194–202.

Winter, D. D. (2007). Understanding identification with the natural environment. *Peace and Conflict: J. Peace Psych.*, 13, 247–250.

Box 9

The swan grazing conflict in chalk rivers

Kevin A. Wood,[1,2] Richard A. Stillman,[2] Francis Daunt[1] and Matthew T. O'Hare[1]
[1]*Centre for Ecology and Hydrology, Midlothian EH26 0QB, UK*
[2]*School of Applied Sciences, Bournemouth University, Dorset BH12 5BB, UK*

The crystal clear waters of the chalk rivers of southern and eastern England are dominated by a keystone plant species, water crowfoot *Ranunculus penicillatus* ssp. *pseudofluitans*, which supports an ecosystem of high conservation value, including abundant invertebrates and fish (Berrie, 1992). This ecosystem has supported the development of economically valuable sport fisheries on many chalk rivers. In recognition of its keystone role in chalk rivers, water crowfoot is protected under the EU Habitats and Species Directive (92/43/EEC). Within the last 30 years, fisheries and conservation interests have become concerned that the foraging by mute swans *Cygnus olor* on water crowfoot degrades the habitat of invertebrates, fish and other animals (Wood *et al.*, 2014). Mute swans are a native species and a natural part of the chalk river ecosystem, but the population in Britain has almost doubled since the 1970s. Some sport fisheries organisations have reported declining membership and income due to swan grazing. Thus, a conflict has arisen between conservationists and anglers on one side and those who seek to protect the swans on the other. Moreover, management of this grazing conflict is complicated not only because mute swans are popular with the public, but also they are protected under both the EU Wild Birds Directive (79/409/EEC) and the Wildlife & Countryside Act 1981. However, there has been little research into swan grazing damage to plants or management options to alleviate the conflict. The few management options that have been explored by stakeholders, such as egg removal and behavioural modifications, have not been successful in preventing grazing damage (Maudsley, 1996; Parrott and McKay, 2001; Watola *et al.*, 2003). Furthermore, early in the conflict there were incidences of illegal shooting of swans, which did nothing to solve the conflict but illustrate the potential problems which can occur when certain groups feel that no action is being taken towards resolution (Trump *et al.*, 1994).

In 2008 we began a four-year research programme into the swan grazing conflict in the River Frome catchment (Dorset, England), a typical chalk river. At the start of the programme we arranged meetings with the different stakeholders to discuss what research was needed to resolve the swan grazing conflict. The stakeholders included representatives from local angling clubs, conservation organisations and land managers directly affected by the grazing conflict, as well as local groups interested in birds. Authorities with statutory responsibilities for the chalk river ecosystem, such as Natural England and the Environment Agency, also participated in these discussions. The stakeholders identified key knowledge gaps, such as the effects of grazing on the plant community, and suggested sites where field studies could be undertaken. These discussions allowed us to identify research needs and ensure that our questions were relevant to the stakeholders involved in the swan grazing conflict. In conjunction with these stakeholders we developed two key objectives. First, we proposed to quantify the effects of swan grazing on the plant community. Second, we proposed to evaluate three management strategies: population control, 'sacrificial' feeding areas (adjacent terrestrial areas where swans could feed away from water crowfoot), and river flow modifications, which stakeholders had suggested would be acceptable options to attempt to alleviate grazing damage. We maintained contact with these stakeholders throughout the programme, which allowed us to share our results with those directly involved in the conflict. We provided the stakeholders with quantitative information regarding the conditions under which grazing conflicts were likely to occur (Wood *et al.*, 2012, 2013a) and how habitat management might be used to reduce these conflicts. Our research was also able to help stakeholders avoid other management options which we predicted would be ineffective, such as controlling the swan population through egg destruction, translocations and culling (Wood *et al.*, 2013b). Nevertheless, there has been no clear, viable management solution developed that would fully resolve the conflict. Therefore, the swan grazing conflict remains currently unresolved, but stakeholders continue discussing mutually acceptable methods of preventing grazing damage. Our experiences show that by involving stakeholders in the design stages, researchers can provide useful scientific information of direct relevance to stakeholders. Moreover, meaningful participation by stakeholders keeps the parties in conflict engaged in seeking a viable resolution to the swan grazing conflict.

References

Berrie, A. D. (1992). The chalk stream environment. *Hydrobiologia*, 248, 3–9.

Maudsley, M. J. (1996). *Swans and Agriculture: A Scoping Study of the Impact of Swans on Agricultural Interests in Britain*. MAFF commissioned R&D Project VCO108. Cambridge: ADAS.

Parrott, D. and McKay, H. V. (2001). Habitat preferences and nest site selection by mute swans: an investigation into the potential for managing swan distribution in leisure fisheries. In *Advances in Vertebrate Pest Management – Volume II,* eds. H. J. Pelz, D. P. Cowan and C. J. Feare, pp. 263–282. Furth: Filander Verlag.

Trump, D. P. C., Stone, D. A., Coombs, C. F. B. and Feare, C. J. (1994). Mute swans in the Wylye Valley: population dynamics and habitat use. *Int. J. Pest Manage.,* 40, 88293.

Watola, G. V., *et al.* (2003). Analyses of two mute swan populations and the effects of clutch reduction: implications for population management. *J. Appl. Ecol.*, 40, 565–579.

Wood, K. A., Stillman, R. A., Clarke, R. T., Daunt, F. and O'Hare, M. T. (2012). Understanding plant community responses to combinations of biotic and abiotic factors in different phases of the plant growth cycle. *PLoS ONE*, 7, e49824.

Wood, K. A., Stillman, R. A., Coombs, T., McDonald, C., Daunt, F. and O'Hare, M. T. (2013a). The role of season and social grouping on habitat use by mute swans (*Cygnus olor*) in a lowland river catchment. *Bird Study*, 60, 229–237.

Wood, K. A., Stillman, R. A., Daunt, F. and O'Hare, M. T. (2013b). Evaluating the effects of population management on a herbivore grazing conflict. *PLoS ONE*, 8, e56287.

Wood, K. A., Stillman, R. A., Daunt, F. and O'Hare, M. T. (2014). Chalk streams and grazing mute swans. *Br. Wildl.*, 25, 171–176.

© Catherine Young.

CHAPTER TEN

Conservation conflicts: ethical issues

NIGEL DOWER
University of Aberdeen

Ethics concerns the nature and justification of moral values and their application to decisions about how to act. How should we act? What kind of life should we live? What are the elements of a 'good' life? Moral values cover various norms of behaviour, such as telling the truth, keeping promises, not causing harm, acting justly, respecting rights and acting for others. These moral norms generally are accepted at least partly because their observance contributes to the realisation of well-being. So we also need, at least, an account of what human well-being consists of – things like pleasure, happiness, health, knowledge, love, relationships, and appreciation of nature. Different thinkers have different views about what the norms are and what the 'goods' are, and how to interpret and justify them. Nevertheless, the ethical point of view is generally contrasted with the self-interested point of view, because it is concerned with what is good, just and right from the point of view of everyone affected by our actions. Therefore, here I present my personal reflections as a philosopher, following 30 years of thinking about environmental issues.

One key issue in environmental ethics that arises in conservation conflicts is this: for whom is conservation being done – just current human beings, or other groups such as future human beings or non-human animals? Another issue is what is the nature of well-being of human beings? It makes a big difference whether what is important to human well-being is having (more) wealth and possessions or having quality in human relationships, which includes our relationship with nature. Being clearer about these questions sharpens our understanding of environmental conflicts by clarifying the underlying values of each position, the nature of differences (deep or superficial) and whether there is a reasonable possibility of compromise.

Practical and theoretical conflicts

Conflicts over the environment fall into two broad categories – practical conflicts about what ought to be done, and theoretical conflicts about the reasons why something ought to be done. There are different views about what ought to be done in respect to a given environment (e.g. whether to make certain changes

Conflicts in Conservation: Navigating Towards Solutions, ed. S. M. Redpath, R. J. Gutiérrez, K. A. Wood and J. C. Young. Published by Cambridge University Press.

to it or not). It might be small scale, such as Donald Trump's application in 2007 to develop the Menie estates in Scotland, a mobile sand dune system protected as a Site of Special Scientific Interest, into a golf course complex. In contrast, it might be a much larger scale, such as the controversial decision of the Icelandic government in 2002 to authorise the Kárahnjúkar Hydropower Plant which involved damming a vast area of wilderness in eastern Iceland to provide electricity for an aluminium plant (Saving Iceland, 2011). Such conflicts often emerge in response to planning applications for development of particular areas or when new government policies are made. Yet usually conflicts exist not because of a planning application or enquiry, but because different groups of people have *ongoing* differences of view about what ought to happen to an environment.

There are different kinds of reasons that one has for policies one supports. That is, conflicts are theoretical because they involve different understandings of environmental ethics and of how humans relate to the natural world. For example, contrast the different rationales of John Muir, founder of the Sierra Club, who wanted to preserve wilderness for its own sake, and Gifford Pinchot, pioneer of sustainable forestry, whose interest was the long-term interests of humans.

These are ethical conflicts insofar as the reasons given for different practical responses involve different general values as goals – economic well-being, long-term human interests, the flourishing of nature. They are not just conflicts between different people with different *private* interests. Private interests often lie behind policies, but what *justifies* them is an appeal to the public good.

There is also the difference between conflicts over values and factual understanding. People take different sides in conservation conflicts because they may have different values (e.g. they value biodiversity protection or value livelihoods; see Box 16), or because they may have different understandings of the facts. Two parties may accept that the same values – environmental and non-environmental – apply to a particular issue, but take different sides in a conflict because they differ on their factual assessments.

Environmental ethics

Environmental ethics has three dimensions: first and foremost, what are the objects of moral concern – humans, higher animals, all animals, plants, species, ecosystems? Second, where are these objects located – where one lives locally, one's country, the world? Third, what is the timescale of our concern – the immediate future, our own lifetime and that of our close descendants or the indefinite future? (Rolston, 1988; Light and Rolston, 2002; Attfield, 2003.)

These dimensions can be combined in different ways. Assuming that someone adopts an ethical point of view at all (as opposed to a totally self-interested or hedonistic approach), there is at one extreme someone who cares for the environment in his own locality only as it affects his fellow human beings in the immediate and fairly near future, with little regard for the rest of the world. Such people might care, for instance, about the natural environment as something which, where they live, is their 'field of significance' and that of those with whom they live, but not care about climate change. At the other extreme, some people might care for all life on the planet and do so in regard to the indefinite future. Their position would in effect be a commitment to sustainability in a strong sense, namely 'we ought to sustain the conditions of life for life as a whole for as long as we can'.

In the middle there are many other combinations. For instance, one might care for humans and non-human life in one's own community and want these conditions to continue into the indefinite future, but not actually be much concerned about life elsewhere in the world – except insofar as what happens elsewhere in the world may have a negative impact on life in one's own area. Thus, one may have only *instrumental* reasons for being concerned about what happens globally, but not because one thought that life, human or non-human, elsewhere in the world really matters. Thus, one's position would be a communitarian one concerning the source of ethical obligation in one's community, rather than a global ethic or cosmopolitan one. Again, one might indeed care for the conditions of life for humans everywhere in the world, but neither accept obligations towards future generations nor accept that forms of non-human life were objects of moral concern.

Anthropocentric and non-anthropocentric values

I first consider different views about intrinsic value. Judgements about what should be done on the basis of these values may also be influenced by how one answers the questions 'locally or everywhere' and 'now or indefinitely'. In most situations the practical action may be much the same, but in some other situations there may be differences depending on the answers to these other questions.

It is common in environmental ethics to distinguish three main approaches to valuing the environment: anthropocentrism, biocentrism and ecocentrism.

Anthropocentrism

From an anthropocentric point of view, the environment is *for* humans in the sense that it is human interests that are served by the environment. What do we mean by human interests? What do humans want from the environment?

First, we want a steady supply of resources needed for life and life's comforts and enjoyments. These things, whether food, water or other resources, come either from one's environment locally or come from other places through trade. Second, we want the environments we live in to be health-sustaining, whether it is the water we drink, the air we breathe, or the soil from which our food comes. Insofar as goods come from elsewhere, we want these to be health-sustaining as well. Third, we want our environments to be positive 'fields of significance', as we perceive them and live in them, and as they enter into the experiences of life (see Cooper, 1992). This could be called the aesthetic dimension, although 'aesthetic' is possibly too narrow a term to cover the 'experiential' dimensions to what we value in our environments, including a spiritual dimension. This kind of concern expresses itself naturally in the context of local environments such as a concern about litter on streets or beaches, or the effects of a wind farm on natural landscape. However, it can also have a wider global dimension. A person's engagement with the world may be such that he values the environment as a whole as the 'home of humans', an approach linked to what is called 'biophilia' (Wilson, 1984). If we value environments for these kinds of reasons, then we will welcome changes that maintain or advance them and resist changes that threaten them.

Even before we bring in non-human considerations, we can see how these different sources of human-centred valuing of the environment can bring radically different approaches to conservation conflicts. A planning application for a wind farm illustrates this (Box 10). A key argument for the wind farm, apart from its contribution to the local economy, is its contribution to reducing CO_2 emissions, a clear environmental argument. Those who take it seriously take the human interests we are concerned about to be global and long-term. Those who oppose a wind farm development may do so because of another environmental consideration: the preservation of an area of natural beauty and environmental tranquillity, enjoyed by countless people, either residents or visitors to the area, that is part of a community's heritage (its 'generational continuum'; see Thompson, 2001).

Biocentrism

Biocentrism accords an independent value to other living things, irrespective of whether their existence benefits humans. Whatever we recognise as having this independent value, we as humans have a duty to act both to promote their conditions of life and not to destroy or harm other living things without sufficient cause.

It is worth noting here, because it is particularly relevant to the way conservation conflicts unfold, that environmentalists for example that want to preserve habitat of certain birds do so because they *value* these birds (e.g. Box 14). Although this may be seen as a human-centred consideration, that is misleading. If bird-watchers want to preserve habitat because they want to preserve the basis of their enjoyment and that of fellow bird-watchers, this is one source of concern. If they want to preserve habitat for biocentric reasons (e.g. because they feel humans have no right to destroy the habitats of other species without good cause) that is another. If they have positive feelings about preserving it, this feeling is a *consequence* of their judgement of value and not its *source*. Among naturalists both sources of concern for habitats often go together, but they need not.

As with the anthropocentric perspective there are multiple ways of advancing the biocentric position, some more limited/weak, some more extensive/strong. A thoroughgoing position might be called biocentric egalitarianism as illustrated in *Respect for Nature* (Taylor, 1986). In this view all life has equal moral status or equal right to life, from higher animals, through trees and plants, to microbes. The ethical consequences of this are fairly dramatic. Taylor does, of course, acknowledge that humans like other living things do have a right to life. This means they may impinge on other living things for self-preservation (e.g. getting food) and self-protection (which covers warding off tigers and microbes). However, Taylor argues that we infringe the rights of other living things if we harm or destroy them for other reasons (e.g. for the sake of luxuries).

The biocentric position need not take this strong position, and in practice most people attracted to biocentrism do not. Even if one takes the perspective that all life has a value, one may accept there is a hierarchy involved, such as an ordering of humans, mammals, birds, fish, trees, other plants and insects. The point remains that in acknowledging this independent status one is claiming that humans should not destroy or damage other living things or their habitats without sufficiently good cause. Few environmentalists would, in practice, regard environmental changes as only justified if they are for the sake of self-preservation or self-protection, but they may regard many types of typical modern developments as unjustified land changes that damage or destroy lives and habitats, especially if there are alternative ways of pursuing such development.

Ecocentrism

Unlike the individualist assumptions behind biocentrism, ecocentrism is a holistic approach. It is the whole ecosystem that has an important and overarching value. What gives an ecosystem its value? Various characteristics are suggested: integrity, stability, beauty, dynamism, resilience, diversity, and so on. While some environmentalists certainly have a value preference for those ecosystems or areas of nature that are relatively unaffected by human interests, hence the special value often attributed to 'wilderness' (Elliott, 1982), the idea of a well-functioning ecosystem does not preclude human activities being part of it, nor does it preclude the idea that ecosystems that have been impoverished can be restored by appropriate human interventions. Nevertheless, in the context of most conservation conflicts the appeal to ecocentric values is generally an appeal to arguments for preserving well-functioning and diverse ecosystems against the kinds of changes driven by certain conceptions of human progress.

There is a particular way of marking the distinctiveness of the ecocentric perspective in conservation conflicts, especially if the ecocentrist formulates and applies an overarching ethical principle. One such principle from Leopold (1949: 224–225) states: 'A thing is right when it tends to preserve the integrity, stability, and beauty of the biotic community. It is wrong when it tends otherwise'. If this principle is seen as *the* ethical principle or as the *primary* ethical principle then it leads to a downgrading of humans. This is seen by many as unacceptable because all concerns for human well-being and for the ethical principles that govern human relationships with each other are seen as unimportant or secondary. However, if it is seen as an important *corrective* to an overbearing concern for or privileging of human welfare, the principle is less challenging and still extremely significant – although it leaves much work to be done over assessing how the balances are to be struck in decision-making. Another of Leopold's telling statements is that 'a land ethic changes the role of *Homo sapiens* from conqueror of the land-community to plain member and citizen of it' (Leopold, 1949: 204). It is this gestalt shift (i.e. radical change of perspective) that is of the last significance.

Arguments of this kind are not merely likely to come up against those from a human-centred focus on development, they may also come against arguments drawn from other more individualist biocentric perspectives. Arguments from consideration of ecosystem balance may favour culling certain species of animal or plant, which may be resisted by those who regard the deliberate destruction of living things as wrong (e.g. Box 4; Box 9). This tension between the holistic and the individualistic perspectives was famously highlighted by Callicott (1980).

Futurity

All purposive action is future-oriented, but the question is: how far into the future? It is clear that most human beings have a natural concern for their own future (prudence) and also for those who they love because love is, beyond mere sentiment, caring for the good, current and continuing, of the one loved. Moral concern for the future can be seen as the generalisation of these concerns of prudence and love. If morality is about not undermining (non-maleficence) and to some extent promoting (beneficence) the good of others generally, then morality is by implication concerned with at least the future of the next two generations of humans. Passmore (1974), an early environmental philosopher, once gave the argument that in practice we need not care for more than the next two generations because current actions will not make any difference beyond that point. The analogy he gave was of a ripple in the pond eventually disappearing. While in practice he was right about many of our activities impacting the environment, issues such as climate change and species loss seem now major exceptions to this.

Should that concern extend beyond the next two human generations? Any theory of ethics which makes ethical obligation depend on mutual reciprocity or a convention or contract has difficulty in understanding moral relations between members of generations if members of the future generations are not around to affect us. This is because such theories work on the basis that I will not harm you and yours provided you do not harm me and mine.

By contrast, the crucial point accepted by many environmental philosophers is this: if it is in our power to affect the well-being of other humans (or non-humans), then that is morally relevant, in principle, to decision-making. It does not matter whether some future state which we could affect is one affecting the next two generations of humans or not. It simply matters that we could affect it.

Even without bringing in distant future people and non-humans, it must be apparent that serious concern about human impacts in the next hundred years, as with climate change, comes from a very different perspective to concerns about preserving particular areas of the natural environment. The latter are often driven by much shorter timescales and motivations. Yet what really lies at the centre of such tension is not merely a present versus future concern, but also a tension between a local and a global perspective.

Local and global

What characterises many environmental disputes is a tension between concerns for the local versus global environment (Cooper, 1992). For any person or group of people there is a sense in which 'their' environment is that which surrounds them, is perceived in their active engagement in it as a 'field of significance' for them, and is primarily the location in which they move and have their being. Insofar as their environment is good for them, then they want to preserve it and insofar as it is bad for them, they want to change and improve it. The question is whether in addition they have (moral) reason in general to uphold other people's desire to sustain their environments, if good, or change them, if bad. Or, to put it another way, do they have moral reason to advocate and work for protecting the environment as a whole or the global environment (Dower, 1994)? Again, as with considerations about the future, the issue is raised about how to think about the nature of morality. Should one defend a view of morality that dismisses or makes light of our moral obligations to all other peoples (and by extension to other kinds of living things anywhere), or advocate some form of global or cosmopolitan ethic or responsibility (Dower, 2007)? Such a global ethic is resisted or rejected for a variety of reasons, such as communitarianism or relativism. Communitarianism claims that serious moral obligation arises from a well-established community with shared ethical traditions, which the world is clearly not. Relativism claims that ethical values vary from society to society so there are no universal values upon which to found a global ethic. Certain theories of international relations claim that states are entitled to pursue their own agendas and priorities, with little or no regard for what is good from a global point of view (see Dower, 2007). Whether or not a cosmopolitan approach to environmental problems is accepted, the approach is implicit in many environmentalists' moral concern about global impacts of human activity. Certainly sometimes sources of conflict in conservation conflicts lie in these rather different ethical bases for caring for the environment. Concern for climate change action may come from a cosmopolitan perspective in a way that concern for protecting an area of natural beauty or the habitats of particular species may not.

Ethical convergence and divergence

Having identified some sources of environmental values that may lie in the background of different perspectives in conservation conflicts, I now want to emphasise a contrasting point; namely, that as often as not thinkers with different theoretical positions will agree on what they want to be done or decided. Such areas of convergence may be short-term, contingent and pragmatic, arising merely in a particular case, or long-term and grounded in wider commonalities of value. Thus those who hunt as a sport may be in alliance with a group wanting to preserve an area of countryside in order to stop certain development whereas on most other issues they would be diametrically opposed.

Often battle lines on conservation conflicts are drawn between certain groups of anthropocentrists and other groups of anthropocentrists who have a *general* convergence and common cause with the non-anthropocentrists. The former group then are people who either have no or little interest in environmental issues, or, if they do have an interest, think that really little needs to be done to change the ways we behave. Given their reading of the 'facts' of our environmental situation and their understanding of what is important to human well-being, the arguments of the green environmentalists are seen as ill-founded in fact and in value.

However, the latter group of 'green' anthropocentrists may actually agree with most of the things that non-anthropocentrist thinkers favour. (For the importance of recognising convergence, see Norton, 1991.) This may be because they have another reading of the 'facts' (e.g. about climate change or the effects of industrial pollution) or have a more serious commitment to thinking in the long term about impacts on future generations. Or they may have a view of human well-being that places much less emphasis on material prosperity and much more emphasis on various other forms of well-being in meaningful activities, social relations and relations with the natural world (Fromm, 1979). In the last respect the ecocentric perspective and such a form of anthropocentrism actually tend to converge. Being 'plain members of the biotic community' is after all a *relational* fact, and it is this relating to the natural world that is an important part of human well-being for many anthropocentrists.

The conservation conflict which arose over the application to turn a sand dune into a golf course illustrates this well. Those who favoured it saw economic benefits, and if they thought of environmental impacts at all did not see destroying an SSSI as being of great significance. Those who wished to preserve it wished to preserve an area of wildness which had value both for anthropocentric and non-anthropocentric reasons, which was reflected in its SSSI status (and as well as other reasons such as concern for the rights of local people).

Environmental politics

Environmental politics refers to the processes in which environmental decisions are made, particularly where in a given place, region or country different groups of people bring different priorities and agendas into the decision-making process. There are four kinds of question here, as explored below. My perspective assumes that politics is, as Aristotle argued, in a certain sense subservient to ethics.

How are decisions actually made?

This question concerns the diverse ways in which environmental decisions get made. Whether or not one thinks it ought to be otherwise, politics is often about the exercise of power. The relative power of an organisation or individual in influencing decisions in a conservation conflict has to do with the skill of their advocacy, their general public standing and extent of their followers, how financially strong they are, whether and to what extent powerful organisations and individuals exercise 'influence' behind closed doors, and so on. This belongs to political sociology. However, it is no doubt important to understand political power if one wishes to exercise or counter it.

What are the accepted norms that are meant to govern procedures in resolving conservation conflicts?

This question relates to the rules that govern public enquiries or planning applications in which parties can take part, and how decisions are properly made. What these procedures are and the associated norms vary considerably from country to country, region to region, and on the type of conflict involved. It is important that the norms and procedures are properly applied so the outcome is perceived as legitimate.

What ought to be done to achieve an outcome one wishes?

When, in what is now referred to as the Chipko movement, women in India decided to hug the trees in their area to prevent the loggers from cutting down the trees that constituted their home, they were certainly not following accepted procedures, yet many environmentalists throughout the world approved of their action (Weber, 1988). A large march or rally may not be a formal part of a planning decision that affects the environment, but it may be seen as a legitimate way of voicing a perspective. Such intervention may be judged as morally appropriate because it contributes to what one wants. When Greenpeace attempts to stop whalers, many endorse what they do because they approve of trying to stop whaling.

What should the norms be that apply to the resolution of conservation conflicts?

Some people, often those with formal power, generally disapprove of direct non-violent action as a way of bringing influence to bear on decisions, yet, they might make exceptions as in the case of the Chipko movement because the end goal was deemed sufficiently important to justify unusual means. However, for those people who think that non-violent direct action has a legitimate place in decision-making, then this would be one's general ethical position on such matters. This is one reason why many environmentalists, while wholly sympathetic to Greenpeace's goal of stopping whaling, find it difficult to endorse what amounts to violent direct action because they are uneasy about allowing violent methods (as opposed to nonviolent ones) to have a legitimate influence on decision-making.

Complexity of ethical considerations

What the above discussion illustrates is that there is a whole range of ethical issues that are to some extent separate from those of environmental ethics discussed earlier about how decisions ought to be made when different parties with different views about the facts and environmental values engage with each other. I add just two further thoughts.

The range of relevant considerations: criteria of reasonableness

The range and relative strengths of ethical arguments that are brought to bear on a conservation conflict, and form the basis for decision-making, may be much broader than the range of ethical arguments that, for any given party, appear to be appropriate to the case. These arguments, as noted earlier, are not merely various kinds of environmental arguments; they are also other kinds of arguments, like the arguments for economic growth or the regeneration of an area.

For any given individual, faced with an environmental issue, the various ethical arguments the individual accepts are used to try and reach a practical conclusion about what ought to be done. Insofar as ethical enquiry is a rational enquiry, each individual uses reason to apply values and accepts the facts of the situation. Whether the individual *presents* these arguments to others in just this form or in other perhaps tailored forms is again an exercise of reason. However, this is a different kind of exercise of reason because it is the assessment of what will be most convincing.

When it comes to those who weigh evidence and make decisions in hearings, they are exercising reason in yet a third way. What is reasonable here is recognising the ethical perspectives of the various parties involved and recognising that there are appropriate procedures for reaching decisions that properly reflect these arguments as well as the status of the groups presenting them. The assessments of decision-makers are not merely from their point of view. It is what others, as third parties, can endorse both as observers and also as participants who have perhaps contributed their own specific lines of argument about what ought to be done.

The democratic paradox

I have already made the point that any party to a procedure that leads to decisions that affect them may acknowledge that the procedure, an election, a planning application, an enquiry or some other conflict resolution mechanism, if it is a fair procedure, may lead to outcomes other than what one they would have preferred. However, if a decision is reached in a democratic manner, is what is done right because it was democratically decided, or is it still wrong because that is what one judged it to be before the result (and one has not changed one's mind meantime over the primary considerations)? This arises in all areas of public decision-making.

There is no simple answer, certainly for any environmentalist who is also a citizen – that is, someone who acknowledges civil norms and values as well as environmental ones. The force of moral norms comes from various separate sources. Yet in a way this was already recognised earlier in this chapter where we saw that the ethical considerations that enter conservation conflicts come from diverse sources, both within the broad camp of environmental ethics and beyond them in other areas of ethical enquiry. If we take seriously the personal, social and political values that arise in human relations as well as those that arise directly from our human embodiment in the physical environment, we recognise the complexity of the moral life.

Conclusion

Conservation conflicts arise because different people want to do different things. This may be because of conflicting interests or different understandings of the facts. But often it is because they have different ethical values and goals. Clarifying what these values are helps make discussions of issues informed, reasonable and respectful. It is also helpful to recognise that there are ethical issues involved in how decisions involving conflicts should be made as well.

References

Attfield, R. A. (2003). *Environmental Ethics: An Overview for the Twenty-First Century.* Cambridge: Polity.

Callicott, B. (1980). Animal liberation: a triangular affair. *Environ. Ethics*, 4, 314–338.

Cooper, D. E. (1992). The idea of environment. In *The Environment in Question*, eds. D. E. Cooper and J. A. Palmer, pp. 165–180. London: Routledge.

Dower, N. (1994). The idea of the environment. In *Philosophy and the Natural Environment*, eds. R. A. Attfield and A. Belsey, pp. 143–156. Cambridge: Cambridge University Press.

Dower, N. (2007). *World Ethics: The New Agenda.* Edinburgh: Edinburgh University Press.

Elliot, R. (1982). Faking nature. *Inquiry*, 25, 81–93.

Fromm, E. (1979). *To Have or To Be.* London: Abacus.

Leopold, A. (1949). *A Sand County Almanac.* Oxford: Oxford University Press.

Light, A. and Rolston, H. (2002). *Environmental Ethics: An Anthology.* Oxford: Blackwell.

Norton, B. G. (1991). *Toward Unity Among Environmentalists.* Oxford: Oxford University Press.

Passmore, J. (1974). *Man's Responsibility for Nature.* London: Duckworth.

Rolston, H. (1988). *Environmental Ethics.* Philadelphia: Temple University Press.

Saving Iceland. (2011). http://www.savingiceland.org/2011/12/time-has-told-the-karahnjukar-dams-disastrous-economical-and-environmental-impacts/. Accessed 18 June 2013.

Taylor, P. (1986). *Respect for Nature.* Princeton: Princeton University Press.

Thompson, J. L. (2001). Planetary citizenship: the definition and defence of an ideal. In *Governing for the Environment*, eds. B. Gleeson and N. Low, pp. 135–146. Basingstoke: Palgrave.

Weber, T. (1988). *Hugging the Trees: The Story of the Chipko Movement.* New York: Viking.

Wilson, E. O. (1984). *Biophilia.* Cambridge, MA: Harvard University Press.

Box 10

Raptors and wind farms in Scotland

Aly McCluskie

Royal Society for the Protection of Birds, 2 Lochside View, Edinburgh Park, Edinburgh ED12 9DH, UK

Recently, there has been substantial expansion of the renewable energy industry in Scotland, particularly the development of wind farms. Renewable energy development is partly a consequence of governmental goals to generate 100% of energy from renewable sources and to achieve carbon neutral energy supplies as well as the availability of government subsidies. However, the locations and even existence of wind farms remain a source of conflict because of potential environmental impacts. This case study is unusual because a species is not at the centre of conflict; rather, species are being used, often with scant evidence, to justify a general opposition to wind farms.

Preferred sites for wind farm developments are often geographically coincident with remote upland areas used by raptors. Raptors and their habitats are protected by national and international legislation. Conflicts arise between those who aim to develop wind farms and those who seek to protect raptors and their habitats. However, the conflict dynamic is not as simple as development versus conservation; it is among those who have fundamentally different views on landscape, development and climate change.

Relationships between raptors and people in Scotland cannot be viewed without considering another longstanding environmental conflict between raptor conservationists and bird hunters (Box 3). As a consequence, some species of raptors are routinely killed to enhance hunting interests, driving populations of raptors to local extinction. The species that have historically been the most persecuted, such as golden eagle *Aquila chrysaetos* and hen harrier *Circus cyaneus* are among those that are potentially threatened by wind farm developments.

The main potential impacts of wind farms on raptors are collision mortality, disturbance and loss of habitat. The greatest of these risks is likely to be collision. While there are scant data on the processes that determine collision risk, there is consensus that birds sometimes collide with turbines, and raptors are one of the main bird groups at risk of collision (e.g. Gove *et al.*, 2013). Numerous studies show that birds either collide directly with turbines or are thrown to the ground after being caught in vortexes caused by rotating turbine blades. Mortality losses are generally low, but some areas such as Navarre and Tarifa in Spain (e.g. 124 raptors were recovered annually after fatal collisions across 20 wind farms at Tarifa; Ferrer *et al.*, 2012), Altamont Pass in the United States (1127 estimated annual raptor mortalities; Smallwood and Thelander, 2008) and Smøla in Norway (39 white-tailed eagle *Haliaeetus albicilla* collisions detected between 2005 and 2010; Dahl *et al.*, 2013) sustain high collision mortality. As raptors occur at relatively low densities, and are long-lived with a low reproductive output, additive mortality from collision could have large negative effects on population dynamics (Carrete *et al.*, 2009). These population-level effects will be greater where persecution also occurs.

Superficially, the conflict might suggest that more politically conservative stakeholders would object to renewable developments whereas more liberal stakeholders might be in favour. However, some unlikely partnerships have formed, with the green lobby uniting with large energy conglomerates to support development and private shooting estate owners allying with raptor conservationists in opposition.

Many of the direct issues with raptors can be resolved by careful site selection of wind farms. Clear evidence exists that topography is an important factor determining the potential for collisions (Ferrer *et al.*, 2012). A Scottish 'birds and wind farms sensitivity map' has been produced (Bright *et al.*, 2008) based on the locations and movements of sensitive species, and is of great value in 'positive planning'. However, despite this map, there can also be unpredictable consequences of wind farm development. At one Scottish wind farm, where there were no environmental objections to construction based on prior knowledge and environmental impact assessment, three hen harriers were killed, almost certainly by colliding with turbines, shortly after operations began.

Resolution to the conflict may potentially be found through scientific investigation that elucidates the extent of actual impacts. Research at the early stages of wind farm development could remove or reduce the uncertainty currently surrounding deployment of wind farms. Reducing uncertainty through research could lead to a less-aggressive and more useful debate among protagonists.

The Scottish Windfarm Birds Steering Group has recently been established to facilitate dialogue among industry, conservationists and government, and to provide independent scientific advice and research; this group has the potential to resolve many issues. However, even with effective communication of the results of scientific studies, it remains likely that the most entrenched protagonists will remain firmly in their position because their positions are defined not by logic but by predisposition.

References

Bright, J., Langston, R., Bullman, R., Evans, R., Gardner, S. and Pearce-Higgins, J. (2008). Map of bird sensitivities to wind farms in Scotland: a tool to aid planning and conservation. *Biol. Conserv.*, 141, 2342–2356.

Carrete, M., Sánchez-Zapata, J. A., Benítez, J. R., Lobón, M. and Donázar, J. A. (2009). Large scale risk-assessment of wind-farms on population viability of a globally endangered long-lived raptor. *Biol. Conserv.*, 142, 2954–2961.

Dahl, E. L., *et al.* (2013). White-tailed eagles (*Haliaeetus albicilla*) at the Smøla wind-power plant, Central Norway, lack behavioral flight responses to wind turbines. *Wild. Soc. Bull.*, 37, 66–74.

Ferrer, M., *et al.* (2012). Weak relationship between risk assessment studies and recorded mortality in wind farms. *J. Appl. Ecol.*, 49, 38–46.

Gove, B., Langston, R. H. W., McCluskie, A., Pullan, J. D. and Scrase, I. (2013). *Wind Farms and Birds*. Report prepared by BirdLife International on behalf of the Bern Convention RSPB/BirdLife in the UK.

Smallwood, K. S. and Thelander, C. (2008). Bird mortality in the Altamont Pass wind resource area, California. *J. Wildl. Manage.*, 72, 215–223.

CHAPTER ELEVEN

A view from sociology: environmental movement mobilisation over old-growth temperate rainforests in British Columbia

DAVID B. TINDALL
University of British Columbia
JOANNA L. ROBINSON
York University
and
MARK C. J. STODDART
Memorial University

Like many of the social sciences, sociology is a multi-paradigm science (Ritzer, 1975). Different approaches explain social phenomena at different scales (from the individual to the world system) and focus on different aspects of social reality (e.g. the distinction between the objective world and the subjective world). Another distinction is between descriptive and normative analysis. Descriptive analyses focus on explanation and understanding cause and effect relationships. Normative analyses focus on moral dimensions of issues and what we 'ought to do'. While sociological approaches often entail elements of both of these approaches, most work tends to emphasise one or the other. Also, there is a distinction between sociological work that has further theoretical explanation as a primary goal and work that is more applied – that is, work that applies past theory and research to practical empirical problems.

In sociology, there are a variety of views about conflict, and the orientation of any given analysis depends upon the theoretical framework and objectives of the researcher. Thus, the approach that a sociologist might take regarding conflict depends on where her work is situated with regard to these different considerations. Some sociologists might focus primarily on explaining social conflict, such as someone studying the causes of a revolution, while others might focus on trying to resolve it, such as those supporting a land management planning process. In some instances sociologists might actually be interested in facilitating conflict, such as those who work to mobilise collective action among members of an oppressed group. Some sociologists might focus on the

Conflicts in Conservation: Navigating Towards Solutions, ed. S. M. Redpath, R. J. Gutiérrez, K. A. Wood and J. C. Young. Published by Cambridge University Press.

mechanisms that generate conflict (descriptive analysis) – such as the factors that might underlie a conflict over clear-cut logging, while others might focus primarily on the moral dimensions of conflict (e.g. how can gender inequality be reduced in forest-dependent communities). Some sociologists might focus primarily on 'objective' indicators of conflict (e.g. the size of a social protest, and its political outcomes), while others might focus on subjective dimensions, such as how conflict is socially constructed (e.g. such as perceptions about the social values that underlie the conflict). Some sociologists might focus on interpersonal conflict between individuals, such as those focusing on conflict resolution among community members, while other sociologists might examine inter-state or inter-region conflict, such as political economists who might be interested in inter-state conflict over natural resources.

Conventional views about conflict predominantly take the position that conflict is bad, and that an absence of conflict is desirable. Many sociologists question these popular assumptions. From a sociological perspective, society is constituted by various groups who compete for resources, status and power. In many instances, particular groups (in terms of class, race, or gender) dominate other groups and exploit them in various ways (Collins and Annett, 1975). Sometimes stability in society is more a reflection of the complete dominance of ruling classes, rather than a reflection of inherent tranquillity. By contrast, the emergence of conflict can be an indicator of dissatisfaction with the status quo and the increasing power of dominated groups. Indeed, social conflict is one of the major drivers of social change. With respect to natural resources and natural resource management, conflict is often seen as a negative thing to be avoided. However, it is often through conflict that harmful environmental practices, policies and legislation are changed for the better.

Here we briefly consider several sociological approaches to conflict, primarily from the social movements, and the environmental sociology literatures. We then consider conflicts over forestry in British Columbia by focusing on a particular environmental movement. Our approach is mostly rooted in descriptive analysis, and motivated by theoretical interests.

Literature

Under the *Structural Functionalist* approach, the dominant paradigm in sociology until the mid-twentieth century (Ritzer, 1996), social institutions and structures were analysed in terms of the positive functions they served. More recently, this perspective has been heavily criticised for its limited focus on history, conflict and social change as well as its conservative bias; many critics noted that particular structures might benefit some groups, but disadvantage other groups. Subsequently, *Conflict Theory* became a dominant perspective in sociology. Conflict theorists envision society as made up of multiple groups that compete for power, status and resources. Several versions of conflict theory have been developed (Coser, 1957; Ritzer, 1996), although many have their roots in Marxism and Critical Theory.

Some of the criticisms of conflict theory are 'mirror image' to those of functionalism. Conflict theory has been criticised for underemphasising the importance of order and stability (Ritzer, 1996). Indeed, many commentators have noted the necessity of some level of cohesion even within contending groups.

Social conflict, social movements and the environment

Sociological theories of conflict provide a structure to understand the causes and consequences of social conflict, including the structural conditions for its emergence, how it functions and its capacity to cause harm (Wagner-Pacifici and Hall, 2012). They also provide a basis for understanding how conflict is shaped by social relations, institutional arrangements and the uneven distribution of power (Piven and Cloward, 1977; Piven, 2006; Tarrow, 2011).

Those who perceive an imbalance in the way resources and power are distributed may mobilise in response to these perceived grievances and engage in disruptive behaviour, ranging from protest and uprisings to wide-scale violence and social disorder (McAdam *et al.*, 2001; Tilly, 2003; Tilly and Tarrow, 2007; Collins, 2009). Despite the general negative view of conflict, there are scholars who argue that conflict can have beneficial social outcomes, including the establishment of conflict resolution strategies, the strengthening of social relations and positive social or political policy outcomes (Habermas, 2000; McAdam *et al.*, 2001; Tilly, 2003; Tilly and Tarrow, 2007; Collins, 2009). Theories of social movements and contentious politics help us understand social conflict, by highlighting the interplay among symbolic, institutional and relational mechanisms, including relationships between states and social movements (Meyer and Staggenborg, 1996).

Understanding conflict over environmental resources

While social conflict has generally been understood in relation to the fair distribution of power and material resources, in recent decades, environmental destruction and risk have become integral parts of discourse on conflict and conflict resolution (Beck, 1999). In sociology, the growing environmental crisis has largely been understood as a conflict between environmental protection on the one hand and economic expansion on the other. Early eco-Marxist conceptions of the problem were framed in light of the material conflict between economic expansion, supported by the cooperation of the state, capital and labour – and environmental protection, which remained largely ignored in political economy (Schnaiberg, 1980). For example, Schnaiberg and Gould (2000) argue that the demand for constant economic expansion, including capital accumulation and job creation, comes at the expense of environmental protection; environmental concern thus remains external to economic decisions. This demand for growth leads to a permanent cycle of production and consumption, what is called the 'Treadmill of Production', or the continuous withdrawal of energy and resources from ecosystems that leads to increased outputs in the form of pollution and waste (Schnaiberg and Gould, 2000). This process results in a conflict between people within environmental movements on the one hand, and those who favour economic growth on the other hand, including governments, business and labour.

At the same time, adherents of the Treadmill of Production thesis contend that it is a model of both conflict and cooperation. Schnaiberg (1980), for example, saw the consolidation of capitalism as an eventual threat to the stability of the state, to worker well-being and to environmental protection. The expansion of global capitalism, he argued, would necessarily lead to downward pressure on both wages and the environment and thus the environmental crisis could potentially facilitate cooperation between the state and labour around a more sustainable economy. The desire to slow down or reverse the Treadmill of Production provides the common ground for diverse groups, previously divided, to work together for the betterment of the environment and society.

Critics of the Treadmill of Production theory argue that the conflict among people in society over the environment is a false dichotomy and point out that the economy does not have to be a destructive force, but can be re-designed to increase environmental protection (Beck, 1999; Mol, 2001; Mol and Sonnenfeld, 2000; Mol *et al.*, 2009). Ecological Modernisation theorists point to the modernisation process as a means of reversing the ecological crisis through innovation and investments in a clean energy economy (Mol and Spaargaren, 1993; Spaargaren, 2000). Thus, environmental problems such as pollution and resource depletion will spur technological innovation that leads to their solutions. Ecological Modernisation theory and Risk Society theory hold that environmental movements are central actors in causing governments to change, through collaboration and engagement.

Forests in British Columbia

Mass arrests in logging protest RCMP round up more than 250 men, women and children in B.C.

Police charged more than 250 opponents of Clayoquot Sound logging yesterday in the largest mass arrest in B.C. history. The protesters, from children to pensioners and women with babies in their arms, were trying to block the Kennedy River Bridge, the main access point for logging crews into the old-growth forests of Clayoquot Sound. Yesterday's display of civil disobedience, the biggest since the beginning of the blockades at the bridge, had been aimed at halting logging operations for what would have been the fourth day since blockades began five weeks ago. 'I think the arrests today are going to catalyze people in this province and bring them out in even greater numbers over the next few weeks,' said William Thomas, a spokesman for the environmental group Friends of Clayoquot Sound. The blockade is in opposition to the B.C. government's decision in April to allow logging on about two-thirds of the sound, described by activists as the largest remaining temperate rain forest on the continent.

Wilson, Deborah (1993). 'Mass arrests in logging protest RCMP round up more than 250 men, women and children in B.C.' *The Globe and Mail* [Toronto, Ont] 10 Aug 1993: A.1.

British Columbia is a large province in western Canada (95 million hectares) – larger than most individual European countries. About 60% of British Columbia is covered in forests (55 million hectares plus an additional 3.7 million hectares with stunted or scattered trees). Historically, forestry (in terms of industrial timber production) has been the largest and most important industry in British Columbia. The relative economic importance of forestry has declined by historical standards, but it is still significant. For example, in 2009 forest product exports were $7.6 billion (CDN), which made up about 30% of total B.C. exports (British Columbia, 2010).

Another important aspect of forestry in British Columbia is land ownership. Most of the province (e.g. land and forests) is publicly owned. The province owns 94% and the federal government owns 1% of the area of B.C., while private owners hold 4% of the provincial land base. This is much higher than other areas outside of Canada, and has important political consequences. For example, by contrast, of the 514.2 million acres of US forests that are classified as timberland, federal, state and local governments own 22% (USEPA, 2014).

Contention over forest management: the forest industry in British Columbia and the rise of the environmental movement

Historically, representatives of the forest industry had relatively high levels of political influence. Marchak (1983) has documented how many smaller communities were dominated economically and politically by forestry companies. As described by the Treadmill of Production model, capital from forestry profits tended to be reinvested in more sophisticated harvesting equipment. Greater reliance on technology displaced forestry workers and increased the pace of harvesting. The Ministry of Forests was under continual pressure to make cheap timber available for companies, facilitate the creation of jobs and provide revenues for the province through stumpage fees. Consistent with the views of Schnaiberg and his colleagues, the province constantly finds itself in a conflict between protecting the use values of forests (e.g. protecting animal habitat, salmon streams, fresh water, recreation opportunities) and fostering the exchange values of forests (harvesting of timber for the production of lumber in a global commodity market). As the Treadmill of Production model predicts, exchange values have historically been favoured over use values.

Consequently, in the past several decades the forest industry in British Columbia has been criticised for many reasons, including overuse of clear-cutting as a harvesting technique, insufficient replanting, lack of development of secondary industry related to forestry, export of raw logs, lack of environmental protection, negative impacts on wildlife, loss of wilderness, expropriation and destruction of First Nations land (e.g. M'Gonigle and Parfitt, 1994; Marchak *et al.*, 1999). Much of this critique has come from the environmental movement.

British Columbia has one of the largest and most vibrant environmental movements in Canada, and has been the centre of a number of significant movement events. Part of the reason why British Columbia has been a centre for environmental politics is the fact that it is a relatively large geographical area, consisting of numerous ecoregions, with abundant natural resources, large forests, mountains and an extensive ocean coastline. Development of large-scale industry and urbanisation occurred later than in many other parts of North America, thus there was more 'nature' for environmentalists to contest.

Concern about forestry in British Columbia coincided with the rise of the environmental movement in British Columbia; environmental activists have been some of the most vocal critics of the forest industry. Aboriginal peoples have also been key players in environmental politics, but a detailed account of their roles in this conflict is beyond the scope of this chapter (see Tindall *et al.*, 2013). A variety of environmental campaigns have been launched over time in British Columbia, including campaigns over fisheries, forestry and wilderness preservation.

In the early 1990s a conflict arose over logging in Clayoquot Sound, which is a relatively untouched area of temperate old-growth rainforests on the west coast of Vancouver Island. This conflict was ostensibly about the potential logging of pristine rainforests in Clayoquot Sound, one of the few remaining intact temperate rainforests in the world. Environmental groups used a variety of tactics to raise awareness of the issues, sway public opinion and influence government decision-makers. Tactics ranged from making public presentations, creating information brochures and videos to protesting on the lawns of the provincial legislature, and blockading logging roads near Clayoquot Sound. The campaign culminated in 1993 with over 850 people arrested for engaging in civil disobedience, the largest instance of civil disobedience in Canadian history, by blockading logging roads leading into the Clayoquot Sound region (Tindall, 2013). Subsequent to the mobilisation over Clayoquot sound, several environmental organisations redeployed their efforts towards protecting the forests lands of the mid-coast of British Columbia, in an area that has come to be known as 'the Great Bear Rainforest'.

Opportunities, strategies and tactics of environmental activists

Environmental activists have undertaken many activities in their efforts to preserve old growth forests in British Columbia, and relatedly, to protect ecological values and make forest management more sustainable. They have interacted with a variety of organisations and individuals in their campaigns, including different levels of government, other NGOs, private-sector forestry (and other natural resource) companies, and First Nations communities. To varying degrees these groups have been in conflict; in some instances, they have collaborated.

Strategies used by environmental organisations to meet their objectives include engaging in direct action and civil disobedience to gain media attention and change public opinion, as well as to pressure governments to create new legislative policies and compel the forestry companies to change certain practices. Some tactics were intended to mobilise resources for the movement while others were designed to create or respond to political opportunities. And, of course, directly or indirectly, most of these tactics were enlisted in an effort to meet movement goals.

Framing

Social movement actors rely on 'frames' to gain attention, mobilise support and sway public opinion. 'Frames' are the interpretive schemas individuals use to perceive, identify and label events in the world. 'By rendering events or occurrences meaningful, frames function to organize experience and guide action' (Snow *et al.*, 1986: 464). In the context of social movements, scholars have pointed to the importance of 'frame alignment', which is the linking of individual values and interests with those of social movement organisations, as a prerequisite to movement participation.

In the conflict over forest conservation/preservation in British Columbia, environmentalists emphasised 'clear-cutting' (felling 100% of trees on a site) as the root of the problem. Video images, postcards and posters showed dramatic images of 'clear-cuts', and these images found their way into the print media.

The public now readily associates the word 'clear-cut' with environmental destruction. Subsequently, an image of clear-cutting can be effectively communicated with the use of a single word. Various commentators (e.g. Moore, 1995; Kimmins, 1997) have noted that there is a tendency for people to associate the visual quality of landscapes with the health of ecosystems (Sheppard and Harshaw, 2000) and because clear-cuts look 'ugly' to most people they are considered to be a threat to ecosystem health. This is a type of frame amplification because problems associated with clear-cutting are underscored, but they are only a subset of more complex arguments.

Books, newspapers, videos, calendars, posters

The movement has produced a number of coffee table books on particular geographical areas (Haida Gwaii/Queen Charlotte Islands, Carmanah, Clayoquot, the Great Bear Rainforest) and topics (Clear-cut), posters ('Big Trees', 'Ancient Forests'), calendars (scenic pictures of wilderness, charismatic megafauna like the Spirit Bear, other iconic images like the spotted owl). These were used as venues for visually oriented communication of binary frames of the forest as pristine/healthy versus clear-cut/devastated.

Media attention

Getting media attention is an important strategy for environmentalists (Gamson and Wolfsfeld, 1993; Gamson, 2007). Media access is important for communicating ideas, and mobilising resources and support. In the short to medium term, media attention serves as a type of scorecard for the effectiveness of the movement, and in some instances as a barometer of the level of support for the movement. Sometimes the ultimate audience of the movement is the general public, but in other instances it can be governments or other actors. Some specific tactics such as protests, or particular stunts, are used to gain media attention which, in turn, are important for influencing public opinion.

Legal and peaceful protests

Protests are a common tactic of many social movements. They can serve a variety of purposes including gaining media attention and support from movement members and participants. Further, to the extent that movement organisers can mobilise large numbers of people to participate, protests can send a message to governments and other actors about the amount of support there is for the movement's goals and objectives.

Talks

In this particular movement, several activists spent a considerable amount of time travelling around the province, as well as other parts of the country, giving talks and slide shows to various audiences, from small selective groups to larger public audiences. These events provided opportunities to disseminate information, and to engage in frame bridging. During the campaigns of the 1990s these talks were frequently accompanied with a slide show displaying various key images (giant, ancient trees versus clear-cuts) for framing purposes.

Meeting with politicians

By definition, social movements are conceived as phenomena that largely occur outside of institutionalised politics (della Porta and Diani, 2009). Still, leaders in social movements regularly interact with elected politicians and government leaders. In some instances, this involves attempts to persuade politicians to adopt particular policies or legislation; in others, it is to arrange agreements between different parties (e.g. over the protection of particular geographical areas). Sometimes, unpredicted alliances can form such as when environmentalists persuaded a politically conservative British Columbia Premier (Gordon Campbell) to support the plan to protect the *Great Bear Rainforest* (British Columbia Ministry of Agriculture and Lands, 2006). Both the Treadmill of Production Model and the Ecological Modernisation Model outline the role of environmental non-governmental organisations (ENGOs) in trying to influence government policy and legislation, although the Treadmill of Production Model depicts this in the context of conflict, while the Ecological Modernisation approach emphasises collaboration. The British Columbia case demonstrates that both types of relationships between states and movements can coexist, and sometimes evolve from one to the other.

Direct action

In some situations, social movement participants adopt *direct action* as a tactic, wherein activists directly intervene in the normal flow of business (e.g. by blocking logging roads or climbing or sitting in trees that are about to be cut) in the hope of disrupting regular forestry operations (Doherty *et al.*, 2004). Direct action escalates the conflict, increasing the likelihood of media and government attention to the conflict. Civil disobedience in support of a cause that is seen as worthy and just can also lead to widespread public support. Similarly, harsh counteraction by governments and others can increase public support when the moral dimensions of the movement's goals resonate widely with the general public. However, when direct action is seen as illegitimate or a threat to the public, public support for a movement can wane. Blockading logging roads as part of the Clayoquot Sound protests constituted one of the largest instances of civil disobedience in Canadian history, and received media attention on an international scale. The protests led to a variety of positive outcomes in terms of meeting the movement's goals (Tindall, 2013).

At the same time, some direct actions have had negative consequences for the environmental movement in British Columbia. For example, during protests in the 1980s and 1990s there were claims of 'tree spiking'. Tree spiking refers to the practice of hammering steel spikes or nails into living trees. Generally, this does not do serious damage to the tree, but it can damage equipment (such as chain saws) when trees are felled and also sawmill equipment when the timber is processed. It can potentially lead to serious injuries to forestry workers. While it is not entirely clear who engaged in these actions, or how widespread they were, these actions resulted in hostility toward the environmental movement on the part of some communities and some segments of the general public.

Another aspect of direct actions is that they can sometimes lead to 'radical flank' effects. This refers to situations where the existence and actions of 'radical groups' lead governments (and others) to interact and/or negotiate with relatively more moderate social movement groups that they might otherwise ignore (Haines, 1984). In British Columbia the existence and actions of groups like the Friends of Clayoquot Sound and Greenpeace apparently led the provincial government to take other groups, like the Sierra Club, more seriously.

Stunts

Environmental groups often used a variety of stunts to gain media attention, or disrupt situations. Greenpeace is the exemplar for using this type of tactic. Berman and Leiren-Young (2011) provide an insider's account of some of the stunts that Greenpeace (and some other groups) undertook during the campaigns in the 1990s to protect old-growth forests in British Columbia. A typical stunt involved displaying large and potentially embarrassing banners in a strategic public location such as when Greenpeace infiltrated ceremonies for British Columbia house in London, England with bogus Canadian Mounted Police who then brandished banners protesting logging (Berman and Leiren-Young, 2011).

Boycotts and other market campaigns

The Political Process Model examines the political opportunities and constraints to social movements, and the extent to which movement participants are successful in responding to these. In British Columbia an important shift occurred in the 1990s when movement participants moved from focusing mostly on domestic protests and domestic publicity-focused campaigns to an international market campaign (Berman and Leiren-Young, 2011). This was partly because of the political environment that the movement faced. Both the provincial government and forest companies were beginning to ignore environmental protests, and thus demonstrations and other tactics used to gain public support were having diminishing returns. At this time, activists switched tactics, and approached large companies in Europe and the United States who were customers of British Columbia forest companies about their purchase of dimensional timber, pulp and paper products. These companies were told that the environmental movement (including the allied movements in Europe and the US) would initiate consumer boycotts of their products if they continued to purchase British Columbia forest products whose source was old-growth temperate rainforests. This tactic was highly successful, with a number of high-profile companies either cancelling their contracts for British Columbia forest products or reducing their purchases.

Certification

In the 1990s, organisations in the environmental movement and others (such as foresters, and environmentally concerned woodworkers) began to explore the possibilities for independently certifying forests and forest products that met certain ecological and social criteria (Bartley, 2007). In 1993 they created the Forest Stewardship Council (FSC). This is an international body 'that sets standards for responsible forestry, accredits auditors or "certifiers" to inspect operations in the field, and grants the use of its logo on qualified products' (Bartley, 2007: 234).

As Bartley discusses, the motivation to establish an independent certification scheme was an effort to find positive alternatives to tropical timber boycotts in Europe and North America. Certification processes, however, soon included wood products from temperate forests, and have played a significant role in environmental politics in British Columbia (Pechlaner and Tindall, 2013).

Conclusion

In this chapter we have discussed how conflict is viewed in sociology, and have described some specific perspectives within the sub-fields of social movement scholarship and environmental sociology. Many sociologists see society as consisting of groups that contend for status, power and resources. Cleavages between groups can be based on various things, including differences in social class, race, ethnicity, gender, region, values, culture and related interests (Box 11).

Social movement scholars have been interested in the factors that explain the mobilisation of individuals and groups for collective action to respond to grievances and struggle for change. They have also been interested in the factors that make such mobilisations more or less successful, and in outcomes, both intended and unintended. For example, according to Brym and Fox (1989), collective action is more likely to occur, and to be successful, to the extent that the members of a contending group: (1) are bound together in dense social networks; (2) are highly socially polarised from advantaged groups; (3) are relatively unrepresented by existing groups (or parties); (4) have relatively high levels of shared social identity; (5) compared with dominant/contending groups, have relatively high access to material resources (property, money, jobs), normative resources (communications media, educational institutions) and/or coercive resources (police, armed forces); and (6) are relatively large in number and have a large number of support bases.

Environmental sociologists are interested in many topics pertaining to conflict over natural. Notably, in the context of this discussion, they have paid special attention to how political economic structures and processes are related to environmental and social outcomes, including the extent to which such structures promote or hinder sustainability. Several dominant frameworks consider the political economy of environmental issues. The Treadmill of Production predicts an inherent conflict between exploiting the natural environment for economic uses (exchange values), and protecting for non-economic uses (use values). Conclusions drawn from the Treadmill of Production are generally pessimistic about the outcome of this tension. A contrasting framework, Ecological Modernisation, can be viewed as more optimistic about the possibilities of sustainability. Adherents see market mechanisms as being drivers of changes to promote sustainability. Both of these perspectives see social movements as being a relatively key player in conflicts over natural resources. Treadmill of Production scholars see social movements as being important in resisting pressures from business interests to minimise environmental concerns. Ecological Modernisation theorists see environmental organisations as a key third party in the environmental policy process.

In the view of many sociologists, conflict often plays an important role in societies by driving or otherwise motivating social change. In the case of the wilderness preservation movement in British Columbia, environmental movement organisations developed a number of strategies and tactics that allowed them to influence the general public, and to put pressure on governments and companies to make important changes. Through the creative use of social conflict, the environmental movement was also able to provoke the provincial government and forestry corporations to engage in more collaborative interactions around forestry management and the creation of protected areas. The strategic tactics used by the environmental movement likely resulted in the preservation of greater areas of old-growth temperate rainforests in British Columbia than would otherwise have been protected.

References

Bartley, T. (2007). How foundations shape social movements: the construction of an organizational field and the rise of forest certification. *Soc. Probl.*, 54, 229–255.

Beck, U. (1999). *World Risk Society*. Cambridge: Polity Press.

Berman, T. and Leiren-Young, M. (2011). *This Crazy Time: Living Our Environmental Challenge*. Canada: Knopf.

British Columbia (2010). *The State of British Columbia's Forests*. Third edition. British Columbia: Ministry of Forests, Mines, and Lands.

British Columbia Ministry of Agriculture and Lands (2006). *Province Announces A New Vision for Coastal BC*. British Columbia: Ministry of Agriculture and Lands.

Brym, R. J. and Fox, B. (1989). *From Culture to Power: The Sociology of English Canada*. Toronto: Oxford University Press.

Collins, R. (2009). *Violence: A Microsocial Theory*. Princeton, NJ: Princeton University Press.

Collins, R. and Annett, J. (1975). *Conflict Sociology: Toward an Explanatory Science*. New York: Academic Press.

Coser, L. A. (1957). Social conflict and the theory of social change. *Br. J. Sociol.*, 8, 197–207.

della Porta, D. and Diani, M. (2009). *Social Movements: An Introduction*. Malden, MA: Blackwell Publishing.

Doherty, B., Plows, A. and Wall, D. (2004). Covert repertoires: ecotage in the UK. *Soc. Movement Stud.*, 3, 199–220.

Gamson, W. A. (2007). Bystanders, public opinion, and the media. In *The Blackwell Companion to Social Movements*, eds. D. A. Snow, S. A. Soule and H. Kriesi, pp. 242–261. Oxford: Blackwell Publishing.

Gamson, W. A. and Wolfsfeld, G. (1993). Movements and media as interacting systems. *Ann. Am. Acad. Polit. Soc. Sci.*, 528, 114–125.

Habermas, J. (2000). *The Inclusion of the Other: Studies in Political Theory*. Cambridge, MA: MIT Press.

Haines, H. H. (1984). Black radicalization and the funding of civil rights: 1957–1970. *Soc. Probl.*, 32, 31–43.

Kimmins, H. (1997). *Balancing Act: Environmental Issues in Forestry*. Second edition. Vancouver: UBC Press.

M'Gonigle, M. and Parfitt, B. (1994). *Forestopia: A Practical Guide to the New Forest Economy.* Madeira Park, BC: Harbour Publishing.

Marchak, M. P., Aycock, S. L. and Herbert, D. M. (1999). *Falldown: Forest Policy in British Columbia.* Vancouver: David Suzuki Foundation and Ecotrust Canada.

Marchak, P. (1983). *Green Gold: The Forest Industry in British Columbia.* Vancouver: University of British Columbia Press.

McAdam, D., Tarrow, S. and Tilly, C. (2001). *Dynamics of Contention.* New York: Cambridge University Press.

Meyer, D. S. and Staggenborg, S. (1996). Movements, countermovements, and the structure of political opportunity. *Am. J. Sociol.*, 101, 1628–1660.

Mol, A. P. J. (2001). *Globalization and Environmental Reform: The Ecological Modernization of the Global Economy.* Cambridge, MA: MIT Press.

Mol, A. P. J. and Sonnenfeld, D. A. (2000). *Ecological Modernisation around the World: Perspectives and Critical Debates.* London: Routledge.

Mol, A. P. and Spaargaren, G. (1993). Environment, modernity and the risk-society: the apocalyptic horizon of environmental reform. *Int. Sociol.*, 8, 431–459.

Mol, A. P. J., Sonnenfeld, D. A. and Spaargaren, G. (2009). *The Ecological Modernisation Reader: Environmental Reform in Theory and Practice*, London: Routledge.

Moore, P. (1995). *Pacific Spirit: The Forest Reborn.* West Vancouver, BC: Terra Bella Publishers Canada.

Pechlaner, G. and Tindall, D. B. (2013). Changing contexts: environmentalism, aboriginal community and forest company joint ventures, and the formation of Iisaak. In *Aboriginal Peoples and Forest Lands in Canada*, eds. D. B. Tindall, R. L. Trosper and P. Perreault, pp. 260–278. Vancouver: University of British Columbia Press.

Piven, F. F. (2006). *Challenging Authority: How Ordinary People Change America.* Lanham, MD: Rowmanand Littlefield.

Piven, F. F. and Cloward, R. A. (1977). *Poor People's Movements: Why They Succeed, How They Fail.* New York: Vintage Books.

Ritzer, G. (1975). Sociology: a multiple paradigm science. *Am. Sociol.*, 10, 156–167.

Ritzer, G. (1996). *Sociological Theory.* Third edition. New York: McGraw-Hill Education.

Schnaiberg A. (1980). *The Environment: From Surplus to Scarcity.* New York: Oxford University Press.

Schnaiberg A. and Gould, K. A. (2000). *Environment and Society: The Enduring Conflict.* West Caldwell, NJ: Blackburn Press.

Sheppard, S. R. J. and Harshaw, H. W. (2000). *Forests and Landscapes: Linking Ecology, Sustainability and Aesthetics.* IUFRO Research Series, No. 6. Wallingford: CABI Publishing.

Snow, D. A., Rochford, Jr. E. B., Worden, S. K. and Benford R. D. (1986). Frame alignment processes, micromobilization, and movement participation. *Am. Sociol. Rev.*, 51, 464–481.

Spaargaren, G. (2000). Ecological modernisation theory and the changing discourse on environment and modernity. In *Environment and Global Modernity*, eds. G. Spaargaren, A. P. J. Mol and F. Buttel, pp. 41–73. London: Sage.

Tarrow, S. (2011). *Power in Movement: Social Movements and Contentious Politics.* Third edition. New York: Cambridge University Press.

Tilly, C. (2003). *The Politics of Collective Violence.* New York: Cambridge University Press.

Tilly, C. and Tarrow, S. (2007). *Contentious Politics.* Boulder: Paradigm.

Tindall, D. (2013). Twenty Years After the Protest, What We Learned from Clayoquot Sound. *The Globe and Mail*, 12 August.

Tindall, D. B., Trosper, R. and Perreault, P. (2013). *Aboriginal Peoples and Forest Lands in Canada.* Vancouver: UBC Press.

United States of America Environmental Protection Agency (2014). Forestry. http://www.epa.gov/oecaagct/forestry.html

Wagner-Pacifici, R. and Hall, M. (2012). Resolution of social conflict. *Annu. Rev. Sociol.*, 38, 181–199.

Box 11

Cormorants, fisheries and conflicts across Europe

Mariella Marzano[1]
Forest Research, Northern Research Station, Roslin, Midlothian EH25 9SY, Scotland, UK

Various factors such as international conservation legislation and abundant food supplies have contributed to the increase of great cormorants *Phalacrocorax carbo*, large fish-eating birds that breed throughout Europe (van Eerden *et al.*, 2012). An interdisciplinary research coordination network (INTERCAFE) funded through the EU COST Action Programme was established to address long-standing animosity between commercial and recreational fisheries interests and conservationists over public policy that many feel has facilitated the growing numbers of cormorants. INTERCAFE functioned from 2004 to 2008 and coordinated 70 researchers from the EU and other countries to synthesise current knowledge about ecological databases and analyses, conflict resolution and management, and linkages between science and policy and best practice (Marzano and Carss, 2012). Local perspectives were considered paramount and thus over 170 external stakeholders attended one of nine meetings. The locations for meetings reflected geographical area, habitat types, fishery type, stakeholder groups and current and potential mitigation actions (http://www.intercafeproject.net). The perspectives of contributors spanned many disciplines including fisheries, ecology, ornithology, anthropology, sociology, political science, economics, policy and law. Effective conflict management requires an understanding of the nature of the conflicts before proposing solutions (Carss *et al.*, 2009). The aim of INTERCAFE was to provide an overview of key factors in the cormorant–fisheries conflicts, to encourage people to discuss and exchange ideas, knowledge and experience and to provide examples of the problems that people experience with potential mitigation solutions. Tools exist (e.g. protection through nets, harassment, shooting) and are helpful in certain local situations, but many believe that problems could be better addressed through a pan-European management plan such as widespread, coordinated population reduction.

Pond fisheries are one example where conflicts over cormorant impacts on fisheries mask a multitude of issues facing fisheries (Seiche *et al.*, 2012). Artificial production of carp *Cyprinus carpio* in freshwater ponds is a historically common activity across much of Central and Eastern Europe. Many fish ponds are considered part of the cultural heritage in the regions where they occur, having been an essential part of the landscape, culture and economy for hundreds of years. While initially pond construction involved massive habitat modification, centuries of naturalisation and management have turned these areas into a mosaic of interconnected semi-natural wetlands. These 'new' landscape features are now both familiar to local people and a symbol

[1] Crown copyright.

of a long history of fishing and water management. Indeed, pond fish farming is often the centre of regional identity. Many local communities that have fish ponds take great pride in stewardship for them. However, the market for carp is declining. Pond fisheries face increasing competition from fish importing industries and other economic and social pressures, which make fish farming an increasingly tenuous activity. The increase in great cormorants adds to these pressures because they prey on fish within these ponds. The predation levels currently experienced by fish farmers are thought to cause serious financial losses. In addition, valuable time and resources are spent scaring or shooting cormorants to manage them locally. In many of the traditional fish farming areas of Europe, such control efforts lead to conflict with others dealing with active conservation and regulation. The conflicts are symbolic of the plight of carp pond farmers in terms of economic loss and cultural heritage issues against environmental policies that protect almost all wild bird species across Europe.

Cormorants are shot in carp pond regions in several countries and more are shot in France than in any other European country, with some 30,000 birds being killed there each winter. Nevertheless, many fish farmers do not think it is their role to try and solve the wider cormorant problem, which is the increase in cormorant numbers, expansion of their geographical range and changes in their migratory patterns. They also feel that this issue is not given sufficient recognition by regulatory authorities, in contrast to other environmental issues. In many ways these farmers feel constrained by national and EU legislative frameworks, but they also look to their governments and to the EU for a solution to the cormorant problem, which they see as originating beyond their borders. It is unclear how a European solution (such as widespread shooting) would work in practice (Marzano *et al.*, 2013). Meanwhile, blame for the decline in fisheries is often directed towards the 'environment' or 'wildlife' rather than the other factors discussed above that are also contributing to the decline of fish farming because there is explicit conservation legislation protecting the environment and wildlife. In contrast, farmers feel that preservation of rural communities and their economic livelihood is not given similar protective consideration. INTERCAFE's efforts to coordinate research on cormorant–fisheries interactions across Europe as well as its attempts to include the current and future needs of local stakeholders and policy makers has contributed significantly to a process of opening communication and developing networks of trust with stakeholders that can aid future conflict management.

References

Carss, D. N., Bell, S. and Marzano, M. (2009). Competing and coexisting with cormorants: ambiguity and change in European wetlands. In *Landscape, Process and Power: A New Environmental Knowledge Synthesis*, ed. S. Heckler, pp. 19–121. New York: Berghan.

Marzano, M. and Carss, D. N. (2012). Essential social, cultural and legal perspectives on cormorant-fisheries conflicts. INTERCAFE COST Action 635 Final Report IV (ISBN 978-1-906698-11-9). http://www.intercafeproject.net/COST.html

Marzano, M., Carss, D. N. and Cheyne, I. (2013). Managing European cormorant-fisheries conflicts: problems, practicalities and principles. *Fish. Manag. Ecol.*, 20, 401–413.

Seiche, K., *et al.* (2012). *Cormorant-fisheries conflicts at Carp ponds in Europe and Israel – an overview*. INTERCAFE COST Action 635 Final Report V (ISBN 978-1-906698-10-2).

van Eerden, M., van Rijn, S., Volponi, S., Paquet, J.-Y. and Carss, D. N. (2012). *Cormorants and the European Environment; exploring cormorant status and distribution on a continental scale.* INTERCAFE COST Action 635 Final Report I (ISBN 978-1-906698-07-2).

© Catherine Young.

CHAPTER TWELVE

Peace research and conservation conflicts

PAUL ROGERS
Bradford University

I here outline the development of peace research as an area of study, analyse its current approach as an interdisciplinary field of work and discuss whether approaches prevalent within peace research are of value in relation to the resolution of conservation conflicts. Peace research is a discipline which endeavours to analyse and understand the roots and structures of conflicts, to prevent them before they develop, resolve them when they have commenced and engage in post-conflict peace-building.

Historical context

Prior to the Second World War, pioneering researchers studying the causes of interstate wars included Quincy Wright, Pitrim Sorokin and Lewis Fry Richardson. Their work attracted little attention until the early post-war period when peace research was boosted by the failure of conventional international politics to prevent two massive world wars in the space of 30 years, and it was further boosted by the advent of the nuclear arms race and the risk of a global nuclear catastrophe.

By the mid-1950s, peace research centres were being established principally in Scandinavia and North America, but with a persistent interest in Japan. A notable initiative was the establishment in 1957 of the *Journal of Conflict Resolution,* based at the Centre for Conflict Resolution at the University of Michigan, with the founding editors presenting two reasons for the need for such a journal:

> The first is that by far the most important practical problem facing the human race today is that of international relations – specifically the prevention of global war. The second is that if intellectual progress is to be made in this area, the study of international relations must be made an interdisciplinary enterprise, drawing its discourse from all the social sciences and even further. (Boulding, 1973: 3)

The interdisciplinary call was reflected in the backgrounds of North American peace researchers including the mathematician Anatol Rapoport, the economist Kenneth Boulding, the sociologist Elise Boulding and the psychologist Herb

Conflicts in Conservation: Navigating Towards Solutions, ed. S. M. Redpath, R. J. Gutiérrez, K. A. Wood and J. C. Young. Published by Cambridge University Press.

Kelman. By the late 1960s there were also peace research institutes established in Stockholm and Oslo and a number of smaller centres elsewhere in Western Europe, with much of the focus being on the dangers of the nuclear arms race. This orientation persisted until the end of the decade, but by the early 1970s peace researchers were also focused on issues of north–south relations and global environmental limits to growth (Meadows *et al.*, 1972), as well as the problem of proxy wars indirectly involving the superpowers during the late Cold War era, especially in Vietnam, the Horn of Africa and Central America.

The early 1980s saw the onset of a particularly dangerous phase of the Cold War. In addition to the hawkish attitudes of some western governments, there were developments in nuclear strategy that increased the risk of a nuclear exchange. These included NATO's preparedness to use nuclear weapons first (most likely matched, if officially denied, by the Soviet Union), the belief that a limited nuclear war was possible and the deployment of highly accurate intercontinental ballistic missiles which might be used to 'disarm' a first strike.

Appeasement studies

In response to the dangers of nuclear war, anti-nuclear movements were re-invigorated in many western countries and this was paralleled by a renewed interest in international security by peace researchers. Some peace research centres, such as the Stockholm International Peace Research Institute (SIPRI), had long specialised in arms control and disarmament processes, but they were joined by academics from a number of universities. Their research on nuclear issues often provided resources for anti-nuclear activists and they therefore incurred the enmity of conservative politicians who were prone to term the subject 'Appeasement Studies'. There was also a marked tendency to scrutinise research outputs closely in case poor-quality research could be identified and used to damage peace research as a whole.

Being subject to such unusual scrutiny served two useful purposes. One was that academics working in this field had to achieve singularly high levels of research competence – if everything they published was intensely examined, the standards had to be particularly good. The other was that the field attracted many able young scholars who were personally committed to research on key international security issues and sought a higher impact in a wider community than was usual in more conventional academic disciplines.

It is significant that during this period, which lasted until the end of the 1980s and the fall of the Berlin Wall, antipathy to peace research came primarily from politicians, whereas it was not uncommon for peace researchers to be invited to lecture at senior defence colleges where officers were eager to debate alternative views. In this context, those peace researchers who had an interdisciplinary background, especially if it involved the natural sciences, could readily command respect, especially if they were knowledgeable on the more technical aspects of nuclear strategy.

The relevance of this experience to the current subject is that if one is working in an area of controversy, the research has to be of a particularly high standard if it is to be accepted. It is especially useful if a researcher goes beyond exploring the significance of his/her research and seeks to visualise it from the perspective of those who may have a very different outlook. This requires a well-developed ability to maintain a degree of detachment that may be difficult when opinions are sharply divided. During the Cold War it was not uncommon for some peace researchers to attempt to understand the Soviet world view, not in terms of the 'blue/red' studies undertaken by the western military and intelligence agencies which sought to win conflicts, but to help to understand Soviet attitudes with a view to avoiding conflicts.

After the Cold War

During the 1980s, peace researchers continued to work on other themes beyond the East/West confrontation, and interest in issues such as global north–south relations, environmental and resource conflicts and the theory and practice of conflict resolution proved to be attractive fields of research. This was significant when the Soviet Union collapsed because it left many of the more traditional international relations scholars finding it difficult to adapt to the radical changes in international relations as their narrowly focused ethnocentric world was torn apart. What made this even more troublesome was that the view expressed in the early 1990s that the world was entering a relatively peaceful phase of a 'New World Order' now that the east/west confrontation had ended was quickly proved wrong.

The early and mid-1990s saw the Gulf War, intensive and prolonged conflicts in the Balkans and the Caucasus and the appalling violence of the Great Lakes region of eastern Africa, as well as continuing conflicts in Sri Lanka, Afghanistan and Northern Ireland. This was markedly different to an orderly world, but most international security specialists were so fixed in the old East/West rivalry that they had difficulty coming to terms with the numerous smaller yet often devastating conflicts. By contrast, the work of peace researchers on issues such as peacekeeping and conflict resolution seemed more relevant.

Characteristics of peace research

Over the past 20 years the peace research community has retained its normative orientation, seeking to undertake original research that is likely to have a direct and positive effect in terms of aiding and enhancing the development of more peaceful societies. This is by no means different from many other applied fields of study, including much of the medical sciences, engineering and environmental science. It is an outlook illustrated by the mission statement of one UK peace research centre (Bradford, 2011: 1):

> We combined empirical, theoretical and applied research with sustained engagement at international, regional, national and local levels to analyse, prevent and resolve conflicts and develop peaceful societies. We aim for an enabling environment for international research excellence involving diverse and critical approaches.

Rogers and Ramsbotham (1999) discuss seven characteristics of peace research. Together, these seven characteristics illuminate the kind of mission statement quoted above.

Underlying causes

Peace research goes beyond the absence of war to focus on the reasons why societies may be more peaceful. As such, it explores a range of inequalities whether stemming from race, class or gender and does so at levels ranging from the international right down to the community and the individual. Peace researchers commonly talk of 'structural violence' (Galtung, 1969) as a condition in which there are structural inequalities that work against cooperative and equitable relationships within and between communities, and may lead to physical violence and even interstate war.

Interdisciplinary focus

As mentioned above, the early development of peace research involved scholars from several disciplines and this continues, so that the larger research centres, whether university departments or institutes, will typically draw on specialists in politics, international relations, history, economics, sociology and anthropology, as well as the natural sciences. Such diversity may lead to differences over methodological approaches and theoretical frameworks, but also to a richness of understanding that transcends the limitations.

Non-violent transformations

A common orientation among peace researchers is towards non-violent transformation of otherwise violent interactions. This does not necessarily imply that only pacifist orientations are acceptable; more that in a progression from military to non-military solutions peace research works vigorously to analyse and understand the mechanisms which move situations towards the latter. Thus research on war may focus not on how victory is achieved, but on the underlying causes, the actual spurs to the outbreak of war, factors enhancing the early achievement of ceasefires and conditions likely to encourage an irreversible move towards post-conflict peace-building. Even so, the emphasis on non-violence may not be absolute and there is recognition of the possible relevance of force in narrow circumstances such as a requirement for humanitarian intervention. The emphasis in peace research is on non-violent transformation and the many examples studied include the non-violent transformations in East European states at the end of the Cold War and many of the instances of liberation movements at the end of the colonial era.

Multi-level analysis

Such analysis, whether at the individual, community, state or interstate level, aims to overcome the compartmentalisation of conflict analysis. Two reasons for this are the extent to which a particular level provides analytical insights that may aid the understanding of other levels, and the more practical aspect, such as in the post-Cold War environment, many conflicts operate at very different levels. Wars in Iraq and Afghanistan after the 9/11 attacks were, in part, intranational, with interconfessional conflict in Iraq causing the majority of deaths and injuries, but initiated by and involving an international dimension.

Global outlook

As indicated above, with the interest in the 1970s in the 'Limits to Growth' debate, peace research seeks to emphasise a global dimension, but this also relates to diverse conceptions of peace and non-violent social change that come from the understandings within different cultures. In terms of this aspect of a global outlook, peace research attempts to avoid the western ethnocentric analysis that has tended to dominate the study of international relations, especially in the Cold War era.

Analytical and normative

The normative element in peace research does not mean that peace research is not properly analytical: 'While there has been a tendency to ground peace studies in quantitative research and comparative empirical study, the reality is that most scholars have been drawn to the field by ethical concerns and commitments. Deterministic ideas have been largely rejected, whether in realist or Marxist guises, with large-scale violence not seen as inevitable features of the international system, but as consequences of human actions and choices' (Rogers, 2013: 60).

Theory and practice

This follows on directly from the previous characteristic in that peace research not only may be rooted in theoretical and empirical analysis but also is frequently policy-relevant. While there may be a distinction between research and activism, the results from peace research may frequently lead to engagement with governments and intergovernmental organisations, the military, business communities and civil society. Indeed, this can be seen as form of testing of theoretical insights.

Peace research and conservation conflicts

With these characteristics of peace research in mind, what might it have to offer the analysis of conservation conflicts and what, in turn, can peace research learn from such analysis? One of the most relevant aspects of peace research in relation to conservation conflicts is a common understanding among many peace researchers of the nature of international threats to human security, and this stems from the concern, dating back to peace research in the 1970s, about the implications of global environmental limits to human activity. In its most basic representation it proposes main drivers of international conflict (namely socio-economic divisions and environmental constraints) over the next 30 years. It can be argued that an appreciation of these drivers has a direct relevance for the analysis of conflicts over conservation of biodiversity (Abbott *et al.*, 2006).

One of these drivers has, at first sight, little to do with conservation, but is in reality highly relevant. This is the growing wealth/poverty divide and the concentration of most of the fruits of recent decades of economic growth in about one-fifth of the world's population. It is not that the poor are getting poorer or that a few countries are forging ahead, but it is a matter of the relative marginalisation of the majority of the global population, with a trans-global elite of some 1.5 billion people accruing between 80% and 85% of the income and wealth (Davis *et al.*, 2006).

This concentration is exacerbated by marked improvements in education and literacy across much of the global south with even the gender gap slowly narrowing. The problem is that while such improvements are important, they have led to higher expectations of life chances which stand little or no likelihood of being fulfilled, which results in deep frustration and bitterness as the nature of the rich/poor divide becomes transparent. In its more extreme manifestation this may lead to social revolts, such as Maoism in Nepal and neo-Maoist Naxalites in India. It also underpins many of the frustrations that underlie the Arab Awakening, especially in countries such as Tunisia and Egypt where many tens of thousands of young graduates have little prospect of employment.

Under the current liberal market system there is no sign of this divide being narrowed, and even the shock of the 2007–08 financial crisis failed to engender support for serious reforms. Moreover, it is likely to be exacerbated by environmental limitations to economic growth. These take two main forms, one being intense competition and even conflict over resources such as strategic minerals and oil and gas, and the other being the potentially disruptive effects of climate change. Ecologists working in areas of conservation conflict are invariably aware of the likely impacts of climate disruption, and yet it is a phenomenon that is still widely denied by fossil fuel lobbies and some liberal market economists.

Although there are some differences between the results of climate modelling and the interpretation of climate data, two aspects of climate disruption are particularly important – acceleration and asymmetry. It is becoming clear that changes in the global climate, when measured over decades, are accelerating (Kirby, 2013) with the decade 2001–2010 showing the most marked changes. Moreover, it is already clear that some parts of the world are experiencing much faster rates of change than others, not the least of which is the near-Arctic. It is also highly likely that a long-term trend in temperature changes will show faster rates of increase over the tropical and sub-tropical land masses than over oceans, leading to a risk of a marked decline in the ecological carrying capacity of the croplands upon which the majority of the world's people depend for their food. These regions also experience the greatest disparities in wealth and the highest incidence of marginalisation. They are therefore going to be least able to cope with problems of food supply and responses to severe weather events.

The combination of socio-economic divisions and environmental constraints as a source of sustained fragility and instability is recognised not just by peace researchers but in other areas of study. It is also widely recognised by military and security experts of 'think tanks' in the West, but the prevailing attitude of military planners may be to acknowledge the threats and to respond primarily by developing a security architecture that is considered appropriate to maintaining the security and well-being of their own states and alliances, rather than addressing the underlying problems. The third trend is the persistence of a security architecture that stops far short of addressing the causes of instability arising from the first two trends. It is a security paradigm that may be termed 'liddism' – keeping the lid on a boiling pot rather than turning down the heat. For example, there may be considerable support offered to a weak state facing civil disorder to ensure that it can maintain public order and overall stability, primarily through enhancing the capabilities of police, paramilitary and arms units. It is an approach that may appear to achieve short-term results but neglects the underlying economic, social and environmental issues that may be at the root of the instability.

The human impact on the homeostasis of the global ecosystem will increasingly be exacerbated by persistent socio-economic divisions. This needs to be met by a transition to more emancipated forms of ultra-low carbon living, which will amount to a third societal transition as radical as the first two – the Neolithic and Industrial Revolutions. Although issues of conservation conflicts are usually specific to times and places, there is value in recognising this wider dimension as it is likely to have a generic impact on such conflicts and how they might be resolved.

Peace research may also be able to contribute to conservation conflict resolution through the application of a range of methods and approaches that have been developed over the past half century. They range from standard methods of arbitration to diverse forms of mediation, the latter commonly utilising the methodologies of more established disciplines such as psychology. Such methods operate at many levels, from the individual and community through to the international. The past 30 years have seen a particular growth in mediation in conflicts between states, especially with 'Track 2' processes. In such processes, intermediaries work with people from the affected states that have links with centres of power but whose involvement in the 'negotiations' can be readily denied. Such processes enable mediated solutions to be developed which can then be used in more official and acknowledged dialogue. There may also be a specific value in the experience of non-government organisations that work in the global south and which have accrued considerable experience, often involving the use or adaptation of traditional conflict resolution practices. Such groups in the UK include *Conciliation Resources*, *International Alert* and especially *Peace Direct*. What is significant about their work in this context is that they support groups often involved in neighbourhood conflicts which may have an environmental element such as access to land or natural resources. Traditional practices may have evolved over many generations and have a societal familiarity which aids community confidence in their efficacy. They may therefore be capable of adaptation in order to work in the case of conservation conflicts which may be relatively new to a particular community.

However, there are two other issues relating to the development of conflict resolution that may be relevant to conservation conflicts, one theoretical and one practical. The first is that conflict resolution may be palliative rather than transformative, treating symptoms rather than underlying causes. As such it may succeed in apparently resolving conflict in the short term without exploring solutions that prevent a later re-emergence of conflict. This may be a particular problem for conservation conflicts if long-term pressures of population growth or economic development continue while the search is made for an immediate solution to a problem. The more practical issue relates to this, in that individuals or organisations seeking to resolve conservation conflicts may need to demonstrate success in order to ensure further funding. This can result in overstating claims of success and looking for 'quick-fix' solutions.

Broadly similar issues have been intensively debated within the peace research community and one of the most useful responses comes from Ramsbotham *et al.* (2011: xvii):

> We suggest that peace and conflict research is part of an emancipatory discourse and practice which is making a valuable and defining contribution to emerging norms of democratic, just and equitable systems of global governance. We argue that conflict

resolution has a role to play in the radical negotiation of these norms, so that international conflict management is grounded in the needs of those who are the victims of conflict and who are frequently marginalized from conventional power structures.

As to the value of the study of conservation conflicts to peace researchers, there are two elements to be recognised – one general and one specific. At the general level, although there has long been an interest in issues of environmental limits to growth, the impact of climate disruption still tends to be under-recognised by peace researchers. This is perhaps understandable given the plethora of conflict issues of a more conventional nature, such as the war on terror and intercommunal conflict across much of the Middle East, but environmental limits are likely to increase in importance as underlying drivers of conflict. More knowledge of basic ecological principles would therefore be of considerable value to peace researchers. While some of their ranks are drawn from the life sciences, knowledge of issues such as biodiversity and ecosystem homeostasis remain limited and this constrains the analysis of wider conflict trends related to environmental limits to growth. At the specific level, knowledge of specific conservation conflicts is very limited in peace research yet might have much to offer, especially in the practice of conflict resolution related to other kinds of conflicts.

References

Abbott, C., Rogers, P. and Sloboda, J. (2006). *Global Responses to Global Threats: Sustainable Security for the 21st Century*. Oxford: Oxford Research Group.

Boulding, K. (1973). Editorial. *J. Conflict Resolut.*, 1, 3.

Bradford. (2011). *Peace Research at Bradford, Annual Report 2010*. Bradford: University of Bradford.

Davis, J., Sandstrom, S., Shorrocks, A. and Wolff, E. N. (2006). The world distribution of household wealth. In *WIDER Angle, No 2*. Helsinki: World Institute for Development Economics Research.

Galtung, J. (1969). Violence, peace and peace research. *J. Peace Res.*, 6, 167–191.

Kirby, A. (2013). Unprecedented climate extremes marked last decade, says UN. *Guardian Environment Network*, 3 July.

Meadows, D. H., Meadows, D. L., Randers, J. and Behrands W. H. III. (1972). *Limits to Growth*. London: Earth Island.

Ramsbotham, O., Woodhouse, T. and Miall, H. (2011). *Contemporary Conflict Resolution*. Cambridge and Malden, MA: Polity.

Rogers, P. (2013). *Peace studies. In Contemporary Security Studies*, ed. A. Collins, pp. 54–66. Oxford: Oxford University Press.

Rogers, P. and Ramsbotham, O. (1999). Then and now: peace research – past and future. *Polit. Stud.*, 47, 740–754.

Box 12

Lead shot in Europe: conflict between hunters and conservationists

Julia Newth[1], Ruth Cromie[1] and Niels Kanstrup[2]
[1]*Wildfowl & Wetlands Trust, Slimbridge, Gloucestershire GL2 7BT, UK*
[2]*Danish Academy of Hunting, Skrejrupvej 31, DK-8410 Rønde, Denmark*

Lead shot is a widely used, effective and cheap ammunition. During shotgun shooting, the majority of spent lead shot falls to the ground where it can be ingested by birds because they mistake it for food or grit.

Lead poisoning in birds from lead shot was first recorded in the 1870s and is now well documented as a source of morbidity and mortality in wild birds around the world (Mateo, 2009). Predatory and scavenging birds, primarily raptors, also ingest lead embedded in prey or carrion. Conflict arises between those who wish to continue using lead shot for hunting and those who wish to replace lead shot with non-toxic alternatives for all shooting in order to protect birds and ecosystems from the toxic effects of lead (herein termed 'conservationists', accepting that hunters may be conservationists too).

Research on lead poisoning encouraged the development of shared positions of agreement by hunters and conservationists. In 1991, an International Wetlands Research Bureau (IWRB) conference in Brussels representing experts in lead poisoning, government agencies, NGOs, conservation and hunting organisations, and arms and ammunition manufacturers led to statements supporting phased removal of lead shot used over wetlands and prompted the first of several resolutions of the African–Eurasian Waterbirds Agreement, which precipitated national legislation in many countries.

International hunting bodies, such as the European Federation of Associations for Hunting and Conservation (FACE) and the International Council for Game and Wildlife Conservation (CIC), have also promoted the phasing out of lead in ammunition. However, although there is some acceptance within national hunting communities that lead shot poisons birds, there is disagreement about its scale and potential solutions.

Case study England: Conservationists argue that English legislation introduced in 1999 aimed at restricting the use of lead over wetlands fails to protect birds, including wildfowl, feeding in terrestrial areas. Compliance with legislation has been poor and difficult to enforce (Cromie *et al.*, 2010), and has not reduced lead poisoning (Newth *et al.*, 2012). Proposals that allude to further restrictions on lead shot use over terrestrial habitats have been met by strong resistance from many in the shooting community. Motivations for non-compliance with legislation include: perception that lead poisoning was not a sufficient problem to justify the restrictions; non-lead alternatives were expensive, not widely available and not as effective as lead; and restrictions were not enforced (Cromie *et al.*, 2010). Restrictions were also perceived as a means to phase out lead shot completely and perhaps as 'being used' by conservationists to end bird hunting completely. Conservationists see shooters as obstructing efforts to reduce lead poisoning. These perceptions have contributed to mutual mistrust. Prior to, and since, the introduction of legislation in England, conservationists have not worked with hunters to persuade them of the benefits of non-toxic shot and few advocates within the shooting community have supported the use of non-toxic shot.

Case study Denmark: Denmark began regulating lead shot in 1985 and in 1996 it became illegal to use lead shot for all shooting. Initially, Danish shooters shared similar fears about the cost and efficacy of non-toxic shot. However, some key advocates from this community were crucial in persuading other hunters of the benefits of non-toxic shot (Kanstrup, 2006). The success of steel shot for clay pigeon shooting allayed shooter concerns by demonstrating that there were suitable, safe, relatively cheap alternatives to lead. Research by the Danish Hunters Association also demonstrated the efficacy of steel shot for killing birds. When steel shot embedded in trees was deemed unacceptable to foresters, the development of softer alternatives such as bismuth was prompted, despite the comparatively higher cost. As was the case in England, many Danish hunters were concerned that phasing out of lead shot would lead to the end of hunting, but this has not occurred and the number of hunters and the annual bag have not changed significantly. Furthermore, an initial concern that there was an increased risk of irreparable gun damage by steel shot proved unfounded. Scientific studies conducted by hunters demonstrated the efficacy of alternative shot and have reassured the hunting community. Moreover, the general 'image' of shooting within Danish society has been maintained.

These examples illustrate contrasting approaches to lead poisoning having varying success. Denmark successfully reduced environmental lead contamination by enacting legislation that banned the use of lead shot by garnering support of hunters for use of non-toxic alternatives. Danish success can be linked to a few advocates within the hunting community who persuaded other hunters of the benefits using evidence from hunter-led research. All stakeholders were involved in the transition process in the early stages and thus trust between them was maintained. In England, a country of strong traditions, the absence of hunter advocates for non-toxic shot and support for non-toxic shot polarised stakeholders. Partial restrictions have had poor compliance, largely because shooters are not convinced either of the problem or the benefits of non-toxic shot. The engagement of a trusted third party, respected by stakeholders, is needed to facilitate negotiations to resolve the conflict.

References

Cromie, R. L., *et al.* (2010). Compliance with the Environmental Protection (Restrictions on Use of Lead Shot) (England) Regulations 1999. Report to Defra, Bristol, UK.

Kanstrup, N. (2006). Non-toxic shot – Danish experiences. In *Waterbirds Around the World*, eds. G. C. Boere, C. A. Galbraith and D. A. Stroud, 861 pp. Edinburgh: The Stationery Office.

Mateo, R. (2009). Lead poisoning in wild birds in Europe and the regulations adopted by different countries. In *Ingestion of Lead from Spent Ammunition: Implications for Wildlife and Humans*, eds. R. T. Watson, M. Fuller, M. Pokras and W. G. Hunt, pp. 71–98. Boise, ID: The Peregrine Fund.

Newth, J. L., *et al.* (2012). Poisoning from lead gunshot: still a threat to wild waterbirds in Britain. *Eur. J. Wildl. Res.*, 59, 195–204.

© Adam Vanbergen.

CHAPTER THIRTEEN

Linking conflict and global biodiversity conservation policies

ESTHER CARMEN, JULIETTE C. YOUNG and ALLAN WATT
Natural Environment Research Council Centre for Ecology and Hydrology, UK

Policies come about as a result of a series of decisions based on a dynamic and complex process involving a continuous interplay of discussions, political interests and different people that define the goals and actions of organisations (Keeley and Scoones, 2003). This process of policy development and subsequent implementation can lead to conflict (Pierson, 2005; Saito-Jensen and Jensen, 2010). In some cases, conflict itself can lead to policy change (Castro and Nielsen, 2001; Haro *et al.*, 2005). Although policy processes are complex, and conflict between groups may only be one factor within the policy process (Anderies and Janssen, 2013), a broad perspective on the link between policy and conflict (as illustrated in Chapter 15) is needed for understanding and managing conservation conflicts.

In this chapter we examine conservation conflicts as a potential component of the global biodiversity policy process. We outline the potential links to conflict as biodiversity policies move from a focus on protected areas to diversified approaches that acknowledge wider socio-economic objectives. We also highlight the different layers, such as the ecosystem services framework or the green economy, which have been progressively added to these policies to help practitioners reframe recognised conflicts. We then illustrate some of these issues with the example of the Joint Forest Management (JFM) policy in India before concluding with the need to be more explicit about conflicts in policy development.

An overview of global biodiversity conservation policies

There have been a number of conservation policies adopted at the global level since the 1970s. These policies can be approached using two main integrative dimensions (see Hirsch and Brosius, 2013). The 'horizontal' dimension represents the interplay between conservation objectives and wider socio-economic–political goals, while the 'vertical' dimension represents the hierarchical structure of multiple stakeholders and institutions involved in managing natural resources (Karlsson-Vinkhuyzen, 2012). Both these dimensions link closely to

Conflicts in Conservation: Navigating Towards Solutions, ed. S. M. Redpath, R. J. Gutiérrez, K. A. Wood and J. C. Young. Published by Cambridge University Press.

conservation conflicts. As defined in Chapter 1, conservation conflicts occur when parties clash over differences about conservation objectives and when one party asserts, or at least is perceived to assert, its interests at the expense of another. The horizontal dimension addresses the differences over conservation (and other) objectives, while the vertical dimension addresses the structure of the different parties potentially involved in conservation conflicts. In this section we highlight a few of the key policies agreed at the global level over biodiversity, their horizontal and/or vertical dimensions, and how these dimensions could link to conservation conflicts.

Early conventions tended to focus on the conservation of particular habitats and species, for example the 1971 Convention on Wetlands of International Importance ('Ramsar Convention'), the 1973 Convention on International Trade in Endangered Species of Wild Fauna and Flora (CITES), and the 1979 Convention on the Conservation of Migratory Species of Wild Animals ('Bonn Convention'). The development of the United Nations Environment Programme (UNEP) in 1972 was an attempt at a more comprehensive approach to biodiversity conservation. The United Nations Conference on the Human Environment in 1972 addressed mainly the horizontal dimension, emphasising the need to link economic development and environmental pressures (Mansfield, 2008). Subsequent policy developments have strengthened this horizontal dimension, for example by framing global conservation policies around the concept of sustainable development (Roe, 2008).

In the 1980s and 1990s integration across the vertical dimension became increasingly prevalent, as global policies such as the Rio Declaration on Environment and Development in 1992 recognised the importance of local resource users in taking part in and achieving policy goals. These policy shifts were influenced by a number of different processes. Many citizens increasingly felt frustrated and disconnected from political processes and institutions (Scharpf, 1999), leading to challenges to traditional approaches to representative democracy and the 'participation explosion' in the 1960s (Steelman and Ascher, 1997). In addition, from the 1970s onwards, there were moves in conservation towards community-based approaches, the recognition that protected areas were insufficient for conservation and growing criticism about the negative effects of exclusionary policies on local people, which could lead to conflict (Hulme and Murphree, 2003; Brockington and Igoe, 2006; West *et al.*, 2006).

The appointment of the World Commission on Environment and Development in 1983, and the subsequent Brundtland report (World Commission on Environment and Development, 1987), were the main triggers for the United Nations Conference on Environment and Development in 1992, also referred to as the Rio Summit. During this conference, binding agreements were opened for signature, including one on biodiversity referred to as the Convention on Biological Diversity (CBD), subsequently signed by 168 parties, which came into force in December 1993. The main aims of the CBD are to promote biodiversity conservation, the sustainable use of all its components and the equitable sharing of genetic resources. Another major, and closely linked, international landmark at the Rio Summit of 1992 was Agenda 21 (UNCED, 1992a), a central feature of which is public participation, viewed as 'one of the fundamental prerequisites for the achievement of sustainable development' (paragraph 23.2). Although the Rio Summit led to the formalisation of public participation as a non-binding policy goal, specifying in Principle 10 of the Rio declaration (UNCED, 1992b) that 'environmental issues are best handled with the participation of all concerned citizens, at the relevant level', it was not until 1998 that this was translated into a set of implementing measures with the adoption of the 'UNECE Convention on access to information, public participation in decision-making and access to justice in environmental matters' (the so-called Aarhus Convention). The Aarhus Convention is unique in that it goes further than simply stressing the need for participation. It sets out public participation requirements, including the timely notification of the public, reasonable time frames for participation, free access to all information relevant to the decision-making, an obligation on the decision-making body to take due account of the outcome of the public participation, and prompt public notification of the decision (Article 6). The Convention entered into force in 2001. The CBD and Agenda 21 marked a key development in global conservation policies by expanding protected area approaches to include wider socio-economic–political goals (the horizontal dimension), leading to more 'people-centred' conservation policies (the vertical dimension) involving multiple stakeholders and institutions in the management of natural resources (Miller *et al.*, 2011).

This change in emphasis led to an increased focus on integrated approaches to poverty alleviation, and an emphasis on decentralisation, i.e. the partial transfer of power from national institutions to democratically elected local representatives (Secretariat of the Convention on Biological Diversity, 2000: principle 2; Larson and Ribot, 2004; de Haan and Zommers, 2005; Ribot *et al.*, 2006). The Millennium Development Goals (MDG) in 2000 created targets for governments to integrate sustainable development into national policies and reverse the loss of environmental resources (Pisupati and Warner, 2003; Roe and Elliott, 2004; Bass *et al.*, 2006; Hulme, 2009). In 2002, this horizontal dimension was further strengthened at the World Commission on Sustainable Development, which stressed the development of 'national biodiversity strategies' (United Nations, 2002: 44(d)) and providing 'support to developing countries and countries with economies in transition [...] with a view to conserving and the sustainable use of biodiversity' (United Nations, 2002: 44(m)). In addition, this Commission emphasised the need to promote sustainable community-based use of biodiversity, recognising 'the rights of local and indigenous communities [...], benefit sharing mechanisms on mutually agreed terms' (United Nations, 2002: 44(j)) and promoting the 'participation of indigenous and local communities in decision and policy making' (United Nations, 2002: 44(l)), with the potential to reduce conflict with local resource users and advocates of indigenous peoples' rights. In practice, decentralisation is often linked to a combination of economic and political factors, such as structural adjustment, fiscal reforms and difficulties of enforcing regulations in remote areas, as well as conservation conflicts relating to demands by local groups for recognition of their traditional rights (Conroy *et al.*, 2002; Larson, 2002; Diaw *et al.*, 2008; Pulhin and Dressler, 2009; Coleman and Fleischman, 2012; Larson and Dahal, 2012).

At the start of the twenty-first century, some claimed that conservation had 'fallen off' the global agenda (Roe and Elliott, 2004). This resulted in the strengthening of both the horizontal and vertical links in biodiversity policies. For example, the fifth meeting of the Conference of the Parties to the CBD recognised the positive role of certain human activities on the environment (horizontal dimension) and that 'management should be decentralized to the lowest appropriate level [...], should involve all stakeholders and balance local interests with the wider public interest' (vertical dimensions) (Secretariat of the Convention of Biological Diversity, 2000: Principle 2). Both of these aspects can be seen as having the potential to reduce conflict with local resource users. In 2002, signatories to the CBD agreed to achieve a significant reduction of biodiversity loss by 2010 (Decision VI/26). This marked an important turning point in international environmental agreements, being the first time a large group of governments agreed to a quantitative target for reducing biodiversity loss (Balmford and Bond, 2005).

More recent international biodiversity policy discussions (e.g. Rio+20 in 2012) continue to focus on the horizontal and vertical dimensions, and win–win situations to integrate these dimensions better. Specifically, in the vertical dimension, discussions have stressed the need to engage stakeholders 'from all levels of government' (United Nations, 2012: 42), 'civil society' (United Nations, 2012: 44) and the 'private sector' (United Nations, 2012: 46) and highlighted the contribution to and dependence on biodiversity by 'indigenous peoples and local communities' (United Nations, 2012: 197). This need for integration of stakeholders and their expertise was strengthened in the recent creation of the Intergovernmental Platform for Biodiversity and Ecosystem Services (IPBES), which aims to develop 'the science policy interface [...] for the conservation and sustainable use of biodiversity, long term human wellbeing and sustainable development [...engaging with] indigenous peoples and local communities and the private sector' (UNEP, 2012: 1) to facilitate more informed decisions by national policy makers. This broadening perspective may, if the range of relevant perspectives is ensured, help reduce conflicts which previously emerged during policy implementation while implicitly recognising that the complex mix of factors which influence outcomes often varies from place to place, moving away from a more prescriptive outlook which can result in unintended consequences (Barrett *et al.*, 2005).

With the growing recognition of the need for a multi-institutional perspective across governance levels came the increasing focus on finding win–win solutions. A key theme emerging from Rio+20 was a focus on large-scale processes to create a 'green economy [. . .] to eradicate poverty as well as sustain economic growth, enhancing social inclusion, improving human welfare and creating opportunities for employment [. . .], while maintaining the healthy function of the Earth's ecosystem' (Brand, 2012; United Nations, 2012: 56). A similar approach was the development of the ecosystem services framework which emphasises the relationships between biodiversity, ecosystem goods and services and human well-being, and the benefits societies and individuals gain from the environment (Millennium Ecosystem Assessment, 2005). The ecosystem services framework, similarly to the green economy, and concepts of natural capital, nature-based solutions and so on, may be aiming to reframe conservation conflicts into potential win–win situations: for example, developing eco-tourism opportunities for local people (horizontal dimension), who can in theory reap benefits (vertical dimension) from a species otherwise considered a pest or a threat. Whether such win–win situations exist or whether they run the risk of raising expectations remains to be seen. Thus, the policy process to reframe the long-standing conflict between environmental and economic objectives into potential win–win situations continues alongside the need for cooperation with a broad range of stakeholders (IIED, 2012).

The example of the Joint Forest Management policy in India

As with most conventions and agreements reached at the global level, it is at the regional and national levels that implementation is best examined. In this section we explore the often implicit links between conservation conflicts and policy by exploring the example of forest resources in general and the example of the Joint Forest Management (JFM) policy in India in particular. These examples are chosen as they illustrate both the horizontal and vertical dimensions, and the complex relationships among socio-economic contexts, conservation policies and their outcomes.

Conflicts over forest resources have been particularly widespread (e.g. Box 5; Box 19; Chapter 11). Historically, the control of forest resources has been a strategy to demarcate territory and organise economic activities to enhance state building (Vandergeest and Peluso, 1995). The creation of forest reserves often involved annexing forests and then managing them for exclusive hunting and timber production for state revenue (Larson and Pulhin, 2012). This model included regulation by the state, and often undermined the role of local people in using forest resources (Williams and Mawdsley, 2006). Although the socio-political and ecological contexts have varied, two common outcomes have emerged from this model: conflict between the state and local resources users, and loss of forest cover (Peluso, 1993; Negi *et al.*, 1997). In some situations, violence, widespread protests and more implicit acts of sabotage, such as cutting or burning trees, have ensued (Brosius, 1997; Klooster, 2000; Agrawal and Ostrom, 2001). These conflicts have also often been multi-dimensional, involving private logging companies and plantation owners (Kröger, 2012). Furthermore, many such conflicts have been linked to increasing differentiation between informal and traditional legal rights (Saito-Jensen and Jensen, 2010). As such, conflicts over forest resources have been common (Castro and Nielsen, 2001), multi-faceted, connected to current biodiversity conservation policies across time, varied in spatial scale and involved different interests. However, despite the often long and complex histories of these conflicts, these aspects are commonly overlooked or simplified when policies are developed (see Chapter 4).

Resource management has a long history in India, with laws concerning the protection of forests and wild animals to protect the hunting resources of the elite dating back to the third century BC (Madhusudan and Shankar Ramen, 2003). Towards the end of the nineteenth century, British colonial rule instigated a system of plantations to supply timber for ship-building and railway construction (Karanth and DeFries, 2010). Accordingly, administrative structures were created to manage resources more effectively through a centralised government forest department (Agrawal, 2001). Amid competing demands from other sectors of the government, local timber merchants and other local groups, forests were partitioned, with two-thirds being controlled by the Forest Department for commercial exploitation (Rangan, 1995). This partitioning alienated local people by strictly controlling their activities (Gadgil, 1992). This management policy continued after independence, but by the 1970s its failures were widely recognised (Williams and Mawdsley, 2006), including loss of forest resources for which local people were often blamed (Rangan, 1995) and widespread violent protests (Shiva, 1991). Elsewhere, however, examples of locally initiated protection of forest resources started to emerge despite the lack of formal rights given to local people (Conroy *et al.*, 2002). Perhaps as a result of these local conflicts, together with pressure from external donors, more conservation-focused and participatory policies began to emerge (Larson and Dahal, 2012). This shift was formalised in the 1988 Forest Policy, closely followed by the 1990 Joint Forest Management (JFM) policy to decentralise decision-making to involve local people in more meaningful ways (Sekhar, 2000).

Although the JFM policy implementation is an example of large-scale decentralisation that may have been in part due to conservation conflicts, it is unclear how much of the resulting increased forest cover is directly attributable to the JFM policy (Agrawal and Ostrom, 2001; Capistrano, 2008), or indeed whether the JFM policy has reduced conflict. In some instances, conflicts have arisen during implementation, sometimes leading to a loss of biodiversity (Larson and Pulhin, 2012). One area of contention is a lack of recognition during implementation of existing, informal arrangements to use resources (Larson and Pulhin, 2012). As such, some villages are unwilling to adopt JFM as it may erode locally negotiated agreements to use forest resources, the loss of which could have consequences for marginalised social groups (Beck and Nesmith, 2001). While JFM may help to resolve conflict between some local resource users and the Forest Department, it may also exacerbate local conflicts among or within villages and different user groups (Saito-Jensen and Jensen, 2010).

Conclusions

Global conservation policies since the 1970s have placed more emphasis on integration between conservation and other socio-economic objectives (the horizontal dimension), and engagement of all relevant stakeholders (the vertical dimension). In practice, however, implementation of global conservation policies has often remained dichotomous. Along the horizontal dimension, implementation at the national and local level often places an emphasis either on conservation or development, and rarely both (Berkes, 2007). Indeed, nearly all the international conventions rely on designating areas for conservation, which, although essential to conserve biodiversity at the global and regional scales, are deemed insufficient to conserve the full range of biodiversity, and are unlikely to lessen the impact of conservation conflicts (Millennium Ecosystem Assessment, 2005). In addition, many conservationists believe more attention still needs to be paid to the vertical dimension in practice, with appropriate involvement of all relevant levels of society (Fox *et al.*, 2006; Secretariat of the Convention on Biological Diversity, 2010). For example, decentralisation may only involve an incomplete transfer of power to local levels with administrative tasks transferred to the local level while decision-making either remains with central government or control becomes concentrated within privileged local groups (Larson and Ribot, 2004). This imbalance may itself lead to new conflict and further unsustainable use of resources (Capistrano, 2008; Saito-Jensen and Jensen, 2010).

The links between conservation conflicts and the development and outcomes of global policies are more often than not difficult to define. Although conflicts may contribute to the overall development of policies, they will most often only be a component of the range of factors instigating the development of policies. Similarly, it is difficult to define the exact impact of a policy on conservation conflicts due to changing contexts and other parameters. With conflicts becoming a key challenge to modern conservation, conflicts may need to be better acknowledged, and become a clear focus of future global conservation policies and their implementation. In many ways, focusing on conservation conflicts could integrate the horizontal and vertical dimensions of policies, by tackling some of the root causes of biodiversity loss. So, instead of recognising biodiversity loss and putting in measures to address this loss (e.g. protected areas), policies could directly acknowledge and focus on conservation conflicts as a cause of biodiversity loss, and put in measures to address those conflicts, which in turn may benefit biodiversity.

References

Agrawal, A. (2001). Common property institutions and sustainable governance of resources. *World Dev.*, 29, 1649–1672.

Agrawal, A. and Ostrom, E. (2001). Collective action, property rights and decentralization in resource use in India and Nepal. *Polit. Soc.*, 29, 485–514.

Anderies, J. M. and Janssen, M. A. (2013). Robustness of socio-ecological systems: Implications for public policy. *Policy Stud. J.*, 41, 513–536.

Balmford, A. and Bond, W. (2005). Trends in the state of nature and their implications for human well-being. *Ecol. Lett.*, 8, 1218–1234.

Barrett, C. B., Lee, D. R. and McPeak, J. G. (2005). Institutional arrangements for rural poverty reduction and resource conservation. *World Dev.*, 33, 193–197.

Bass, S., Bigg, T., Bishop, J. and Tunstall, D. (2006). Sustaining the environment to fight poverty and achieve the Millennium Development Goals. *Reciel*, 15, 39–55.

Beck, T. and Nesmith, C. (2001). Building on poor people's capacities: the case of common property resources in India and West Africa. *World Dev.*, 29, 119–133.

Berkes, F. (2007). Community-based conservation in a globalized world. *Proc. Natl Acad. Sci. USA*, 104, 15188–15193.

Brand, U. (2012). Green economy – the next oxymoron? *Gaia*, 21, 28–32.

Brockington, D. and Igoe, J. (2006). Eviction for conservation: a global overview. *Conserv. Soc.*, 4, 424–470.

Brosius, J. P. (1997). Prior transcripts, divergent paths: resistance and aquiescence to logging in Sarawak, East Malaysia. *Comp. Stud. Soc. Hist.*, 39, 468–510.

Capistrano, D. (2008). Decentralization and forest governance in Asia and the Pacific: trends, lessons and continuing challenges. In *Lessons from Forest Decentralization: Money, Justice and the Quest for Good Governance in Asia–Pacific*, eds. C. J. P. Colfer, G. R. Dahal and D. E. Capistrano, pp. 211–232. London: Earthscan.

Castro, A. P. and Nielsen, E. (2001). Indigenous people and co-management: implications for conflict management. *Environ. Sci. Pol.*, 4, 229–239.

Coleman, E. A. and Fleischman, F. D. (2012). Comparing forest decentralization and local institutional change in Bolivia, Kenya, Mexico, and Uganda. *World Dev.*, 40, 836–849.

Conroy, C., Mishra, A. and Ajay, R. (2002). Learning from self-initiated community forest management in Orissa, India. *Forest Pol. Econ.*, 4, 227–237.

de Haan, L. and Zommers, A. (2005). Exploring the frontiers of livelihood research. *Dev. Change*, 36, 27–47.

Diaw, C. M., Blomley, D. and Lescuyer, G. (2008). *Elusive Meanings: Decentralisation, Conservation and Local Democracy*. Abingdon: Earthscan.

Fox, H. E., Christian, C., Nordby, J. C., Pergams, O. R., Peterson, G. D. and Pyke, C. R. (2006). Perceived barriers to integrating social science and conservation. *Conserv. Biol.*, 20, 1817–1820.

Gadgil, M. (1992). Conserving biodiversity as if people mattered: a case study from India. *Ambio*, 21, 266–270.

Haro, G. O., Doyo, G. J. and McPeak, J. G. (2005). Linkages between community, environmental, and conflict management: experiences from northern Kenya. *World Dev.*, 33, 285–299.

Hirsch, P. D. and Brosius, P. (2013). Navigating complex trade-offs in conservation and development: an integrative framework. *Issues Interdiscipl. Stud.*, 31, 99–122.

Hulme, D. (2009). *The Millennium Development Goals: A Short History of the World's biggest Promise.* BWPI Working paper 100. Manchester: Brooks World Poverty Institute, University of Manchester.

Hulme, D. and Murphree, M. (2003). *African Wildlife and Livelihoods: The Promise and Performance of Community Conservation.* London: James Currey.

IIED (2012). *Sharing Solutions for a Sustainable Planet. Fair Ideas Highlights.* London: International Institute for Environment and Development (IIED).

Karanth, K. K. and DeFries, R. (2010). Conservation and management in human-dominated landscapes: case studies from India. *Biol. Conserv.*, 143, 2865–2869.

Karlsson-Vinkhuyzen, S. I. (2012). From Rio to Rio via Johannesburg: integrating institutions across governance levels in sustainable development deliberations. *Nat. Resour. Forum*, 36, 3–15.

Keeley, J. and Scoones, I. (2003). *Understanding Environmental Policy Processes: Cases from Africa.* London: Earthscan.

Klooster, D. (2000). Community forestry and tree theft in Mexico: resistance or complicity in conservation? *Dev. Change*, 31, 281–305.

Kröger, M. (2012). The expansion of industrial tree plantations and dispossession in Brazil. *Dev. Change*, 43, 947–973.

Larson, A. M. (2002). Natural resources and decentralisation in Nicaragua: are local governments up to the job? *World Dev.*, 30, 17–31.

Larson, A. M. and Dahal, G. R. (2012). Forest tenure reform: new resource rights for forest-based communities? *Conserv. Soc.*, 10, 77–90.

Larson, A. M. and Pulhin, J. (2012). Enhancing forest tenure reforms through more responsive regulations. *Conserv. Soc.*, 10, 103–113.

Larson, A. M. and Ribot, J. (2004). Democratic decentralisation through a natural resource lens: an introduction. *Eur. J. Dev. Res.*, 16, 1–25.

Madhusudan, M. D. and Shankar Ramen, T. R. (2003). Conservation as if biological diversity matters: preservation versus sustainable use. *Conserv. Soc.*, 1, 49–59.

Mansfield, B. (2008). Global environmental politics. In *The Sage Handbook of Political Geography*, eds. K. R. Cox, M. Low and J. Robinson, pp. 235–246. London: Sage.

Millennium Ecosystem Assessment (2005). *Ecosystems and Human Well-being: Biodiversity Synthesis.* Washington, DC: World Resource Institute.

Miller, T. R., Minteer, B. A. and Malan, L. C. (2011). The new conservation debate: the view from practical ethics. *Biol. Conserv.*, 144, 948–957.

Negi, A. K., Bhatt, N. P., Todaria, N. P. and Saklani, A. (1997). The effects of colonialism on forests and the local people in the Garhwal Himalaya, India. *Mount. Res. Dev.*, 17, 159–168.

Peluso, N. (1993). Coercing conservation? The politics of state resource control. *Global Environ. Chang.*, 3, 199–217.

Pierson, P. (2005). The study of policy development. *J. Pol. Hist.*, 17, 34–51.

Pisupati, B. and Warner, E. (2003). *Biodiversity and the Millennium Development Goals.* Colombo, Sri Lanka: IUCN.

Pulhin, J. M. and Dressler, W. H. (2009). People, power and timber: the politics of community-based forest management. *J. Environ. Manage.*, 91, 206–214.

Rangan, H. (1995). Contested boundaries: state policies, forest classifications and deforestation in the Garhwal Himalayas. *Antipode*, 27, 343–362.

Ribot, J. C., Agrawal, A. and Larson, A. M. (2006). Recentralizing while decentralizing: how national governments reappropriate forest resources. *World Dev.*, 34, 1864–1886.

Roe, D. (2008). The origins and evolution of the conservation-poverty debate: a review of key literature, events and policy processes. *Oryx*, 42, 491–503.

Roe, D. and Elliott, J. (2004). Poverty reduction and biodiversity conservation: rebuilding the bridges. *Oryx*, 38, 137–139.

Saito-Jensen, M. and Jensen, C. B. (2010). Rearranging social space: boundary-making and boundary work in a joint forest management project, Andhra Pradesh, India. *Conserv. Soc.*, 8, 196–208.

Scharpf, F. W. (1999). The choice for Europe: social purpose and state power from Messina to Maastricht. *J. Eur. Publ. Pol.*, 6, 164–168.

Secretariat of the Convention on Biological Diversity. (2000). *Conference of the Parties to the Convention on Biological Diversity (COP 5), Decision 6*. Nairobi, Kenya.

Secretariat of the Convention on Biological Diversity. (2010). *Global Biodiversity Outlook 3*. Montreal, Canada.

Sekhar, N. U. (2000). Decentralized natural resource management: from state to co-management in India. *J. Environ. Plann. Manage.*, 43, 123–138.

Shiva, V. (1991). *Ecology and the Politics of Survival: Conflicts over Natural Resources in India*. New Delhi: Sage.

Steelman, T. A. and Ascher, W. (1997). Public involvement methods in natural resource policy making: advantages, disadvantages and trade-offs. *Policy Sci.*, 30, 71–90.

UNCED (1992a). Agenda 21. United Nations Conference on Environment and Development, Rio de Janerio, Brazil, 3 to 14 June 1992. United Nations Division for Sustainable Development.

UNCED (1992b). Rio Declaration. United Nations Conference of Environment and Development, Rio de Janerio, Brazil, 3 to 14 June 1992. United Nations Division for Sustainable Development.

UNEP (2012). Functions, operating principles and institutional arrangements of the intergovernmental science-policy platform on biodiversity and ecsosystem services. t.-s. A. Second session of the plenary meeting for IPBES, 2012. Panama City, Panama.

United Nations (2002). Plan of Implementation of the World Summit on Sustainable Development. United Nations Division on Sustainable Development.

United Nations (2012). Rio + 20: Outcomes document – The Future We Want. Rio de Janeiro, June 2012.

Vandergeest, P. and Peluso, N. (1995). Territorialization and state power in Thailand. *Theor. Soc.*, 24, 385–426.

West, P., Igoe, J. and Brockington, D. (2006). Parks and peoples: the social impact of protected areas. *Annu. Rev. Anthropol.*, 35, 251–277.

Williams, G. and Mawdsley, E. (2006). Postcolonial environmental justice: Government and governance in India. *Geoforum*, 37, 660–670.

World Commission on Environment and Development (1987). *Our Common Future*. United Nations Environment Programme.

Box 13

Conservation conflicts in Natura 2000 protected areas

Juliette C. Young[1], Allan Watt[1], Andrew Jordan[2] and Peter Simmons[2]
[1]*NERC Centre for Ecology and Hydrology*
[2]*University of East Anglia*

A traditional approach towards biodiversity conservation has been creating protected areas (Mulongoy and Chape, 2004), now estimated to cover 12.9% of the global terrestrial area (Jenkins and Joppa, 2009). In the European Union, the main policy mechanism for protected areas is the Natura 2000 network, which aims to 'enable the natural habitat types and species' habitats concerned to be maintained or, where appropriate, restored at a favourable conservation status in their natural range' (European Commission, 2000, Habitats Directive, Article 3(1)). To achieve this aim, the network includes Special Protection Areas (SPAs) established under the 'Birds Directive' (79/419/EEC) and Special Areas of Conservation (SACs) to comply with the 'Habitats Directive' (92/43/EEC). The network now covers 17% of EU territory, making it the largest network of protected areas in the world (European Commission, 2010). However, the establishment of Natura 2000 has been problematic because of resistance by private land-owners and managers who perceive the designations of SACs and/or SPAs to be impositions as most Natura areas are established on private land.

Conflicts over Natura 2000 began in 1992 as a result of the selection and designation of sites. The top-down, scientifically driven selection of Natura 2000 sites led to widespread resistance to the network by local residents who owned or managed many of the sites, and who perceived the network to be a threat to their economic or social interests. One extreme example was the 'Groupe des 9' in France, who questioned the legitimacy of implementation in France and ultimately caused the Directive to be suspended temporarily in 1996 (Alphandery and Fortier, 2001). In Finland, the network caused major conflicts between landowners and environmental authorities, leading to hunger strikes by forest owners in the Karvia region (Bergseng and Vatn, 2009) that ultimately affected country-wide attitudes towards biodiversity conservation. Delays have meant that the network took longer to implement than planned, and the marine network is still lagging. Also, there have been concerns over the quantity and quality of designated sites (e.g. Dimitrakopoulos *et al.*, 2004) and the ability of the network to allow species to adapt to environmental change (Sutherland *et al.*, 2010).

The European Commission issued a document in 2000 to help EU member states manage Natura 2000 sites (European Commission, 2000). This document expanded Article 6 of the Habitats Directive, which states that EU member states are required to 'establish the necessary conservation measures' for management plans and statutory, administrative or contractual measures that are consistent with ecological requirements of species when sites are designated as SACs. This document also acknowledged that Natura 2000's success relies on the active involvement of those who live or depend on those areas. More importantly, the document suggested 'important considerations' including the need to consult landowners and other stakeholders during implementation, including the development of management plans. The aim of a management plan has been to achieve the conservation objectives of the Directive. Management plans were not necessarily designed to manage conflicts. Although many conflicts will be present as a result of SACs, there has been only limited exploration of potential for management plans as conflict management tools in the context of Natura 2000.

Multi-disciplinary independent research funded by the UK Natural Research Environment Council involving sociologists, political scientists and modellers from the Centre for Ecology and Hydrology and the University of East Anglia analysed stakeholder perceptions of conflict and its management in three case studies in Scotland where conflicts were present and where management plans had been developed to ensure protection of species. Using quantitative and qualitative data derived from semi-structured interviews, many conditions were identified as necessary to enable Natura 2000 management plans to act as tools for conflict management and to assist in achieving a socially acceptable network of protected areas. Researchers found that management plans could help management of biodiversity conflicts in Natura 2000 protected areas, but considerations included (i) determining if management plans were the best option, (ii) understanding, acknowledging and addressing conflicts present, (iii) acknowledging the importance of leadership, (iv) integrating scientists, decision-makers and local knowledge into management plans, and (v) taking long-term action based on the management plan.

Following these considerations, management plans could be based on a combination of both top-down and bottom-up initiatives (see Box 4 and Young *et al.*, 2012) to ensure that perceptions and knowledge of local stakeholders managing the protected areas are taken into account. To prevent future conflicts from occurring, there is, however, a need for long-term investment in research, adaptive monitoring and evaluation in order to identify and address future conflicts linked to the management of Natura 2000 protected areas.

References

Alphandery, P. and Fortier, A. (2001). Can a territorial policy be based on science alone? The system for creating the Natura 2000 network in France. *Sociol. Rural.*, 41, 311–328.

Bergseng, E. and Vatn, A. (2009). Why protection of biodiversity creates conflict – some evidence from the Nordic countries. *J. Forest Econ.*, 15, 147–165.

Dimitrakopoulos, P. G., Memtsas, D. and Troumbis, A. Y. (2004). Questioning the effectiveness of the Natura 2000 Special Areas of Conservation strategy: the case of Crete. *Global Ecol. Biogeogr.*, 13, 199–207.

European Commission (2000). Managing NATURA 2000 sites: the provisions of Article 6 of the 'Habitats' Directive 92/43/EEC.

European Commission (2010). Options for an EU vision and target for biodiversity beyond 2010. COM(2010) 4 final.

Jenkins, C. N. and Joppa, L. (2009). Expansion of the global terrestrial protected area system. *Biol. Conserv.*, 142, 2166–2174.

Mulongoy, K. J. and Chape, S. (2004). Protected areas and biodiversity. UNEP-WCMC Biodiversity Series No. 21.

Sutherland, W. J., *et al.* (2010). The identification of priority policy options for UK nature conservation. *J. Appl. Ecol.*, 47, 955–965.

Young, J. C., Butler, J. R. A., Jordan, A. and Watt, A. D. (2012). Less government intervention in biodiversity conservation: risks and opportunities. *Biodivers. Conserv.*, 21, 1095–1100.

PART III

Approaches to managing conflicts

CHAPTER FOURTEEN

Modelling conservation conflicts

JOHANNES P. M. HEINONEN and JUSTIN M. J. TRAVIS
University of Aberdeen

Modelling enables theory and empirical evidence to be brought together to build representations of how real-world systems work and how they are likely to respond to external influences. Models can take many forms, such as simple verbal or written descriptions, flow diagrams, sets of mathematical equations or computer programs. Usually the process begins with the development of a verbal or written description of a real-world system (i.e. a 'conceptual model'), which subsequently can be translated into a mathematical or computational format (i.e. an 'implemented model'). This implemented model can then be given appropriate inputs such that outputs, predicting the dynamics of the system of interest, are generated (Edmonds and Hales, 2003; Wilensky and Rand, 2007; Fig. 14.1). The outputs can then be compared to understanding or empirical data related to the behaviour of a natural system and this comparison can result in modification of the conceptual model. This iterative process can make a major contribution to our understanding of how systems work and what may be the crucial drivers of a system (Edmonds, 2000; Fig. 14.1).

It has been argued that the most important goal of modelling is to understand general mechanisms, not to generate specific predictions using models (Grimm, 1999). However, where sufficient, empirically verified, knowledge and understanding of a system exists, models can provide an excellent means for testing how a complex system may respond to different drivers for a natural resource system, and assess the likely responses of a system to alternative possible future management (Frederiksen *et al.*, 2001; Bunnefeld *et al.*, 2011). Importantly, even in cases where knowledge of a system is too limited for modelling to provide robust quantitative predictions, models can still be developed that yield useful qualitative predictions of expected trends and system dynamics (such as population cycles or the risk of extinction), particularly about influential mechanisms of the system.

Models can help us understand where and why ecological conflicts occur. They enable us to identify the main drivers of conflict by simplifying the system to key components that still replicate patterns in the real conflict system. To date, dynamic models (i.e. models which simulate behaviour or events in time)

Conflicts in Conservation: Navigating Towards Solutions, ed. S. M. Redpath, R. J. Gutiérrez, K. A. Wood and J. C. Young. Published by Cambridge University Press.

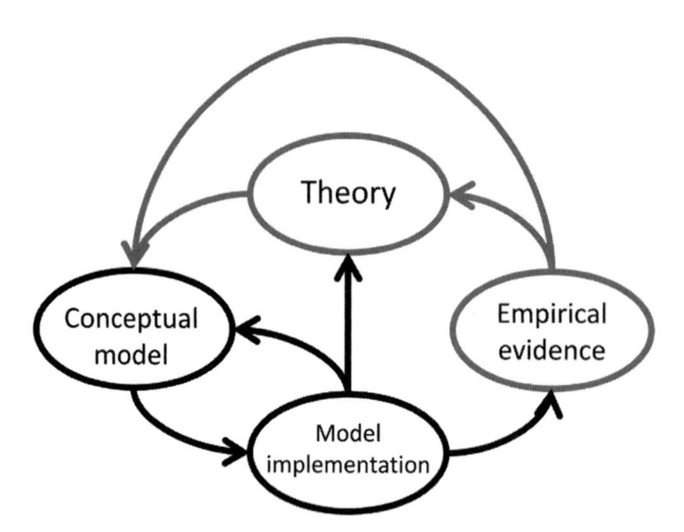

Figure 14.1 Illustration of the model development cycle: the iterative feedback between theory development, making predictions that are tested against empirical evidence and then theory, and model refinement to improve predictive capability is key to effective modelling. Model implementation and analysis of output can lead directly to an updated understanding and thus improved theory. It can also motivate the collection of new empirical data (e.g. ecological or socio-economic) to test the current understanding as encapsulated in the model and its output.

have been used very little in this context. One exception is the socio-ecological HUMENTS model used to explore impacts of poaching and agriculture on wildebeest in the Greater Serengeti ecosystem in East Africa (Holdo *et al.*, 2010). Statistical models have also been used to identify high crop cover as an important indicator of where human–elephant conflicts occurred in one region of Kenya (Sitati *et al.*, 2003).

There has been recent emphasis in ecological modelling on 'predictive systems ecology' (i.e. developing predictive representations of highly complex real-world systems that include humans; Evans *et al.*, 2013). This offers opportunities to integrate ecological and socio-economic knowledge and to test, *in silico*, alternative future policy or management interventions. As highlighted by modellers in other fields, the process 'requires understanding that all models are wrong and humility about the limitations of our knowledge. Such humility is essential in creating an environment in which we can learn about the complex systems in which we are embedded and work effectively to create the world we truly desire' (Sterman, 2002: 501).

While models will always be wrong, they become less wrong as they develop with improved theory, understanding, data availability and technology. However, a set of a priori trade-offs between model properties exist (Matthewson, 2011) and require careful consideration. In particular, in modelling biological populations, a trade-off exists between generality (defined as the number of target systems to which a model applies) and precision (defined as how finely parameters are specified in the model equations; Matthewson, 2011). This has implications for ecological conflicts modelling because trade-offs impose restrictions on what can be achieved. If all required features cannot be optimised in a single model, a multi-pronged approach utilising different but complementary models that optimise different features might provide better information for conflicts than a single modelling approach.

In this chapter we first seek to identify key aspects of conservation conflicts that will influence the modelling process. We then evaluate some existing modelling approaches. Finally, we draw conclusions about the requirements and goals for modelling conservation conflicts.

Modelling conservation conflicts – some key aspects

There are conceptual and practical aspects to consider in modelling conservation conflicts. Conceptual aspects include the disciplinary approaches that shape and direct the model-building and interpretation, dealing with variation, scale of representation and uncertainty. Practical aspects include data limitations and difficulties and bias in interpretation of results.

Conceptual aspects

Conservation conflicts arise because different stakeholders have different views about the system (Chapter 1). For modelling, this requires the incorporation of heterogeneity of attitudes in decision-making by humans, resulting in variability both in the decisions and actions a stakeholder might take. For example, gamekeepers in UK upland areas managed for red grouse may be more inclined to take and support actions that will reduce numbers of hen harriers on grouse moor, while conservationists may be more biased towards actions that will promote hen harrier population growth (Box 3). However, the tendency towards these actions may also vary between individuals within these two groups (Redpath *et al.*, 2004).

Variation is not merely between people. Biological populations are also heterogeneous in nature (Matthewson, 2011): for example, individual African elephants *Loxodonta africana* are not equally predictable in their crop-raiding (Sitati *et al.*, 2003), nor do all lynx *Lynx lynx* exhibit the same tendency to prey on sheep (Stahl *et al.*, 2002). Therefore, both social and ecological heterogeneity must be modelled, otherwise there is risk of misrepresenting the overall system.

The spatial scale and spatial distribution of processes are other important considerations. For example, a change in management that results in an overall economic benefit may still result in an economic loss for a local community or a specific group of stakeholders. This is an example of scale dependence in the socio-economic outcome (see Chapter 15). Space is also important because conflicts are not evenly distributed (Woodroffe *et al.*, 2005; Young *et al.*, 2010). For example, African elephant crop-raiding incidents have been found to be highly clustered at the 1 km^2 scale but less so at the 25 km^2 scale (Sitati *et al.*, 2003). Hence distributions need to be considered at more than one spatial scale.

Uncertainty must also be modelled. According to Milner-Gulland (2011), uncertainties include:

- observation uncertainty (occurs when field data is collected),
- parameter uncertainty (in terms of knowledge about parameter values),
- unpredictability in environmental variation and other system processes (process uncertainty), and
- uncertainty about the representation of processes of the ecological system such as density dependence (structural uncertainty).

In conflict systems, these uncertainties also apply to representation of stakeholders in models; the uncertainty in observation or perception of the situation by the stakeholder is one example (Osborne and Rubinstein, 1994). A few other forms include uncertainty in successful implementation of chosen actions, and uncertainty in compliance with regulations (Bunnefeld *et al.*, 2011). Ignoring uncertainty carries the risk of choosing management strategies that are prone to fail (Bunnefeld *et al.*, 2011). Related to our earlier discussion of the importance of heterogeneity, a key uncertainty is how much heterogeneity there is in, for example, attitudes that exist within and between stakeholder groups.

Practical limitations

Uncertainty is also a practical limitation. Observation uncertainty occurs when data are collected, partly because of process uncertainty in nature and limitations in our ability to obtain data of sufficient quality, which can be related to funding limitations or technological constraints. Uncertainty exists, for example, in our knowledge about how stakeholders make decisions; this is a form of structural uncertainty (An and López-Carr, 2012). Resources will always be limited, so it is important to evaluate the trade-off between collecting high-quality data on a few aspects or lower-quality data on many aspects of the conflict.

The issue of limited resources also has implications for trade-offs when model-building. As noted by Odenbaugh (2003: 1498), 'photographically exact models' (Levins, 1968) that are realistic and precise (i.e. models that include all the details of the system being modelled) will be difficult if not impossible to interpret. This is because of the difficulty in identifying the links between parameters, and system dynamics increases with the number of parameters.

Data limitations also determine whether (and with what confidence) parameters can be estimated to produce a realistic model. A general, less-detailed model with few parameters could provide broad insights into system dynamics, while a more precise model requiring potentially more assumptions and certainly more parameter estimates – which might or might not be more realistic – can potentially provide quantitative predictions for how a specific conflict situation may respond to alternative interventions. It is necessary to consider which approach would be more useful for a particular conflict situation given the existing availability of knowledge and data and the ease with which further data can be collected to inform the modelling process.

Possible approaches to modelling conservation conflicts

Individual-based models

Individual-based models (IBMs), or agent-based models (ABMs), have become popular in both natural and social sciences (An, 2012; Heinonen *et al.*, 2012; Box 14). These types of models simulate behaviours or processes of individual entities, with larger-scale patterns emerging from the interactions of these entities (Grimm, 1999; Stillman and Goss-Custard, 2010). For example, individual decisions about land use by farmers will collectively affect the large-scale patterns in land use and habitat (Vitousek *et al.*, 1997; Valbuena *et al.*, 2010), which have potential impacts on species of conservation concern.

Because these types of models enable multiple characteristics of individuals to be modelled explicitly, they also enable variation in individual characteristics (heterogeneity) to be modelled. This allows for modelling systems that consist of heterogeneous groups of entities, such as biological populations (Matthewson, 2011) and human societies (Wan *et al.*, 2002). Together with heterogeneity, the emergence of system-level patterns from individual behaviours and interactions enables IBMs to incorporate some of the complexity of real conflict situations (see, for example, descriptions of conflict characteristics in Young *et al.*, 2010). IBMs can also incorporate space and spatial complexity explicitly (Travis *et al.*, 2011), which allows modelling of the scale characteristics of conflicts as described by Young *et al.* (2010) and in the 'key aspects' section above. IBMs also provide an easy means for incorporating informational uncertainty in the decision process, an area receiving increasing attention in both ecological (Bocedi *et al.*, 2012) and socio-economic (Delton *et al.*, 2011) disciplines.

The potential for incorporating unlimited empirical information and complexity also produces a problem for IBMs. There may be a temptation to incorporate unnecessary details to make a model more 'realistic' (Grimm, 1999). This increases the complexity of the model, which increases the number of parameters and requires more research. In turn, this could compromise the robustness of data and model outputs including model accuracy and general applicability. The difficulty lies in finding the appropriate level of detail. Grimm (1999) suggests a scaling-down approach, where one starts with a coarse model and gradually increases the level of detail to address the question of interest. In a conservation conflict, this could involve representing broad stakeholder groups as single units with no variation between individuals, and then increasing the level of variation in attitudes or actions within groups, thus building stakeholder representation from high-level to individual-level. This could provide insight about how general patterns emerge from the individual level, including the influence of individual variation within stakeholder groups on outcomes of potential management strategies.

Game theory

Game-theoretic models involve players, or 'agents', making decisions by selecting choices from a set of strategies (Osborne and Rubinstein, 1994) and searching for optimal strategies (Colyvan *et al.*, 2011; Lee, 2012). For example, the Prisoners' Dilemma game involves two individuals, in separate cells, who each have the option either to confess or not confess a crime (Osborne and Rubinstein, 1994). If both confess (i.e. if they betray each other), each receives a three-year prison sentence; if one confesses and the other does not, the former will be freed and becomes a witness against the latter, who will receive a four-year sentence. If neither confesses (i.e. both cooperate), both will be convicted of a minor offence and receive a one-year sentence. Neither suspect knows what action the other will take, so the best option for either player is to confess; thus the game reveals a double confession as the equilibrium outcome under the set conditions even though there would be gains from cooperation (see Osborne and Rubinstein, 1994). Extensions of the Prisoner's Dilemma game which allow for cooperation between players, such as the Hawk–Dove game and the Stag Hunt game, have also been used in studies of competition and escalation of conflicts and of the evolution of cooperation (Doebeli and Hauert, 2005; Cole and Grossman, 2010; DeDeo *et al.*, 2010).

The game-theoretic framework is useful for conservation conflicts because it can be considered a type of adaptive management: it requires clear goals, strategies for achieving those goals, and monitoring and adapting of strategies based on how other agents affect those goals (Colyvan *et al.*, 2011). Game theory is also concerned with multiple agents (or stakeholders) with conflicting interests, which is consistent with conflict characteristics. Building nested games in models and allowing features such as nature and governments to be both rule-makers and players using different levels of the game allows for a degree of socio-economic and ecological complexity (Colyvan *et al.*, 2011).

Game theory assumes that the players (stakeholders) are expert game theorists who know their own best strategy and what the strategies of other players should be (Colyvan *et al.*, 2011). However, in reality this may not be the case; individuals may be limited in their knowledge in both aspects. Additionally, uncertainty about the game type (e.g. Prisoners' Dilemma, Hawk–Dove) leads to uncertainty about the best penalty and rewards scheme to introduce (Colyvan *et al.*, 2011). Colyvan *et al.* (2011) have suggested using decision theory to represent different games that might be played as different states of the world and assessing the probability that each one is being played.

The use of nested models with different players at different levels may provide a method for nesting local-scale games and strategies within a larger-scale perspective. For example, a game of 'tragedy of the commons' with local stakeholders as players could be nested within a simple cooperative game to represent a situation where stakeholders within a country individually exploit the resource, but on a national scale it is most advantageous for countries to cooperate to conserve the common-pool resource. Some aspects of ecological patterns can be incorporated by varying how nature acts as a player or rule-maker at the different levels, but detailed dynamics of ecological populations or systems are probably not easy to incorporate. In spatial games, where players are located on a two-dimensional grid, the structure of interactions can affect the dynamics that emerge (Doebeli and Hauert, 2005). In these models, game theory integrates with an individual-based modelling approach, with game theory rules determining the interactions between individuals. Thus, different modelling approaches can be combined in order to represent characteristics of real-world systems, such as space and human–human interactions.

Top-down and bottom-up approaches

Game theory is generally a top-down approach because most models assume all individuals to be equally rational and to make decisions based on the same criteria. Hence it is more concerned with population-level descriptions which can be applied to all kinds of populations (Grimm, 1999). However, it can incorporate individual variability and spatial structure, and hence build from the level of actions of individual players, which would make it a bottom-up approach. Individual-based modelling is a bottom-up approach, starting at the bottom (individual) level of the system.

Top-down approaches, dealing with population-level parameters and processes, have been prevalent in ecological and economic modelling. 'Mean field models' describe processes using mathematical equations, and can include processes such as density-dependence and dispersal (Heinonen *et al.*, 2012). They can also incorporate economic processes and links between ecology and economics; an example of this is the bio-economic harvest model for moose and timber developed by Wam *et al.* (2005; Fig. 14.2). This model uses a set of differential equations to model growth of different forest stands over time and the growth of a moose population (the moose are assumed to distribute

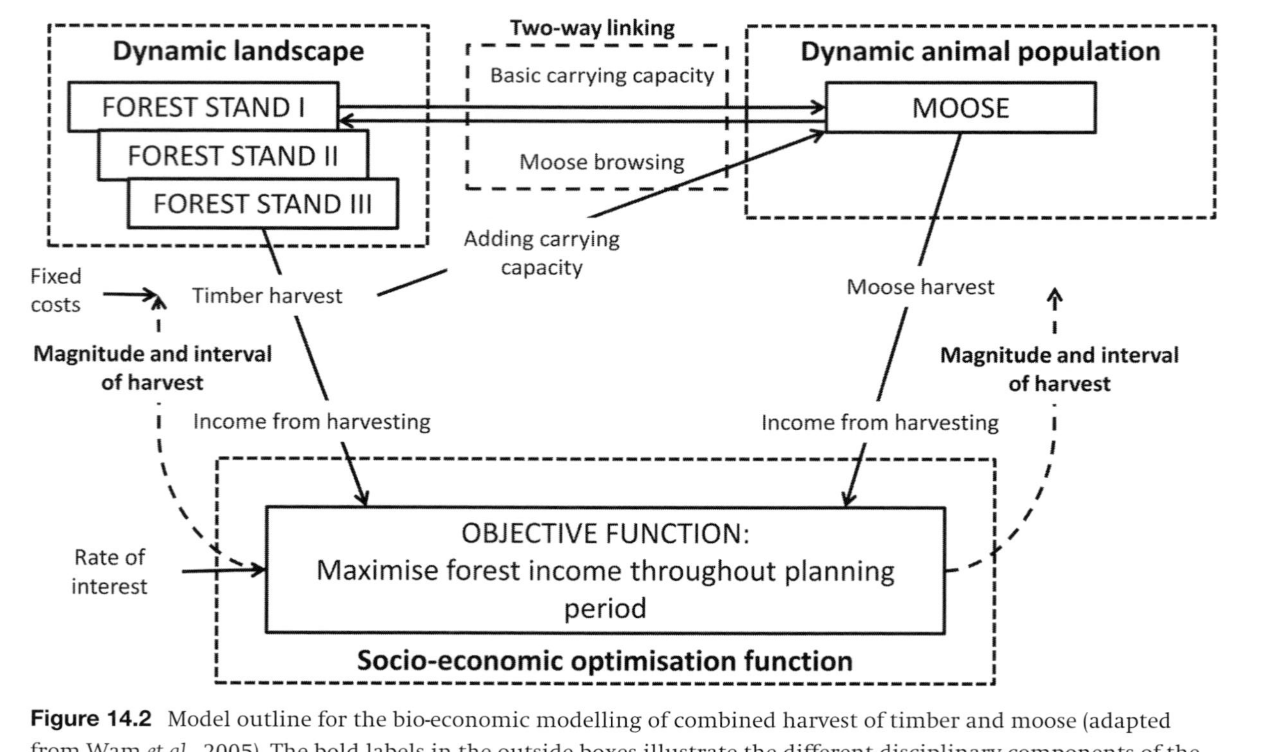

Figure 14.2 Model outline for the bio-economic modelling of combined harvest of timber and moose (adapted from Wam *et al.*, 2005). The bold labels in the outside boxes illustrate the different disciplinary components of the model and the dashed lines and bold text illustrate or highlight the linkages between them (not shown in original figure by Wam *et al.*).

themselves optimally in terms of density – hence foraging opportunities are equalised across stands, with no spatial variation; Wam *et al.*, 2005). Forest stand growth incorporates the influence of damage from browsing by moose. The profit from harvesting and replanting is calculated as a function of growth, yield and value; all factors are stand-specific (incorporating a degree of spatial variability). Profit also comes from moose harvesting and is a function of the meat and recreation value and harvest yield. The model is an optimisation model, and the objective function of the model is to maximise the profit over time. Results showed that the optimal strategy was the combined production of timber and moose, with the most influential factor being the fixed cutting of timber (Wam *et al.*, 2005). Even so, contribution from moose hunting to profit was very low for all scenarios explored (Wam *et al.*, 2005); hence, in a conflict situation, profit from moose hunting might not compensate for the damage to timber by moose browsing, although perhaps a reasonable reduction in moose densities through hunting could be acceptable given the likely associated reduction in browsing damage. This type of a modelling approach could be useful for addressing questions about the effectiveness of harvest regulations and their use in generating profit.

Top-down and bottom-up approaches are two extremes of a spectrum of modelling approaches and have their own strengths and weaknesses (Table 14.1). These should not be seen as either–or choices but as complementary approaches which can refine each other (Grimm, 1999).

Tactics and strategy

It is important to understand whether the goals of conflict modelling are tactical or strategic. A tactical goal can be the provision of information for, or the resolution of, a specific conflict situation. Tactical goals do not focus on general understanding or results that are necessarily relevant to other situations. Tactical conflict modelling would aim to increase precision and realism at the expense of generality (Fig. 14.3). The higher the precision, the more likely the outputs are to be useful in informing or resolving the conflict. Likewise, the greater the realism, the more likely the model is to represent what is actually occurring in the real system, and hence the more likely the outputs are to be useful. This type of modelling incorporates as many relevant parameters as possible and requires accurate measurement of the values of these parameters from the real system (Matthewson, 2011). Hence tactical models demand data of high quantity and quality, and thus require resources in terms of time, work and money. Such resources are always limited, and hence tactical modelling of every specific conflict situation is effectively impossible. This limitation can be stretched to some extent if the stakeholders involved in each conflict are willing and able to invest in the modelling of their particular conflict situation.

Table 14.1 *Comparing bottom-up and top-down modelling approaches.*

Modelling aspect	Bottom-up models	Top-down models
Individual variability	Individuals differ in some characteristics	Individuals in a group are the same
Interactions	Can be represented at individual level and higher	Interactions between higher level groupings (e.g. populations) can be represented
Scale and space	Often spatially explicit and can incorporate multiple levels of scale, with patterns at one level emerging from behaviour at a lower level	At the extreme, have no spatial dimensions and only represent the system at the population scale
Model interpretation	System properties can be traced back to individual behaviour and interactions	System properties are a result of system-level parameters which may have little empirical meaning[1]
Contribution to general theory	General theoretical concepts need to be first translated to individual-level	Relatively straightforward to incorporate general theoretical concepts[2]
Data requirements	Generally high	Generally lower
Model output interpretation	Can be difficult to relate results to specific parameters and behaviour	Relatively easy to quantify influences of different parameters on system dynamics[3]

[1] Odenbaugh (2003).
[2] Grimm (1999).
[3] Travis *et al.* (2011).

In contrast to tactical models, strategic models focus on generality at the expense of either realism or precision (Fig. 14.3). Sacrificing precision may lead to concerns about model validation and the usefulness of model outputs (Matthewson, 2011). Sacrificing realism incorporates unrealistic assumptions and idealisations of systems or parts thereof (Matthewson, 2011). Thus the outputs of such precisely tuned models might not actually provide any real understanding about the dynamics of real conflicts. The trade-off between generality and precision might not be the only model property trade-off in these systems.

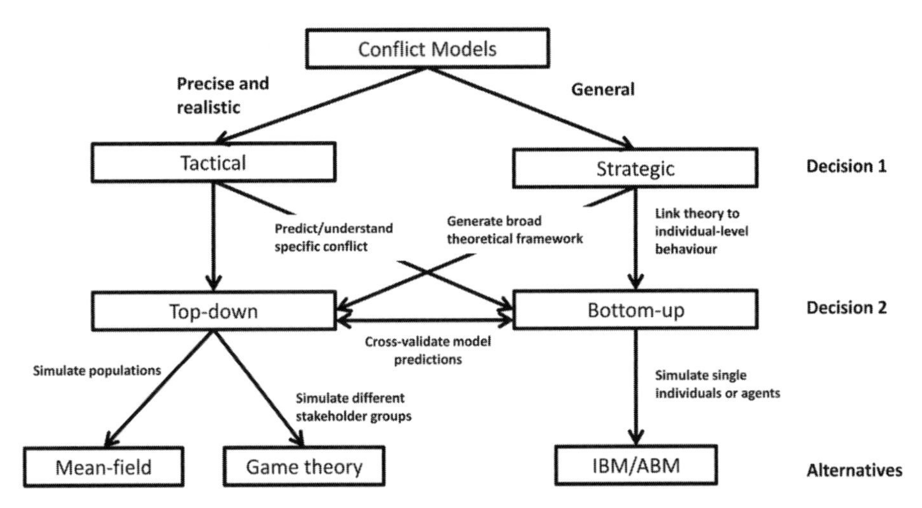

Figure 14.3 Decision tree for modelling approaches. There are two key decisions to be made. First, what is the model for? Is it for making predictions of a specific system, i.e. a tactical model, or is it for generating generic insights, i.e. a strategic model? The second decision is related to modelling philosophy, with the distinction being between top-down (typically more aggregated representation) and bottom-up (where greater complexity is typically incorporated). It is important to recognise that strategic and tactical modelling can both take either top-down or bottom-up approaches. Once the modeller has identified the purpose of their model and decided on the philosophical approach, there can then be alternatives for implementing and analysing the model.

Despite trade-offs, the general trends in system dynamics can, as noted before, give indications as to the mechanisms that drive conflict dynamics. In order for these to be reliable, the model must represent real systems to some degree.

Strategic models aim to inform or resolve conflicts in the long term. This goal requires more general understanding of conflict dynamics so that a broad strategy can be formed which can then be adapted and applied to a range of conflict situations. This leads to a modelling approach with generality and realistic representations of the system, but with imprecise or qualitative predictions (Matthewson, 2011).

Discussion and future directions

Models can help us understand what may potentially happen under different management scenarios. Therefore, they have a clear role in informing conservation conflict management. As discussed above, there are two types of information that models can provide about conflicts: specific information about what can be expected to happen in a particular conflict situation and under different management strategies, and general information that can help to develop understanding of conflict dynamics and prepare long-term strategies for conflict management.

Past approaches to conservation conflict modelling have tended to be one-sided (Liu *et al.*, 2007; An, 2012), focusing either on simulating wildlife populations, with simple or static impacts of human activities (e.g. Starfield *et al.*, 1981; Shea *et al.*, 2006), or simulating human decision-making with economic or sociological interests with a simple incorporation of the wildlife or ecosystem (e.g. Valbuena *et al.*, 2010). More recently, models have included more realistic representations of both human activities and decision-making and of ecological processes (e.g. Wilkinson *et al.*, 2009; Holdo *et al.*, 2010). Socio-economic approaches vary in many aspects, including pre-programmed decision-making rules versus adaptive decision-making (de Almeida *et al.*, 2010; Valbuena *et al.*, 2010) and economic-based versus more social-based approaches (An, 2012).

Conflict models have become more complex in seeking to incorporate ecological, social and economic processes, which potentially yield more reliable understanding and predictions of conflicts. The next step may be a more multi-pronged approach. The combination of top-down approaches and bottom-up approaches can provide a balance of generality, realism and precision. Bottom-up models such as IBMs provide the complexity and flexibility needed to address complex questions which may often be site-specific (Travis *et al.*, 2011; Box 14). For both the strategic and tactical goals, top-down approaches can provide the broad framework from which to devise the questions of interest and verify predictions from models with more detail (Grimm, 1999; Fig. 14.3). In terms of strategy and theoretical understanding, the top-down models produce possible explanations but no testable predictions, while without a general theoretical framework (largely built from top-down models) the bottom-up models produce precise predictions but no links drawn between them to form a coherent picture (Grimm, 1999).

Carefully formulated simulation designs and analysis of model results are important to build a coherent understanding of how the model works and the main drivers in the simulated system (Grimm, 1999). Validation of the model is crucial: a cross-validation can be obtained through comparison with analytical models (Travis *et al.*, 2011), but ultimately the use of real-life systems with high data quality and quantity for model parameterisation is desirable. Together, bottom-up and top-down approaches, for both tactical and strategic purposes, can help advance the field of model-building to understand and inform conservation conflict management and ultimately contribute towards their resolution.

Acknowledgements

We thank Jos Milner, Marius Wenzel, Chloe Denerley and Rosalind Bryce for comments at earlier stages of the development of this chapter. Johannes Heinonen has been funded by a scholarship from the College of Life Sciences and Medicine of the University of Aberdeen during the work leading to this chapter.

References

An, L. (2012). Modeling human decisions in coupled human and natural systems: review of agent-based models. *Ecol. Model.*, 229, 25–36.

An, L. and López-Carr, D. (2012). Understanding human decisions in coupled natural and human systems. *Ecol. Model.*, 229, 1–4.

Bocedi, G., Heinonen, J. and Travis, J. M. J. (2012). Uncertainty and the role of information acquisition in the evolution of context-dependent emigration. *Am. Nat.*, 179, 606–620.

Bunnefeld, N., Hoshino, E. and Milner-Gulland, E. J. (2011). Management strategy evaluation: a powerful tool for conservation? *Trends Ecol. Evol.*, 26, 441–447.

Cole, D. H. and Grossman, P. Z. (2010). Institutions matter! Why the herder problem is not a prisoner's dilemma. *Theory and Decision*, 69, 219–231.

Colyvan, M., Justus, J. and Regan, H. M. (2011). The conservation game. *Biol. Conserv.*, 144, 1246–1253.

De Almeida, S. J., Ferreira, R. P. M., Álvaro, E. E., Obermayr, R. P. and Geier, M. (2010). Multi-agent modeling and simulation of an *Aedes aegypti* mosquito population. *Environ. Modell. Softw.*, 25, 1490–1507.

DeDeo, S., Krakauer, D. C. and Flack, J. C. (2010). Inductive game theory and the dynamics of animal conflict. *PLoS Comput. Biol.*, 6, e1000782.

Delton, A. W., Krasnow, M. M., Cosmides, L. and Tooby, J. (2011). Evolution of direct reciprocity under uncertainty can explain human generosity in one-shot encounters. *Proc. Natl Acad. Sci. USA*, 108, 13,335–13,340.

Doebeli, M. and Hauert, C. (2005). Models of cooperation based on the Prisoner's Dilemma and the Snowdrift game. *Ecol. Lett.*, 8, 748–766.

Edmonds, B. (2000). The use of models – making MABS more informative. In *Multi-Agent-Based Simulation: Second International Workshop*, MABS 2000 Boston, MA, USA, July, Revised and Additional Papers, eds. S. Moss and P. Davidsson, pp. 15–32. Berlin: Springer.

Edmonds, B. and Hales, D. (2003). Replication, replication and replication: some hard lessons from model alignment. *J. Artif. Soc. S.*, 6, 11.

Evans, M. R., *et al.* (2013). Predictive systems ecology. *Proc. R. Soc. B*, 280, 20131452.

Frederiksen, M., Lebreton, J.-D. and Bregballe, T. (2001). The interplay between culling and density-dependence in the great cormorant: a modelling approach. *J. Appl. Ecol.*, 38, 617–627.

Grimm, V. (1999). Ten years of individual-based modelling in ecology: what have we learned and what could we learn in the future? *Ecol. Model.*, 115, 129–148.

Heinonen, J. P. M., Travis, J. M. J., Redpath, S. M. and Pinard, M. A. (2012). Combining socio-economic and ecological modelling to inform natural resource management strategies. In *International Environmental Modelling and Software Society (iEMSs) 2012 International Congress on Environmental Modelling and Software. Managing Resources of a Limited Planet: Pathways and Visions under Uncertainty*, Sixth Biennial Meeting, Leipzig, Germany, eds. R. Seppelt, A. A. Voinov, S. Lange and D. Bankamp, http://www.iemss.org/society/index.php/iemss-2012-proceedings. ISBN: 978-88-9035-742-8.

Holdo, R. M., Galvin, K. A., Knapp, E., Polasky, S., Hilborn, R. and Holt, R. D. (2010). Responses to alternative rainfall regimes and antipoaching in a migratory system. *Ecol. Appl.*, 20, 381–397.

Lee, C.-S. (2012). Multi-objective game-theory models for conflict analysis in reservoir watershed management. *Chemosphere*, 87, 608–613.

Levins, R. (1968). Ecological engineering: theory and technology. *Q. Rev. Biol.*, 43, 301–305.

Liu, J., *et al.* (2007). Complexity of coupled human and natural systems. *Science*, 317, 1513–1516.

Matthewson, J. (2011). Trade-offs in model-building: a more target-oriented approach. *Stud. Hist. Philos. Sci.*, 42, 324–333.

Milner-Gulland, E. J. (2011). Integrating fisheries approaches and household utility models for improved resource management. *Proc. Natl Acad. Sci. USA*, 108, 1741–1746.

Odenbaugh, J. (2003). Complex systems, trade-offs, and theoretical population biology: Richard Levin's 'Strategy of model building in population biology' revisited. *Philos. Sci.*, 70, 1496–1507.

Osborne, M. J. and Rubinstein, A. (1994). *A Course in Game Theory*. Cambridge, MA: MIT Press.

Redpath, S. M., *et al.* (2004). Using decision modeling with stakeholders to reduce human–wildlife conflict: a raptor–grouse case study. *Conserv. Biol.*, 18, 350–359.

Shea, K., Sheppard, A. and Woodburn, T. (2006). Seasonal life-history models for the integrated management of the invasive weed nodding thistle *Carduus nutans* in Australia. *J. Appl. Ecol.*, 43, 517–526.

Sitati, N. W., Walpole, M. J., Smith, R. J. and Leader-Williams, N. (2003). Predicting spatial aspects of human–elephant conflict. *J. Appl. Ecol.*, 40, 667–677.

Stahl, P., Vandel, J. M., Ruette, S., Coat, L., Coat, Y. and Balestra, L. (2002). Factors affecting lynx predation on sheep in the French Jura. *J. Appl. Ecol.*, 39, 204–216.

Starfield, A. M., Shiell, J. D. and Smuts, G. L. (1981). Simulation of lion control strategies in a large game reserve. *Ecol. Model.*, 13, 17–28.

Sterman, J. D. (2002). All models are wrong: reflections on becoming a systems scientist. *Syst. Dynam. Rev.*, 18, 501–531.

Stillman, R. A. and Goss-Custard, J. D. (2010). Individual-based ecology of coastal birds. *Biol. Rev.*, 85, 413–434.

Travis, J. M. J., Harris, C. M., Park, K. J. and Bullock, J. M. (2011). Improving prediction and management of range expansions by combining analytical and individual-based modelling approaches. *Method. Ecol. Evol.*, 2, 477–488.

Valbuena, D., Verburg, P. H., Bregt, A. K. and Ligtenberg, A. (2010). An agent-based approach to model land-use change at a regional scale. *Landscape Ecol.*, 25, 185–199.

Vitousek, P. M., Mooney, H. A., Lubchenco, J. and Melillo, J. M. (1997). Human domination of Earth's ecosystems. *Science*, 277, 494–499.

Wam, H. K., Hofstad, O., Nævdal, E. and Sankhayan, P. (2005). A bio-economic model for optimal harvest of timber and moose. *Forest Ecol. Manag.*, 206, 207–219.

Wan, H. A., Hunter, A. and Dunne, P. (2002). Autonomous agent models of stock markets. *Artif. Intell. Rev.*, 17, 87–128.

Wilensky, U. and Rand, W. (2007). Making models match: replicating an agent-based model. *J. Artif. Soc. Sci.*, 10, 2.

Wilkinson, D., Bennett, R., McFarlane, I., Rushton, S., Shirley, M. and Smith, G. C. (2009). Cost–benefit analysis model of badger (*Meles meles*) culling to reduce cattle herd tuberculosis breakdowns in Britain, with particular reference to badger perturbation. *J. Wildlife Dis.*, 45, 1062–1088.

Woodroffe, R., Thirgood, S., and Rabinowitz, A. (eds) (2005). The future of coexistence: resolving human–wildlife conflicts in a changing world. In *People and Wildlife: Conflict or Coexistence?* pp. 388–405. Cambridge: Cambridge University Press.

Young, J. C., *et al.* (2010). The emergence of biodiversity conflicts from biodiversity impacts: characteristics and management strategies. *Biodivers. Conserv.*, 19, 3973–3990.

Box 14

Balancing shorebird conservation and shellfish harvesting in UK estuaries

Richard A. Stillman and Kevin A. Wood
School of Applied Sciences, Bournemouth University, Dorset BH12 5BB, UK

The intertidal areas of UK coasts are important habitats for shellfish species, such as cockles *Cerastoderma edule* and mussels *Mytilus edulis*. Commercial harvesting of shellfish is worth an annual £250 million to the UK economy, providing both food and employment (DEFRA, 2013). These shellfish are also the principal food resource for overwintering shorebirds such as oystercatchers *Haematopus ostralegus* and knots *Calidris canutus*. Shorebirds are key components of UK coastal biodiversity and are protected under the European Union Wild Birds Directive (2009/147/EEC), which legally obligates the UK government to maintain healthy shorebird populations. Thus, many estuaries have been designated estuarine Special Protection Areas (SPAs) which must be managed to sustain overwintering shorebird populations. Additionally, shorebirds are popular with bird watchers and, therefore, benefit tourism in coastal areas. Thus, a conflict has occurred since the 1970s between those desiring to conserve shorebirds and those desiring to harvest shellfish commercially (Tinker, 1974). On one side, conservationists and bird watchers want strict limits to be placed on shellfish harvesting and disturbance in order to protect the birds. On the other side, the fishermen and their representatives want to maximise shellfish harvesting. Statutory authority for the management of estuarine fisheries, and thus, the responsibility for conflict resolution, is held by local government organisations. These organisations set an annual Total Allowable Catch (TAC) of shellfish in each estuary in an attempt to avoid overfishing. Each year meetings are held with stakeholders to discuss the TAC for the subsequent year; both fisheries and conservation interests lobby the government organisation to increase or decrease the TAC, respectively.

One attempt to balance conservation and economic interests was the 'rule of thirds', whereby the statutory authority designates one-third of the total annual shellfish stock for fisherman, one-third for shorebirds, and one-third for subsequent shellfish recruitment. However, this approach lacks an evidence base, as these values are independent of the shorebird's food requirements. Hence, in some years there may be too little food left for the birds which leads to their starving, while in other years too much may be left for the birds which lowers the economic potential of the fishery (Atkinson *et al.*, 2010). Despite these potential flaws, the rule of thirds continues to be used by managers for some UK estuaries. During the 1990s, the organisations responsible for shellfishery management tasked researchers with evaluating whether the TACs were predictable scientifically, by considering how much food the birds actually required relative to what was available in estuaries. Given their key role in setting TACs, and the pressure from stakeholders, these government organisations have driven the search for evidence-based management of shellfisheries.

For example, in the Dutch Wadden Sea, a policy was introduced whereby enough shellfish to meet 70% of the bird population's energy requirement had to remain unharvested, with the assumption that the remaining 30% could be acquired from non-harvested species. However, research which related survival to food availability showed that shorebirds required between 2.5 and 7.7 times more food than would be predicted from their energy needs, as not all available food can be found and individual birds are often excluded from food through competition with others (Goss-Custard *et al.*, 2004). Since the mid-1990s individual-based models (IBMs) have been developed which account for these factors when predicting the quantity of shellfish that a bird population needs to survive the winter. These IBMs have been used in the process of setting TACs for some UK shellfisheries, most recently the Burry Inlet in Wales (Stillman *et al.*, 2010). However, specialist computational knowledge is required to run the models and they have typically been applied on a site-by-site basis, limiting more widespread use. Despite recent attempts to make IBMs more user-friendly, model complexity is still perceived as a barrier by stakeholders to the successful management of the shellfish conflict. Stakeholders would prefer a simplified approach which they can use to set TACs. Such a simplified model has been developed (Stillman and Wood, 2013a) and has been used to inform management of the Dee Estuary in Wales (Stillman and Wood, 2013b). The main role of researchers in this conflict has been to provide managers with decision-support tools, such as models that accurately predict the food requirements of the shorebirds. This allows managers to set TACs which enhance the economic potential of the shellfishery without threatening the conservation of shorebirds. While stakeholders are keen to implement the new decision-support tool, it remains unclear whether this will help to resolve the conflict. Currently, neither fisherman or conservationists are satisfied, as both feel insufficient shellfish are devoted to their concern. UK shellfish stocks remain historically low due to overfishing, pollution and climate change; unless these problems can be addressed, the conflict between fishermen and conservationists over dwindling shellfish stocks may intensify.

References

Atkinson, P. W., Maclean, I. M. D. and Clark, N. A. (2010). Impacts of shellfisheries and nutrient inputs on waterbird communities in the Wash, England. *J. Appl. Ecol.*, 47, 191–199.

DEFRA. (2013). https://www.gov.uk/shellfisheries-several-orders-and-regulating-orders. 13th June 2013.

Goss-Custard, J. D., *et al.* (2004). When enough is not enough: shorebirds and shellfishing. *Proc. R. Soc. B Biol. Sci.*, 271, 233–237.

Stillman, R. A. and Wood, K. A. (2013a). *Towards a Simplified Approach for Assessing Bird Food Requirements on Shellfisheries. A Report to the Welsh Government*. Poole: Bournemouth University. 34 pp.

Stillman, R. A. and Wood, K. A. (2013b). *Predicting Oystercatcher Food Requirements on the Dee Estuary. A report to Natural Resources Wales*. Poole: Bournemouth University. 29 pp.

Stillman, R. A., *et al.* (2010). Assessing waterbird conservation objectives: an example for the Burry Inlet, UK. *Biol. Conserv.*, 143, 2617–2630.

Tinker, J. (1974). Why shoot oystercatchers? *New Sci.*, 64(918), 125.

CHAPTER FIFTEEN

Defining scales for managing biodiversity and natural resources in the face of conflicts

JOHN D. C. LINNELL
Norwegian Institute for Nature Research

Researchers are documenting a wide diversity of conflicts that emerge among stakeholders about biodiversity conservation (Redpath *et al.*, 2013). This body of evidence challenges the often stated assumption that all biodiversity has positive benefits towards human well-being (Maier, 2013) as different stakeholders may have very different views on the costs and benefits of different situations. The reality is that while much biodiversity conservation (hereafter 'conservation') benefits many humans, there can be real economic or social costs for conservation. The extent to which a given biodiversity component or conservation action represents a service or a disservice can vary with scale. For example, species that represent 'public goods' in general may represent 'public bads' locally (Bostedt, 1999). Large carnivores are a classic example. Because the costs of economic and social conflicts resulting from their presence are felt locally, attitudes to these species are often significantly less positive in the areas where they occur than in distant areas and cities (Karlsson and Sjöström, 2007; Box 15). However, the opposite situation may also occur. For example, in the harvest of wild ungulate populations the benefits (recreational opportunities, sale of licences and meat) of harvesting a 'public good' often fall to the local landowner while the costs (e.g. compensation for forest damage, vehicle collisions and infrastructure to mitigate vehicle collisions) usually fall on society as a whole (Kenward and Putman, 2011; Langbein *et al.*, 2011; Reimoser and Putman, 2011).

This issue of scale represents a much neglected topic in conservation policy thinking, although it is emerging as a crucial discourse in a wide diversity of political sectors (Cash *et al.*, 2006; Young *et al.*, 2013a). In this chapter I will explore the issue of scale with relevance to conservation and conflict management by considering ecological, social and political dimensions of spatial scale (Cash *et al.*, 2006). I will illustrate these relationships mainly using examples drawn from large mammal conservation in Europe, but with supporting reference from

Conflicts in Conservation: Navigating Towards Solutions, ed. S. M. Redpath, R. J. Gutiérrez, K. A. Wood and J. C. Young. Published by Cambridge University Press.

around the globe. My focus is on conservation within multi-use landscapes and not within protected areas.

Changes in the way scale has been considered in conservation

In attempts to halt biodiversity decline and increase the sustainability of natural resource exploitation there have been many scale-related paradigm shifts by scientists and conservationists who manage natural resources. These shifts have occurred in three areas.

Integrating goals for nature and society

The goals of biodiversity conservation, the sustainable use of its components, and the fair and equitable sharing of benefits are widely accepted in international agreements, including the Convention on Biological Diversity (CBD). The objectives for management are now regarded as a matter of societal choice, as are the importance of building democratic, just, and prosperous societies as a prerequisite for successful nature conservation. This implies that multiple scales need to be considered in socio-ecological systems, including spatial, temporal, jurisdictional, knowledge and network scales (Cash *et al.*, 2006; Kok and Veldkamp, 2011).

Moving management towards the local level

There has been widespread focus on decentralisation and devolution of natural resource management to local levels in an attempt to improve effectiveness and justice (see Chapter 13). For example, the principle of subsidiarity ('decisions within a political system should be taken at the lowest level consistent with effective action'; Jordan and Jeppesen, 2000) is codified in European Union legislation.

Holism

There has been movement away from viewing resources in isolation from each other. Now the dominant paradigm is to focus on whole ecosystems, not only from the point of view of a wider range of ecological processes, but also from that of a wider range of stakeholders and interest groups. The Ecosystem Approach (Smith and Maltby, 2003) that is advocated by the CBD and the parallel field of Ecosystem Management (Brussard *et al.*, 1998) which is emerging among North American management agencies are frameworks designed to embrace this holism.

Despite the dramatic increases in public and political awareness, in scientific knowledge, in practical experience with various management systems, and with the development of conceptual models such as the Ecosystem Approach there is still a great deal of controversy and uncertainty about how these ideas work in practice. At least part of this controversy is because of the belief that some elements in each actually conflict or negatively affect each other, even to the point of working towards mutually exclusive goals. The potential contradiction lies in moving responsibility for resource management down to a local level while simultaneously expanding the view to include the ecosystem, which inevitably will include ecological or social processes that operate at very large (non-local) spatial scales.

Different perspectives on scale in conflict management

In conservation conflicts, there is frequently a need to make management decisions linked to spatial planning or resource use. Because these decisions can be controversial, different groups will often seek to move the decision-making process to a scale that favours them. It is therefore important to examine the issue of the appropriate spatial scale of management from different perspectives.

Biology perspective

In the midst of social and political controversy over management, conservation biologists often introduce biological perspectives to the management process by identifying appropriate biological scales for management of wildlife populations. This task is often not as easy as it would seem. The wide range of potential patterns of distributions and spatial structures that define species make global definitions of populations difficult (Thomas and Kunin, 1999). Conceptually, most ecologists view a population to be a group of interbreeding individuals whose population dynamics are mainly governed by birth and death rather than immigration or emigration. The problem comes in defining this operationally or quantitatively. Lacking theoretical consensus, managers are usually forced to resort to a range of ad hoc approaches including taxonomic, ecosystem, distributional, geographical, demographic, economic and behavioural approaches. Different ecological processes, from decomposition to community dynamics, also occur on variable scales – ranging from a few square metres to continental scales. Therefore, there can be no single correct scale for management. Rather, different processes will need to be managed at different scales. For example, an individual landowner may be able to influence habitat quality for a roe deer by manipulating forest patches of a few hectares, while managing their main predators, lynx and wolves, requires coordination on national or even international scales (Linnell and Boitani, 2012).

Large-scale coordination

Because many threats to biodiversity are related to global processes, or require global responses, a wide range of large-scale legislative instruments have been developed during recent decades. In Europe these include the Bern Convention and Habitats Directive. At the global scale, examples include the CBD and the emerging Intergovernmental Platform on Biodiversity and Ecosystem Services. These instruments have without doubt contributed to slowing, halting and reversing many biodiversity declines around the world. However, they are controversial when applied at a local level, because many people feel that the top-down imposition of legislation reduces their ability to influence their own situation (see Chapter 16). Such feelings of power imbalance are often central to conservation conflicts and evoke issues of environmental justice (Williams, 1999; Hiedanpää, 2002). To address conflicts related to top-down and large-scale processes there has been movement towards more local-level management initiatives, with mixed results.

Community-based conservation

Many community-based conservation and development projects have been initiated since the 1980s (e.g. Box 1; Box 2), especially in the tropics. The rationale of these projects has been to link conservation with rural development by allowing local communities to benefit from the natural resources within their neighbourhoods. However, after more than 30 years there is little evidence that the approach has helped biodiversity conservation in these areas. Many critical reviews point to the weakness and false assumptions of the approach, including (1) lack of individual or institutional capacity at local scales, (2) presence of local-level corruption and the inability of local authorities to resist domination by local and/or global power players (= elite capture), (3) lack of interest in many communities, (4) increased economic well-being often leading to increased environmental impact through heightened consumption or immigration, (5) project goals are too broad and too long-term to be achievable or measurable, (6) local communities are often internally divided, and (7) local communities have often already outstripped the resource base (Kellert *et al.*, 2000; Adams *et al.*, 2004; du Toit *et al.*, 2004). Furthermore, conservation and long-term sustainability are rarely, if ever, more profitable in the short term than 'resource mining', habitat conversion and intensive land use. Therefore, the central problem lies in the false expectation that local communities will automatically give up tangible, lucrative and short-term development opportunities in favour of relatively uncertain, abstract and long-term goals (Adams *et al.*, 2004; du Toit *et al.*, 2004). This is especially true for many biodiversity components that are not resources, especially those that are actually sources of economic losses and may even be dangerous (elephants, large carnivores, etc.; Bostedt, 1999; Box 18). Consequently, there have been attempts to improve these approaches, but there is still little evidence they provide a general model for conservation (Berkes, 2004).

Decentralisation and devolution

In the field of political geography the focus has been on evaluating the success of decentralisation and devolution of control over natural resource management to more local levels, and again the literature mainly illustrates situations in the tropics. The goals have also been to increase environmental management efficiency, and improve equity and social justice (see Chapter 13). Typically, evaluations consider management of specific resources and not biodiversity. Even then, published accounts are rather negative (Larson, 2001; Ribot, 2002; Lane, 2003). Problems identified above, such as lack of capacity, lack of resources, local-level corruption and elite capture, lack of incentives and the unwillingness of local communities to forego alternative development opportunities associated with non-sustainable resource use, also appear in this literature. A common problem is that when the controlling power of central government is removed, the local resources can be easily dominated by local (or global) power elites, making the access to and control of resources even less democratic than it was originally (Lane, 2003). Decentralisation advocates point out that a common problem has been that not enough power has been decentralised, or that responsibility has been decentralised without the power, authority, or resources to manage (Ribot, 2002). Because of the widespread abuse of the limited powers which have been decentralised, few governments are willing to decentralise even greater powers. Many authors note that central government is needed to provide a non-local point of view and to represent a broader range of values than possible with local control (Larson, 2001).

Furthermore, the wider experience with common-pool resources illustrates the practical difficulties of excluding non-local resource users (Ostrom *et al.*, 1999). In tropical forests much of the current overexploitation of wildlife resources (the bushmeat crisis) is conducted by non-locals for non-local markets (Robinson and Bennett, 2000). Even if local communities wanted to exclude these exploiters, it is unclear whether they would have the enforcement capability. Deterring meat hunters is hard enough, but deterring people seeking to exploit exceptionally high-value products like elephant ivory, tiger bones and rhino horn without paramilitary-style resources has proven to be impossible.

Co-management

Throughout Canada and the United States natural resource management issues often involve the rights of indigenous people. In attempts to balance power between the central and local levels, widespread use has been made of the co-management approach (Caulfield, 1997; Decker *et al.*, 2000; Zachrisson, 2004). Co-management consists of a partnership among multiple stakeholders and responsible authorities where authority and responsibility is negotiated and shared. Results of co-management have been mixed (Dion, 2003), but have been more positive than the community-based conservation and decentralisation results from tropical regions (Decker *et al.*, 2000). Unfortunately, it is difficult to assess whether this is because of management design, the context (rich countries, more institutional capacity, more scientific data, lower human densities), or both (Kellert *et al.*, 2000). When extending the concept of co-management to include any form of participatory management there are many more examples of management intended to secure local peoples' access to natural resources (see Borrini-Feyerabend *et al.*, 2004), and even examples of management of conflict species such as large carnivores (Nie, 2003). However, one factor that may have contributed to the relative success of these models may be that these systems usually focus on a single species or geographically defined resource (e.g. a single caribou population). How these co-management systems will work within an ecosystem management context remains to be seen (Zachrisson, 2004).

Balancing the global and the local

A central tension present in conflicts is that stakeholder values and priorities may vary across scales (McNeill and Lichtenstein, 2003). How then should one decide which set of values should be given priority? Callicott (2002) proposes that human societies, and their social values, form a nested hierarchy. He further proposes that simple issues of preference or lifestyle should give way to issues associated with fundamental values and matters of livelihood, and that widely held values (e.g. at the human or national scale) at higher scales should have priority over local values (e.g. at the ethnic or community scale). According to this rationale, the internationally held values in favour of conservation should have preference over local values that conflict with conservation, although a difficult value conflict can arise when local people are very dependent on non-sustainable practices for short-term survival.

To minimise the need for values held at higher levels to overrule those held locally, Callicott (2002) calls for the creative use of win–win solutions, of which there are many good examples (Rosenzweig, 2003). This is especially true for human-dominated landscapes where conservation goals are set at a more pragmatic level of ambition (with respect to the decreasing extent of human influence) than those set for designated wilderness areas.

There is also a growing focus on using participatory processes to negotiate outcomes that are viewed as being acceptable (Chapter 16). Many formats for participation exist, each with their own advantages and disadvantages (Bath, 2009; Maser and Pollio, 2012; Young *et al.*, 2013b). The major problem with participatory approaches is that they work best on small scales. Upscaling to larger areas means widening participation and can only really work in cases where some effective system of representation works, for example through the involvement of stakeholder organisations. This in turn requires that these representatives be viewed as legitimate by their constituents and that they have an internal mandate to negotiate.

Deciding on appropriate management scales

Natural resources differ greatly in the extent to which they can be managed at different levels, with some resources being more suitable for local-level management than others. Characteristics of resources that may be successfully managed locally include (Ostrom *et al.*, 1999; Zachrisson, 2004) situations where:

- resources are of low to medium value;
- resources are not currently depleted;
- resources occur within small, recognisable areas and have predictable behaviour;
- exploitation efficiency is low such that there are intrinsic negative feedback loops;
- the costs and benefits exist at the same scale;
- users have an interest in the sustainability of the resource;
- local norms for resource conservation and punishments are well established;
- resources cause few conflicts with other human activities, and are not heavily influenced by human land use; and
- the natural history of the resource is well-studied.

These characteristics describe resources such as small game, freshwater fisheries and wild ungulates that have been managed successfully at a local level for decades (Kenward and Putman, 2011). The question remains about the general applicability of local-level management to other species or resources with different characteristics. A few issues are often forgotten when 'natural resources management' principles (where most elements are viewed as having positive and direct value) and experience are extrapolated to biodiversity conservation' (where some values may be indirect, abstract and even negative). These are:

- not all ecosystem components are viewed as resources (Bostedt, 1999);
- not all conflicts associated with biodiversity are material or economic (e.g. Peterson *et al.*, 2002). In fact many are social or political (Box 15);
- not all exploitation is compatible with conservation;
- the constituency of conservation has become global and involves many stakeholders;
- the currency of conservation is more than economic (Jepson and Canney, 2003); and
- some ecosystem components operate on very large spatial or long temporal scales.

These characteristics describe large carnivores, old-growth forests, deep-water corals and migratory ungulates. Large carnivores are associated with costly human impacts, use large areas, have little economic value and their management is controlled by international agreements (e.g. Box 1). Old-growth forest conservation automatically represents an opportunity cost. Migratory ungulates move over large areas and although they represent a valued resource, their conservation requires habitat conservation that prevents the development of other, more lucrative, land uses and activities (Bråtå, 2003). Such species and habitats are clearly much less suitable for local-level management.

However, considering the ecosystem paradigm, it is not possible to assert that different resources or species should be managed independently of each other at different levels because all ecosystem components interact with other components, even in human-dominated ecosystems where many processes are suppressed. It is therefore impossible to consider the exploitation of a resource in isolation from the ecosystem. Furthermore, different ecosystem processes occur at different scales.

Reconsidering the large carnivore–large ungulate example, the presence of large carnivores will influence the number of ungulates that can be harvested because conservation of the large carnivore populations depends on the ungulates being managed to sustain sufficient prey. In addition, the political influences on large ungulate management are mainly local or national, but large carnivore issues are usually international. How then should the ecosystem be managed when its interacting components differ in the ecological scale at which they operate? The answer is that there is never a simple answer for deciding whether management should be local or central. Control of any resource can therefore never be totally delegated. All management must be multi-scalar. The real question is how much power for the management of each ecosystem component should be placed at each level (Giampietro, 1994; Vogt *et al.*, 2002; Wiens *et al.*, 2002). Different degrees of power over various species or resources should lie at different levels depending on their ecological, social and economic characteristics. Such a hierarchy could be nested, in that the upper levels place limitations on the next lower layer, and so on (Apostolopoulou and Paloniemi, 2012). The higher levels should simply define overall goals and general limitations, with each successive layer adopting more specific and more locally adapted rules and policies. As long as local levels operate within the framework of the upper level's overall goals, the system should function.

However, three main problems prevent implementation of this hierarchical system. First, setting the overall goals over conservation, especially in situations where the opportunity costs will be borne at the local level, will often be controversial (Callicott, 2002). Present-day thinking calls for a focus on the process behind decision-making, as much as on the decisions themselves (Nie, 2003; Peterson *et al.*, 2004). This question effectively asks how we perceive democracy (Arblaster, 2002) and how it balances the protection of minority views against majority views. Just as every country has its own version of democracy (Arblaster, 2002), it is likely that each situation will require locally adapted decision-making processes (Borrini-Feyerabend *et al.*, 2004). The key is to ensure bottom-up representation and participation.

A second problem lies with coordinating this nested hierarchy of scales (Perry and Ommer, 2003). Within an ecosystem context, it is inevitable that a wide range of institutions will exist at many different administrative levels that need coordination. Degnbol *et al.* (2003) argue for the creation of a new form of professional whose job is to cross scales, and to ensure that information flows in both directions. It is also important that some form of upward and downward accountability exists. Designing and initiating a functional hierarchical system is likely to need inspired leadership from the upper level (Callicott, 2002).

Finally, there is the issue of developing mechanisms to address the fact that costs and benefits often occur at different scales (see Chapter 6). Economists state there is a need to internalise the external costs. In other words, there is a need for economic incentives that reduce the differences between the value (or cost) of biological diversity to the private individual and its value (or cost) to society as a whole (Folke *et al.*, 1996). Because of different costs and benefits at different spatial scales, one might argue that there needs to be revenue flow from the global to the local that far exceeds current rates from activities such as ecotourism.

Conclusion

As both ecosystems and societies change, it is inevitable that we will need to continually respond to changing circumstances. Although challenging, these issues reflect how as a society we make decisions, how our democracies function, and how we take care of the local in an increasingly globalised world.

The paradox is that local participation and influence can probably be best achieved if clear frames are set by central authorities (Peterson *et al.*, 2002). The challenge here, however, is to maintain open two-way communication and dialogue between levels, such that the local level can influence the overarching frames.

Finally, while human management and political structures can be debated and adjusted, we must not lose sight of one clear constraint, which is that the ecological processes that we seek to exploit, manage or conserve operate on scales that humans did not design. This may often cause inconvenience when biological scales do not fit the jurisdiction of political scales that we have constructed, and may not fit into social constructionist paradigms. However, when managing natural resources and conserving biodiversity we must accept that nature sets the ultimate limits on the available management options.

References

Adams, W. M., *et al.* (2004). Biodiversity conservation and the eradication of poverty. *Science*, 306, 1146–1149.

Apostolopoulou, E. and Paloniemi, R. (2012). Frames of scale challenges in Finnish and Greek biodiversity conservation. *Ecol. Soc.*, 17, 451–484.

Arblaster, A. (2002). *Democracy*. Buckingham: Open University Press.

Bath, A. (2009). Working with people to achieve wolf conservation in Europe and North America. In *A New Era for Wolves and People: Wolf Recovery, Human Attitudes, and Policy*, eds. M. Musiani, L. Boitani and P. C. Paquet, pp. 173–200. Calgary: University of Calgary Press.

Berkes, F. (2004). Rethinking community-based conservation. *Conserv. Biol.*, 18, 621–630.

Borrini-Feyerabend, G., Pimbert, M., Farvar, M. T., Kothari, A. and Renard, Y. (2004). *Sharing Power. Learning By Doing in Co-Management of Natural Resources Throughout the World.* Tehran: International Institute for Environment and Development and the Collaborative Management Working Group.

Bostedt, G. (1999). Threatened species as public goods and public bads. *Environ. Resour. Econ.*, 13, 59–73.

Brussard, P. F., Reed, J. M. and Tracy, C. R. (1998). Ecosystem management: what is it really? *Landscape Urban Plan.*, 40, 9–20.

Bråtå, H. O. (2003). The Norwegian system for wild reindeer management – major developments since the 19th century. *Rangifer*, Special Issue 14, 29–36.

Callicott, J. B., (2002). Science, value, and ethics: a hierarchial theory. In *Democracy and the Claims of Nature*, eds. B. A. Minteer and B. P. Taylor, pp. 91–116. Oxford: Rowmanand Littlefield.

Cash, D. W., *et al.* (2006). Scale and cross-scale dynamics: governance and information in a multilevel world. *Ecol. Soc.*, 11, 181–192.

Caulfield, R. A. (1997). *Greenlanders, Whales, and Whaling: Sustainability and Self Determination in the Arctic.* London: University Press of New England.

Decker, D. J., Schusler, T. M., Brown, T. L. and Mattfeld, G. F. (2000). Co-management: an evolving process for the future of wildlife management. *Trans. N. Am. Wildl. Nat. Res.*, 65, 262–277.

Degnbol, P., Wilson, D. C., Grolin, H. A. and Jensen, S. S. (2003). Spatial scale in coastal zone management: current approaches, challenges and possibilities. In: *Rights and Duties in the Coastal Zone*, pp. 1–14. Stockholm: Conference proceedings.

Dion, R. (2003). Twenty-five years of co-management of caribou in northern Quebec. *Rangifer*, Special Issue 14, 307–311.

du Toit, J. T., Walker, B. H. and Campbell, B. M. (2004). Conserving tropical nature: current challenges for ecologists. *Trends Ecol. Evol.*, 19, 12–17.

Folke, C., Holling, C. S. and Perrings, C. (1996). Biological diversity, ecosystems and the human scale. *Ecol. Appl.*, 6, 1018–1024.

Giampietro, M. (1994). Using hierarchy theory to explore the concept of sustainable development. *Futures*, 26, 616–625.

Hiedanpää, J. (2002). European-wide conservation versus local well-being: the reception of the Natura 2000 reserve network in Karvia, SW Finland. *Landscape Urban Plan.*, 61, 113–123.

Jepson, P. and Canney, S. (2003). Values-led conservation. *Global Ecol. Biogeogr.*, 12, 271–274.

Jordan, A. and Jeppesen, T. (2000). EU environmental policy: adapting to the principle of subsidiarity? *Eur. Environ.*, 10, 64–74.

Karlsson, J. and Sjöström, M. (2007). Human attitudes towards wolves, a matter of distance. *Biol. Conserv.*, 137, 610–616.

Kellert, S. R., Mehta, J. N., Ebbin, S. A. and Lichtenfeld, L. L. (2000). Community natural resource management: promise, rhetoric, and reality. *Soc. Natur. Resour.*, 13, 705–715.

Kenward, R. and Putman, R. (2011). Ungulate management in Europe: towards a sustainable future. In *Ungulate Management in Europe: Problems and Practices*, eds. R. Putman, M. Apollonio and R. Andersen, pp. 376–394. Cambridge: Cambridge University Press.

Kok, K. and Veldkamp, T. A. (2011). Scale and governance: conceptual considerations and practical implications. *Ecol. Soc.*, 16, 23.

Lane, M. B. (2003). Decentralization or privatization of environmental governance? Forest conflict and bioregional assessment in Australia. *J. Rural Stud.*, 19, 283–294.

Langbein, J., Putman, R. and Pokorny, B. (2011). Traffic collisions involving deer and other ungulates in Europe and available measures for mitigation. In *Ungulate Management in Europe: Problems and Practices*, eds. R. Putman, M. Apollonio and R. Andersen, pp. 215–259. Cambridge: Cambridge University Press.

Larson, A. M. (2001). Natural resources and decentralization in Nicaragua: are local governments up to the job? *World Dev.*, 30, 17–31.

Linnell, J. D. C. and Boitani, L. (2012). Building biological realism into wolf management policy: the development of the population approach in Europe. *Hystrix*, 23, 80–91.

Maier, D. S. (2013). *What's So Good About Biodiversity? A Call for Better Reasoning About Nature's Value*. New York, NY: Springer.

Maser, C. and Pollio, C. A. (2012). *Resolving Environmental Conflicts*. London: CRC Press.

McNeill, D. and Lichtenstein, G, (2003). Local conflicts and international compromises: the sustainable use of vicuña in Argentina. *J. Int. Wildl. Law Policy*, 6, 233–253.

Nie, M. A. (2003). *Beyond Wolves: The Politics of Wolf Recovery and Management*. London: University of Minnesota Press.

Ostrom, E., Burger, J., Field, C. B., Norgaard, R. B. and Policansky, D. (1999). Revisiting the commons: local lessons, global challenges. *Science*, 284, 278–282.

Perry, R. I. and Ommer, R. E. (2003). Scale issues in marine ecosystems and human interactions. *Fish. Oceanogr.*, 12, 513–522.

Peterson, M. N., Peterson, T. R., Peterson, M. J., Lopez, R. R. and Silvy, N. J. (2002). Cultural conflict and the endangered Florida Key deer. *J. Wildl. Manage.*, 66, 947–968.

Peterson, M. N., Allison, S. A., Peterson, M. J., Peterson, T. R. and Lopez, R. R. (2004). A tale of two species: habitat conservation plans as bounded conflict. *J. Wildl. Manage.*, 68, 743–761.

Redpath, S. M., *et al.* (2013). Understanding and managing conservation conflicts. *Trends Ecol. Evol.*, 28, 100–109.

Reimoser, F. and Putman, R. (2011). Impacts of wild ungulates on vegetation: costs and benefits. In *Ungulate Management in Europe: Problems and Practices*, eds. R. Putman, M. Apollonio and R. Andersen, pp. 144–191. Cambridge: Cambridge University Press.

Ribot, J. (2002). *Democratic Decentralization of Natural Resources: Institutionalizing Popular Participation*. Washington, DC: World Resources Institute.

Robinson, J. G. and Bennett, E. L. (2000). *Hunting for Sustainability in Tropical Forests*. New York, NY: Columbia University Press.

Rosenzweig, M. L. (2003). *Win–Win Ecology*. Oxford: Oxford University Press.

Smith, R. D. and Maltby, E. (2003). *Using the Ecosystem Approach to Implement the Convention on Biological Diversity: Key Issues and Case Studies*. Cambridge: IUCN Publications.

Thomas, C. D. and Kunin, W. E. (1999). The spatial structure of populations. *J. Anim. Ecol.*, 68, 647–657.

Vogt, K. A., *et al.* (2002). Linking ecological and social scales for natural resource management. In *Integrating Landscape Ecology into Natural Resource Management*, eds. J. Liu and W. W. Taylor, pp. 143–175. Cambridge: Cambridge University Press.

Wiens, J. A., Van Horne, B. and Noon, B. R. (2002). Integrating landscape structure and scale into natural resource management. In *Integrating Landscape Ecology into Natural Resource Management*, eds. J. Liu and W. W. Taylor, pp. 23–67. Cambridge: Cambridge University Press.

Williams, R. W. (1999). Environmental injustice in America and its politics of scale. *Polit. Geogr.*, 18, 49–73.

Young, J. C., *et al.* (2013a). Does stakeholder involvement really benefit biodiversity conservation? *Biol. Conserv.*, 158, 359–370.

Young, J. C., Jordan, A., Searle, K. R., Butler, A., Simmons, P. and Watt, A. (2013b). Framing scale in participatory biodiversity management may contribute to more sustainable solutions. *Conserv. Lett.*, 6, 333–340.

Zachrisson, A. (2004). *Co-Management of Natural Resources: Paradigm Shifts, Key Concepts and Cases*. Umeå, Sweden: Mountain Mistra Programme Report no. 1.

Box 15

Wisconsin wolf management: a cauldron of controversy

L. David Mech
US Geological Survey, Northern Prairie Wildlife Research Center, 8711 37th St. SE, Jamestown, ND 58401, USA
The Raptor Center, 1920 Fitch Ave., University of Minnesota, St. Paul, MN 55108, USA

Wisconsin is caught in conflicting crosscurrents of controversy over its recovered wolf (*Canis lupus*) population. Like other states with recovered wolf populations that allow wolf harvest, the primary protagonists are animal welfare advocates versus hunters, trappers and ranchers. Wolves are similar to dogs, which elicits varying conservation values among people that have a strong gender bias (Kellert and Berry, 1987), with women generally more protective of wolves.

Unlike other states, Wisconsin allows hounds for hunting wolves, which makes their management more contentious (Mech, 2013). Concerned for both wolves and hounds, a coalition of humane societies and private citizens unsuccessfully sued to prevent hunting wolves with hounds (*Sportsman's Daily*, 2013).

Another crosscurrent of conflict involved the Anishinaabe tribe that regards the wolf as kin, so opposes hunting it. The Wisconsin Department of Natural Resources (DNR) reduced its wolf-take quota of 201 by 75 in deference to tribal wishes; however, the tribe believed this reduction insufficient (Johnston, 2012). This conflict is beset with misunderstandings, disagreements and changed judgements about the number of wolves Wisconsin could support with minimal conflicts. Wolf recovery and public attitudes evolved over many years. As Wisconsin's wolf population flourished, the Recovery Plan for the Eastern Timber Wolf (1992) prescribed a recovery population goal for the upper Midwest of at least 100 wolves in Wisconsin and Michigan for at least 5 consecutive years combined with at least 1250 in Minnesota. Although this requirement for Wisconsin and Michigan might seem low now, in the early 1990s there were 30–40 wolves in Wisconsin and few in Michigan. Thus 100 in Wisconsin and Michigan, coupled with 1250 in Minnesota, seemed like a high population. The state management plan prescribed a goal of 350 wolves (Wisconsin Department of Natural Resources, 1999).

The early thinking by biologists and citizens was that wolves required wilderness. A wolf-habitat model predicted that only 14,864 km^2 (11% of Wisconsin) had at least a 50% probability of supporting wolves (Mladenoff *et al.*, 1995). However, this model was descriptive, not prescriptive (Mech, 2006), so a refined model indicated that 42,017 km^2 (39% of WI) had at least a 50% chance of supporting wolves (Mladenhoff *et al.*, 2009). As information changed, views on the number of wolves the state could support with minimal impact on people also changed, which probably confused the public.

In addition, over time, the former goals, findings and opinions were forgotten or changed, and new players became involved. As with other wolf-recovery areas, publicity and public involvement became intense because wolves were perceived as endangered. Thus, when Wisconsin's wolf population of 800 was delisted, much of the public rebelled at the thought of wolf reduction. Although wolf numbers far exceeded official recovery levels and Wisconsin's goal of 350, animal welfare groups and wolf preservationists filed three lawsuits between 2007 and 2013, based on legal technicalities rather than on biological issues. These lawsuits resulted in relisting twice, and a third suit is pending. They greatly conflicted with concerns of hunters, trappers, ranchers and rural residents whose dogs and livestock had been killed by wolves. Including the other issues discussed above, these strong divergent factions contributed to the cauldron of crosscurrents pervading Wisconsin wolf management.

Following delisting in 2012, the Wisconsin legislature opened a hunting and trapping season with a quota of 201 wolves, including 85 allocated for harvest by tribes. The tribes took no wolves, but 117 were harvested by the public, not enough to impact the population. In 2013, some 197 wolves have been taken thus far from a quota of 251 wolves.

The Wisconsin recovery goal of 350 wolves has also been challenged, adding to the controversy. Although that goal is about seven times the official federal recovery plan goal, it is so much lower than the actual wolf population that some have questioned the 1999 goal itself as too small (large). The goal was publicly reaffirmed by the DNR's Wolf Science Advisory Committee in 2005 when the wolf population estimate was 435. Thus various factions, including some biologists, now reason that if the state currently supports 800 wolves, then 350 may be too low a population goal. Regardless, to reduce the population from 800 to 350 would require increased quotas.

Because of these crosscurrents, Wisconsin's wolf-management controversy persists as a far more complex issue than that of other states with recovered wolf populations. Thus, it appears that the state's cauldron of controversy will continue to boil, and there seems to be little anyone can do to quell the problems encountered with the success of the biological recovery of wolf populations.

References

Johnston, J. (2012). Some tribal perspectives on Wisconsin's first wolf hunt. *Int. Wolf*, 22(4), 4–7.

Kellert, S. R. and Berry, J. K. (1987). Attitudes, knowledge, and behaviors toward wildlife as affected by gender. *Wildlife Soc. B*, 15, 363–371.

Mech, L. D. (2006). Prediction failure of a wolf landscape model. *Wildlife Soc. B*, 34, 874–877.

Mech, L. D. (2013). The challenge of wolf recovery: an ongoing dilemma for state managers. *The Wildl. Prof.*, 7, 32–37.

Mladenoff, D. J., Sickley, T. A. Haight, R. G. and Wydeven, A. P. (1995). A regional landscape analysis and prediction of favorable gray wolf habitat in the Northern Great Lakes Region. *Conserv. Biol.*, 9, 279–294.

Mladenoff, D. J., Murray, K. C., Pratt, S. D., Sickley, T. A. and Wydeven, A. P. (2009). Change in occupied wolf habitat in the northern Great Lakes Region. In *Recovery of Gray Wolves in the Great Lakes Region of the United States*, eds. A. P. Wydeven *et al.*, pp. 119–138. New York: Springer.

Recovery Plan For The Eastern Timber Wolf (1992). Twin Cities, Minnesota: US Fish and Wildlife Service.

Sportsman's Daily (2013). Wisconsin Judge Allows Dogs In Wolf Hunt. 14 February.

Wisconsin Department of Natural Resources (1999). Wisconsin Wolf Management Plan, Madison, Wisconsin.

© Catherine Young.

CHAPTER SIXTEEN

Mediation and conservation conflicts: from top-down to bottom-up

MARK S. REED and JULIAN SIDOLI DEL CENO
Birmingham City University

Conservation conflicts tend to be highly complex, feed on uncertainty and affect people and organisations at different scales (Chapter 15). Traditionally, resolving these conflicts has been a 'top-down' process, led by governments and their official representatives, supported by scientifically trained specialists, with those affected by the conflict often relegated to the role of data-gatherers and passive recipients of information and instructions. These top-down conflict resolution processes typically seek the 'right' answer, as determined by national politics, laws or academic opinions. They are often framed in Western discourses about the intrinsic value of nature, over-riding older, more utilitarian values as 'backward' and 'damaging' (Zammit-Lucia, 2011). However, rather than resolve conflict, these top-down approaches have often inflamed conflict – for example, prompting violent protest or resettlement of protected areas by evicted communities (Brockington and Igoe, 2006).

In contrast to this, it has been claimed that more bottom-up, participatory approaches to controversial conservation issues have the capacity to avoid, cope with or resolve conflicts. Bottom-up approaches to resolving conservation conflicts, it is argued, have the capacity to build trust and facilitate learning among stakeholders, who are then more likely to support project goals and implement decisions in the long term (Beierle, 2002; Reed, 2008). However, there are also many critics of bottom-up approaches to conservation conflicts. For example, problems with stakeholder representation or participatory process design mean processes fail to achieve their goals or exacerbate conflict (Stringer *et al.*, 2007; Scott, 2011). As such, it has started to be recognised that the outputs (e.g. strategies, plans or other agreements) and ultimate outcomes (e.g. social learning, network forming, preference change, implementation of solutions) of bottom-up approaches to conservation conflicts are highly dependent on the selection of participants, the process design and the context in which they are conducted (de Vente *et al.*, in press).

Top-down and bottom-up approaches represent two opposite extremes, and in reality elements of both approaches are often combined successfully. These

Conflicts in Conservation: Navigating Towards Solutions, ed. S. M. Redpath, R. J. Gutiérrez, K. A. Wood and J. C. Young. Published by Cambridge University Press.

hybrid approaches may, for example, include mediation between conflicting groups in certain locations as part of a more top-down, national process of designating sites for conservation. Bottom-up methods applied to conservation conflicts are often used in combination with each other – for example, 'citizen's juries', participatory mapping and participatory scenario development. One of the most commonly used methods is mediation, often as part of a wider participatory process involving other methods. In this chapter we define mediation broadly as a method for intervening in conflicts that enables the parties to reach their own settlement through the facilitation of a neutral mediator, rather than having one imposed on the parties from above.

This chapter explores mediation approaches to conservation conflicts, to propose good practice principles that may be applied to conservation conflicts in a range of contexts. The first section of the chapter considers top-down discourse of conflict resolution, arguing that the 'right answer' thesis is inadequate to deal with the epistemological gap between facts and values, and ethical questions around utility versus the intrinsic value of nature. The next section considers whether mediation methods (from the anthropology of law and grassroots justice literature) based upon stakeholder participation and mediation reduce the likelihood of conflict, enhance trust and mutual learning, and whether these approaches lead to better and more durable outcomes for the environment, compared to top-down approaches. Finally, the chapter uses this analysis to derive good practice principles for mediating conservation conflicts from the bottom up, and illustrates these with a case study from UK uplands.

Problems of the top-down approach

Traditional top-down approaches to conservation conflicts often face serious problems when it comes to implementation (Cramb *et al.*, 1999; Knill and Lenschow, 2000). Often, these problems can be attributed to the lack of ownership over the process among those who have the power to implement decisions (e.g. state personnel or affected citizens and land owners), leading to low rates of acceptance. This may then lead to these groups delaying or preventing the implementation of decisions to preserve their interests. This section considers the theoretical basis for these challenges, providing a conceptual grounding for more bottom-up, mediation-based approaches that value a plurality of perspectives.

Traditionally, 'justice' is a 'top-down' concept that emerges ultimately from Plato (Jaegar, 1943). The two dominant forms of legal philosophy, natural law and positivism, both rely on a top-down, hierarchical approach, although their source of authority is different. Natural law originates with Aristotle and, at least in its traditional guise, conceives of a divine hierarchy with eternal principles such as 'truth' and 'justice' emerging from God's creation (see Chapter 2). These principles, it is argued, can be attained by god-inspired reason. Positivism is in many ways a reaction against natural law, in particular the conflation of law and morality advocated by most natural law theorists. It rejects a divine hierarchy for a man-made one. Positivism, in its Austinian form, states that law is law because it emerges from the lawgiver, a secular ruler who issues decrees and compels compliance with force and is recognised as such by their subjects. At the heart of both natural law and positivism is a strict, top-down relationship between the lawgiver (divine or otherwise), and the subject of the law, whose role is to obey or to be punished for failing to obey. While critical of both natural law and positivism, Dworkin (1986) also arrives at what is termed the 'right-answer thesis', and posits a 'Herculean' judge who is able to grapple with the diverse range of facts, law and principle and emerge with the right solution.

In recent times, the top-down approach has received sustained criticism from a variety of sources. These include the Critical Legal Studies movement in the USA. Kennedy (1997), for example, taking inspiration from Marxist and feminist discourses, has drawn attention to the hidden motivations and power structures of law. This sustained critique of an overbearing, paternalistic and often patently ill-informed, top-down approach has led to a re-conceptualisation of justice as something that emerges from the discourse of equals; a more bottom-up account of justice where reason and argument, and quite possibly compromise, are the hallmarks of a qualitatively distinct form of dialogue between parties.

A different critique of top-down justice emerges from anthropology. A large number of studies of conflict resolution in different community settings have uncovered a great diversity of approaches to conflict resolution (Nader and Todd, 1978; Falk Moore, 2004), including many non-western methods and approaches (e.g. Pospisil, 1971). These studies demonstrate that many societies have radically different approaches to conflict resolution (Goh, 2002). This in turn has raised larger questions about the validity of the top-down approach and moved the focus onto pluralistic approaches to dispute resolution, notably mediation.

Bottom-up mediation approaches to conservation conflicts

The intellectual roots of mediation can be traced to many disparate sources (Menkel-Meadow, 2000). Fuller (1971) brought intellectual credibility to mediation partly by attempting to develop 'the principles of social ordering' (Winston, 1981). He argued that there were a number of ways in which conflicts could be settled. Some types of conflict require adjudication through the courts or the like, but adjudication may not be necessary or even desirable in other types of dispute. Fuller argues that polycentric disputes where there are multiple parties and multiple interests (which is typically the case in conservation matters) are more appropriately dealt with through mediation. Further, mediation then can be viewed as a bottom-up or 'horizontal' form of justice (Rock, 2006). Others have argued that mediation can empower other voices not heard in traditional conflict resolution (Cobb, 1993); for example, women (Alberstein, 2009).

The history of mediation in environmental and conservation disputes dates back as far as the 1970s (Buckle and Thomas-Buckle, 1986; Newig and Fritsch, 2009). Mediation is often linked to the 'justice from below' movement (Brandt, 1995; McEvoy and McGregor, 2010). In reality, there are a variety of 'grassroots movements' that have sought alternative structures based on differing accounts of justice. These movements have highlighted bottom-up norms and localised settings rather than appealing to universal norms.

One of the difficulties in assessing the outcomes of mediation in conservation conflicts (as in mediation generally), is that there are no universally agreed criteria with which to assess mediation success (Bercovitch, 2007). Sheppard (1984) divides the concept of mediation success into two aspects: process and outcome. While it is possible to measure the number of disputes that are settled, many have argued that it is the 'quality of the settlement' that matters: is a mediation that narrows a significant range of issues a success, a partial success or a failure (Sidoli del Ceno, 2013)?

Multiple hypotheses can be offered regarding the factors that determine the outcomes of bottom-up, mediated approaches to conservation conflicts. Many of the studies that have informed these hypotheses have focused on the social benefits of the process as much as the outcomes for conservation or stakeholders. Broadly speaking, there are four hypotheses, relating to the quality of decisions, the likelihood that decisions are implemented, and the role that local context may play in well-designed mediation processes (Newig and Fritsch, 2009; de Vente *et al.*, in press).

The first hypothesis relates to the outcomes of bottom-up mediation processes, rather than the benefits of the process itself. In this hypothesis, it is suggested that the bottom-up approach leads to more or less beneficial conservation outcomes because of the wider range of information inputs that the decision-making process can draw upon. This is argued to be particularly relevant in local decision-making contexts, where lay people may be able to provide detailed knowledge of the local context (Pellizzoni, 2003). In some cases, lay people may be excluded from the process due to the highly technical nature of the decision being made (Thomas, 1995). However, methods have been developed, such as citizen juries, where lay people can be extensively briefed about technical details, so they can input successfully into decision-making. Conversely, however, in contexts where the decision-making body is highly committed to delivering conservation outcomes, participation of other stakeholders less interested in conservation may compromise desired outcomes (Layzer, 2008). As such, Dryzek *et al.* (2005) argue that the extent to which bottom-up approaches to conservation conflicts lead to outcomes that are beneficial to conservation depends upon the extent to which participants value conservation outcomes.

The second hypothesis also relates to the conservation outcomes of bottom-up mediation processes. It suggests that bottom-up processes are more or less likely to lead to beneficial conservation outcomes because they engage those responsible for implementing decisions fully from the outset (e.g. Bulkeley and Mol, 2003). This is because by including all those who can affect or who are likely to be affected by decisions arising from the mediation process, the decision is more likely to reflect the views of those who have to implement it (Papadopoulos and Warin, 2007; Reed *et al.*, 2009). These assertions assume effective facilitation of the mediation process and good representation of legitimate interests in the process, and if this is not the case, acceptance by third parties who did not participate in the process is likely to remain low. Conversely, however, Coglianese (1997) challenges these claims, arguing that participation may reduce the level of acceptance due to disagreements over who is given the right to participate, which can never be fully resolved. Furthermore, a participatory process may bring disadvantageous aspects of the decision to light, that participants may not have been previously aware of, further reducing acceptance.

The third hypothesis relates to the process rather than the outcomes of engaging in bottom-up mediation processes. It suggests that bottom-up processes are more likely to foster learning among participants and those in their social network, because they take a pluralist approach which seeks and values all perspectives on the conflict. By enabling participants to listen to a wider range of perspectives with less prejudice, mediation may enable learning to occur at a number of levels. This may range from better understanding the conservation challenges on a cognitive level, to deeper learning that can enable participants to re-evaluate the assumptions and values that underlie their positions, leading to changes in attitudes that may shift their positions, so that they are more in line with their values in relation to the environment (Keen and Mahanty, 2006). The new understandings, attitudes and positions that arise from this learning process may then diffuse from those directly participating in the process to those in their social network (Reed *et al.*, 2010).

The previous three hypotheses relate strongly to the design of mediation processes. The fourth hypothesis suggests that both social and conservation outcomes may be dictated by local context. Many studies have emphasised the role that local context can play in determining the outcomes of participatory processes (e.g. Koontz, 2005; Stringer *et al.*, 2007; Blicharska *et al.*, 2011). Most of this research has focused on the socio-economic, cultural and institutional contexts within which mediation is enacted (Delli Carpini *et al.*, 2004). For example, it is argued that bottom-up processes with significant power asymmetries are more likely to suppress the interests of weaker actors than more formalised, top-down processes in which power dynamics are more effectively controlled. These power dynamics may impact the nature of the decision that is made, as well as its acceptance, because those who feel disadvantaged by the process may choose to delay or prevent implementation of the decision; for example, by taking legal action (Cupps, 1977; Turner and Weninger, 2005).

These four hypotheses seek to explain why bottom-up mediation may or may not lead to beneficial outcomes for conservation and to social benefits for those involved in the process. It is clear that if well-designed, then bottom-up approaches to conservation conflicts can increase the likelihood that mediation leads to beneficial outcomes. However, these benefits are often strongly constrained by local contextual factors, and the range (and values) of participants who engage with the process.

Good practice principles for mediating conservation conflicts

Much has been written about the benefits and failings of bottom-up approaches to conservation conflict (Reed, 2008; Redpath *et al.*, 2013). Some point to evidence of bottom-up processes that have failed to reach their goals, exacerbated conservation conflicts or allowed special interest groups to bias outcomes (e.g. Cook and Kothari, 2001; Stringer *et al.*, 2007; Scott, 2011). Others suggest that these processes were always likely to fail due to poor process design, or because the context in which the process took place was not conducive to a bottom-up approach (Reed, 2008; de Vente *et al.*, in press). We argue that if good practice is followed, bottom-up mediation approaches are more likely to succeed than top-down approaches in most contexts. There are two elements to this. First, it is necessary to assess whether or not the context is conducive to a bottom-up mediated approach. Second, if it is, then the process should be designed and executed according to known good practice.

There are many contexts in which a bottom-up, mediated approach to conservation conflicts may not be appropriate, for example: where there is widespread apathy and disengagement among stakeholders, making it difficult to mobilise participation in the mediation process; where there is an autocratic culture (i.e. with little decision-making autonomy for individuals, such as former communist states); where there are significant power imbalances between participants; or where some or all participants do not really have decision-making power (Reed, 2008; de Vente *et al.*, in press).

Assuming that it is appropriate to opt for a mediation approach, then it is essential to design the process effectively (Chapter 17; Chapter 18). It is possible to identify three good practice principles for such a design (Reed, 2008; de Vente *et al.*, in press).

The right people

A diverse group of well-informed people from different backgrounds is likely to provide the most relevant and innovative ideas. There are two tasks that need to be performed: identifying the right people for inclusion in the mediation process; and getting these people engaged in the mediation process. First, it is necessary to identify the most relevant individuals and organisations who can represent the full range of interests in the conflict. If key parties are missing from the mediation process, then they are likely to question the legitimacy of the process and potentially block progress towards implementing outcomes. If possible, once representation has been achieved, it is useful to consider whether the right people have been invited from the organisations that are represented. If these people do not have decision-making power, it will be difficult to reach any agreement as part of the mediation process. Finally, where possible, trying to target a few representatives who are known for being innovative and creative can help the mediation process achieve creative solutions to the conflict. Second, it is necessary to bring these people into the mediation process. A mediation process that has identified all the key players but fails to engage them is likely to lead to biased outcomes with low acceptance and implementation. Partly this is about effective communication, and making involvement attractive and easy for all participants. For some participants, this may be about practical considerations (e.g. avoiding meetings at certain times of day or certain seasons) or financial or other types of incentives (e.g. payments for participation, offering meals and opportunities to network). For many participants, they simply need to believe there is a high probability that engaging in the process will lead to direct benefits (e.g. access to land, compensation, etc.). By working through existing trusted contacts and networks, it may be possible to reach and convince potential participants who may not otherwise have been contacted and involved.

The right atmosphere

Creating an open and respectful environment needs to start at the very beginning and continue throughout people's engagement with the process. There is little point in having the right people engaged in the process if some dominate and others feel powerless to speak. Working with a professional, independent mediator can help create the right atmosphere, enabling everyone to participate. They can manage conflicts as they arise, and they often have tools that enable people to think critically together. Highly skilled mediation is particularly important in situations where there is a breakdown of trust.

The methods that a mediator employs can help create an atmosphere of trust. For example, methods must be adapted to the socio-cultural context of the mediation process e.g. avoiding methods that require participants to read or write in groups that might include illiterate participants. Depending on the power dynamics of the group, methods may need to be employed that equalise power between participants to avoid marginalising the voices of the less powerful. There is evidence that less powerful participants who are marginalised during decision-making can delay or prevent implementation through litigation.

Making it relevant

It is important to make the mediation process as relevant as possible for all participants. Partly this is about the content and focus of the mediation process, and how this focus is derived. Partly it is about the perceived credibility of the process, and the likelihood that it will lead to beneficial change. If participants do not perceive that the process has credibility to affect the issues that concern them, then they will view it as irrelevant and not participate. Negotiating a set of ambitious but achievable goals with all participants from the outset can help demonstrate that their participation is likely to make a real difference. If the goals are developed through dialogue (making trade-offs where necessary) among participants, they are more likely to take ownership of the process and build partnerships with outcomes more likely to be relevant to stakeholder needs, priorities and desire to remain actively engaged.

Once the content and focus of the mediation have been negotiated and agreed by all parties, the approach to the mediation needs to be made as relevant as possible to all participants. For example, language can sometimes be used to erect barriers between groups who each have their own exclusive vocabulary, so it is important to use language that is familiar and impartial to all parties. When dealing with complex conservation conflicts with intangible concepts (such as biodiversity), it may be necessary to focus on aspects that are more tangible to participants (such as indicator species that have known uses for humans).

Case study: mediation of conservation conflict in UK uplands

The Sustainable Uplands project (Reed *et al.*, 2013) took a bottom-up, mediation approach to a conflict in UK uplands from 2005 to 2011. It applied emerging best practice in stakeholder participation (Reed, 2008) and effectively illustrates each of the mediation principles outlined in the previous section. Conflict had arisen over the management of land for game and sheep using managed burning versus conservationists and water companies who were concerned that inappropriate burning and over-grazing were compromising internationally important habitats and water quality. The goal of the research was to mediate an agreed set of management options that could enable each of these groups to meet their goals in the context of climate, social and policy change. The research led to the development of a Peatland Code, which provides payments from private companies (including water companies) to landowners and managers to restore damaged peatland habitats.

The research started with a systematic assessment of who may hold a stake in the conflict, using Reed *et al.*'s (2009) extendable interest–influence matrix during a focus group with a small number of cross-sectoral stakeholders (from the National Park Authority and National Trust), followed by semi-structured interviews with key stakeholders. During the focus group and interviews it was possible to identify the key characteristics of the conflict and refine the focus of the project. This was followed by a Social Network Analysis to understand relationships between stakeholders, including power dynamics, which was subsequently used to select participants for the mediation process, ensuring that key individuals were present with decision-making power, but seeking to avoid unhealthy power imbalances.

The mediation started badly, with an inexperienced mediator exacerbating conflict among participants. Despite expressing their frustration with the workshop, participants agreed to continue with the mediation, provided that the mediation was facilitated more effectively in the future. Subsequently, a more experienced mediator with many years' experience working in conservation conflicts was employed, and the researchers received training and shadowed the mediator till she was confident that they were sufficiently competent. To reduce disparities of power between participants, the mediation was primarily conducted during site visits to the locations where conflict was occurring, empowering land management stakeholders with little formal education to interact more equally with other more formally educated conservation stakeholders.

Conclusion

This chapter has explored the rationale, benefits and common problems associated with top-down versus bottom-up approaches to mediating conflict in conservation. It has argued that the 'right-answer' thesis struggles to deal with the gap between facts and values, and decisions based around utility versus intrinsic value. It has therefore explored the basis for more bottom-up approaches to conservation conflicts based on mediation, and derived good practice principles, namely identifying and involving the right people; ensuring the right atmosphere; and making the process relevant.

Although bottom-up approaches to conservation conflict will always be challenging, and may be highly time-consuming and resource-intensive, designing the process to these good practice principles may increase the likelihood of success. It can further be argued that there is great long-term value in developing a collaborative culture in these types of dispute. Whether success means achieving conservation goals or simply leads to an increase in trust and more positive working relationships, the evidence suggests that there are a range of important outcomes that may come from taking a more bottom-up approach. However, more empirical data are required to test this hypothesis.

References

Alberstein, M. (2009). The jurisprudence of mediation: between formalism, feminism and identity conversations. *Cardozo J. Conflict Resol.*, 11.1, 1–28.

Beierle, T. C. (2002). The quality of stakeholder-based decisions. *Risk Anal.*, 22, 739–749.

Bercovitch, J. (2007). Mediation success or failure: a search for the elusive criteria. *Cardozo J. Conflict Resol.*, 7.2, 289.

Blicharska, M., Isaksson, K., Richardson, T. and Wu, C. J. (2011). Context dependency and stakeholder involvement in EIA: the decisive role of practitioners. *J. Environ. Plann. Manage.*, 54, 337–354.

Brandt, H-J. (1995). The justice of the peace as an alternative: experiences with conciliation in Peru. In *Judicial Reform in Latin America and the Caribbean: Proceedings of a World Bank Conference*. World Bank Technical Paper Number 280, eds. M. Rowat, W. H. Malik and M. Dakolias, pp. 92–95. Washington, DC: The World Bank.

Brockington, D. and Igoe, J. (2006). Eviction for conservation. A global overview. *Conserv. Soc.*, 4, 424–470.

Buckle, L.G. and Thomas-Buckle, S. R. (1986). Placing environmental mediation in context: lessons from 'failed' mediations. *Environ. Impact Assess.*, 6, 55–70.

Bulkeley, H. and Mol, A. P. J. (2003). Participation and environmental governance: consensus, ambivalence and debate. *Environ. Value*, 12, 143–154.

Cobb, S. (1993). Empowerment and mediation: a narrative perspective. *Negotiation J.*, 9, 245–259.

Coglianese, C. (1997). Assessing consensus: the promise and performance of negotiated rule-making. *Duke Law J.*, 46, 1255–1346.

Cook, B. and Kothari, U. (2001). *Participation: The New Tyranny?* London: Zed Books.

Cramb, R. A., Garcia, J. N. M., Gerrits, R. V. and Saguiguit, G. C. (1999). Smallholder adoption of soil conservation technologies: evidence from upland projects in the Philippines. *Land Degrad. Dev.*, 10, 405–423.

Cupps, D. S. (1977). Emerging problems of citizen participation. *Public Admin. Rev.*, 37, 478–487.

de Vente, J., Reed, M. S., Stringer, L. C., Valente, S. and Newig, J. (in press). How does the context and design of participatory decision-making processes affect their outcomes? Evidence from sustainable land management in global drylands. *J. Environ. Manage.*

Delli Carpini, M. X., Cook, F. L. and Jacobs, L. R. (2004). Public deliberation, discursive participation, and citizen engagement: a review of the empirical literature. *Ann. Rev. Polit. Sci.*, 7, 315–344.

Diduck, A. and Sinclair, A. J. (2002). Public involvement in environmental assessment: the case of the nonparticipant. *Environ. Manage.*, 29, 578–588.

Dryzek, J., Downes, D., Hunold, C. and Schlosberg, D. (2005). Green political strategy and the state: combining political theory and comparative history. In *The State and the Global Ecological Crisis*, eds. J. Barry and R. Eckersley, pp. 75–96. Cambridge, MA: MIT Press.

Dworkin, R. (1986). *Law's Empire*. Cambridge, MA: Harvard University Press.

Falk Moore, S. (2004). *Law and Anthropology: A Reader*. Malden, MA: Blackwell Publishing.

Fuller, L. L. (1971). Mediation: its forms and functions. *South. Calif. Law Rev.*, 44, 305–339.

Goh, B. (2002). *Law Without Lawyers, Justice Without Courts: On Traditional Chinese Mediation*. Farnham: Ashgate.

Gray, S., Chan, A., Clark, D. and Jordan, R. (2012). Modeling the integration of stakeholder knowledge in social–ecological decision-making: benefits and limitations to knowledge diversity. *Ecol. Model.*, 229, 88–96.

Jaegar, W. (1943). *Paideia: The Ideals of Greek Culture, Volume II: In Search of the Divine Centre*. Oxford: Oxford University Press.

Keen, M. and Mahanty, S. (2006). Learning in sustainable natural resource management: challenges and opportunities in the Pacific. *Soc. Natur. Resour.*, 19, 497–513.

Kennedy, D. (1997). *A Critique of Adjudication [fin de siecle]*. Cambridge, MA: Harvard University Press.

Knill, C. and Lenschow, A. (2000). *Implementing EU Environmental Policy: New Directions and Old Problems*. Manchester: Manchester University Press.

Koontz, T. M. (2005). We finished the plan, so now what? Impacts of collaborative stakeholder participation on land use policy. *Policy Stud. J.*, 33, 459–481.

Layzer, J. (2008). *Natural Experiments: Ecosystem-Based Management and the Environment*. Cambridge, MA: MIT Press.

McEvoy, K. and McGregor, L. (2008). *Transitional Justice from Below: Grassroots Activism and the Struggle for Change*. Oxford: Hart Publishing.

Menkel-Meadow, C. (2000). Mothers and fathers of invention: the intellectual founders of ADR. *Ohio St. J. Disp. Resol.*, 16, 1–37.

Nader, L. and Todd, H. (1978). *The Disputing Process: Law in Ten Societies*. New York, NY: Columbia University Press.

Newig, J. and Fritsch, O. (2009). Environmental governance: participatory, multi-level – and effective? *Environ. Pol. Gov.*, 19, 197–214.

Papadopoulos, Y. and Warin, P. (2007). Are innovative, participatory and deliberative procedures in policymaking democratic and effective? *Eur. J. Polit. Res.*, 46, 445–472.

Pellizzoni, L. (2003). Uncertainty and participatory democracy. *Environ. Value.*, 12, 195–224.

Pospisil, L. (1971). *Anthropology of Law: A Comparative Theory*. New York, NY: Harper and Row.

Reed, M. S. (2008). Stakeholder participation for environmental management: a literature review. *Biol. Conserv.*, 141, 2417–2431.

Reed, M. S., *et al.* (2009). Who's in and why? A typology of stakeholder analysis methods for natural resource management. *J. Environ. Manage.*, 90, 1933–1949.

Reed, M. S., *et al.* (2010). What is social learning? *Ecol. Soc.*, 15, 1–10.

Reed, M. S., *et al.* (2013). Anticipating and managing future trade-offs and complementarities between ecosystem services. *Ecol. Soc.*, 18, 5.

Rock, E. (2006). Mindfulness mediation, the cultivation of awareness, mediator neutrality, and the possibility of justice. *Cardozo J. Conflict Res.*, 6, 347–365.

Scott, A. J. (2011). Focussing in on focus groups: effective participative tools or cheap fixes for land use policy? *Land Use Pol.*, 28, 684–694.

Sheppard, B. (1984). Third party conflict intervention: a procedural framework. *Res. Org. Behav.*, 6, 226–275.

Sidoli del Ceno, J. (2013). Construction mediation as a developmental process. *Int. Rev. Law*, 1.

Stringer, L. C., Twyman, C. and Thomas, D. S. G. (2007). Combating land degradation through participatory means: the case of Swaziland. *Ambio*, 36, 387–393.

Thomas, J. C. (1995). *Public Participation in Public Decisions. New Skills and Strategies for Public Managers*. San Francisco, CA: Jossey-Bass Publishers.

Turner, M. A. and Weninger, Q. (2005). Meetings with costly participation: an empirical analysis. *Rev. Econ. Stud.*, 27, 247–268.

Winston, K. I. (1981). *The Principles of Social Order: Selected Essays of Lon L. Fuller*. Durham: Duke University Press.

Zammit-Lucia, J. (2011). *Conservation is not about nature*. IUCN Expert Opinion http://www.iucn.org/involved/opinion/?8195/Conservation-is-not-about-nature (accessed 10 November 2013).

Box 16

Living with white sharks: non-lethal solutions to shark–human interactions in South Africa

Alison A. Kock and M. Justin O'Riain

Department of Biological Science, University of Cape Town, Private Bag X3, Rondebosch, 7701, South Africa

White sharks *Carcharodon carcharias* aggregate seasonally at select coastal sites, some of which are human recreational areas (Kock *et al.*, 2013). This overlap and the predatory nature of white sharks result in some sharks biting humans. The global number of white shark bites has increased gradually, as it has in South African waters (Curtis *et al.*, 2012). This increase is largely attributed to increasing human use of the ocean for recreation, as well as improved reporting (Burgess *et al.*, 2010).

Although sharks rarely injure or kill people, when they do it induces strong negative responses from the public, fuelled by sensationalist media coverage. Furthermore, shark bites typically negatively affect local economies and the perceived amenity value of coastal areas. Conservationists counter negative information by stressing the rarity of bites and highlight the natural role of sharks as top predators and contributors of biodiversity (Heithaus *et al.*, 2008). Despite making evidence-based arguments, conservationists struggle to convince people and authorities about the value of sharks. Negative public perceptions and economic impacts compel stakeholders to urge local authorities to kill sharks to reduce attacks.

Currently white sharks are listed in Appendix II of the Convention on International Trade in Endangered Species and classified as 'Vulnerable' by the World Conservation Union (Fergusson *et al.*, 2009). Despite this, white shark control programmes have been implemented at many coastal recreational areas (Dudley, 1997). These control measures, which include the use of large-mesh gill nets and/or baited drum lines, aim to reduce the probability of a shark bite by reducing shark numbers (Dudley, 1997). However, such methods are environmentally costly because they both decrease this threatened species and unselectively kill many other species (Cliff and Dudley, 2011). Thus, alternative strategies are needed to mitigate conflict between water users and conservationists so that conservation of marine ecosystems and white sharks is not jeopardised.

In Cape Town, a novel community initiative known as the 'Shark Spotter' programme has been implemented as a non-lethal alternative (Kock *et al.*, 2012). The proximity of mountains to popular beaches provided natural vantage points from which designated people can detect white sharks and warn water users using a flag system or sirens. This programme was set up by local business owners, negatively impacted by sharks biting potential customers; after two years it was endorsed and funded by local government. The programme was an independent non-profit organisation, funded by local government, NGOs and the public. It has been successful in reducing the spatial overlap between people and sharks, which assumes that by reducing spatial overlap there will be a reduced probability of shark bites. Thus the programme's success has been largely attributed to finding a balance between the desires of water users with those of conservationists (Kock *et al.*, 2012).

Successful abatement of potential interactions provided useful reflection on a recent incident in Cape Town where the victim elected to swim, despite spotters having closed the beach owing to the presence of sharks. The victim, who lost a leg, confirmed he had swum knowing the beach was closed because sharks were present. Despite this, there were more demands from the public for alternatives to the Shark Spotter programme, including lethal control. In response, local authorities have begun testing another non-lethal method, the exclusion net, in a small portion of the affected beach. However, conservationists worried that more bites will induce Cape Town authorities to consider lethal alternatives. This example highlighted the challenges of educating people about the risks of swimming with sharks present and ultimately mitigating conflict over shark conservation. Additionally, this incident emphasised the low tolerance humans have when they are threatened by predators.

The Shark Spotter programme provides an example of successful conflict mitigation related to shark bites, but sustainable resolution requires ongoing investment keeping people and sharks apart.

References

Burgess, G. H., Buch, R. H., Carvalho, F., Garner, B. A. and Walker, C. J. (2010). Factors contributing to shark attacks on humans: a Volusia County, Florida, case study. In *Sharks and Their Relatives: II. Biodiversity, Adaptive Physiology, and Conservation*, eds. J. C. Carrier, J. A. Musick and M. R. Heithaus, pp. 541–565. Boca Raton, FL: CRC Press.

Cliff, G. and Dudley, S. F. J. (2011). Reducing the environmental impact of shark-control programs: a case study from KwaZulu-Natal, South Africa. *Mar. Freshwater Res.*, 62, 700–709.

Curtis, T. H., *et al.* (2012). Responding to the risk of white shark attack: updated statistics, prevention, control methods and recommendations. In *Global Perspectives on the Biology and Life History of the White Shark*, ed. M. L. Domeier, pp. 477–510. Boca Raton, FL: CRC Press.

Dudley, S. F. J. (1997). A comparison of the shark control programs of New South Wales and Queensland (Australia) and KwaZulu-Natal (South Africa). *Ocean Coast. Manage.*, 34, 1–27.

Fergusson, I., Compagno, L. J. V. and Marks, M. (2009). *Carcharodon carcharias*. In IUCN Red List of Threatened Species. Version 2013.2. www.iucnredlist.org. Downloaded 23 November 2013.

Heithaus, M. R., Frid, A., Wirsing, A. J. and Worm, B. (2008). Predicting ecological consequences of marine top predator declines. *Trends Ecol. Evol.*, 23, 202–210.

Kock, A. A., *et al.* (2012). Shark spotters: a pioneering shark safety program in Cape Town, South Africa. In *Global Perspectives on the Biology and Life History of the White Shark*, ed. M. L. Domeier, pp. 447–466. Boca Raton, FL: CRC Press.

Kock, A. A., O'Riain, M. J., Mauff, K., Meÿer, M., Kotze, D. and Griffiths, C. (2013). Residency, habitat use and sexual segregation of white sharks, *Carcharodon carcharias* in False Bay, South Africa. *PLoS ONE*, 8, e55048.

© Adam Vanbergen.

CHAPTER SEVENTEEN

Designing and facilitating consensus-building – keys to success

DIANA POUND
Dialogue Matters

I am writing this chapter as someone who has designed and facilitated over 70 stakeholder processes, reviewed others, conducted research on best practice (e.g. Pound, 2009) and trained and learnt from the experiences of over 1000 people. I do this through *Dialogue Matters*, established in 2000 to focus on dialogue about the environment. We work from local to international scales, low to high levels of tension, and between sectors and cultures. Our work includes conflict resolution between resource users, between users and conservationists, and also topic-based dialogue where the science and policy is contested (e.g. climate change, global water futures and solutions, sustainable fishing practice, managing bovine tuberculosis (TB) in cattle and badgers).

Our approach continues to evolve but is rooted in *Consensus-building*, more often now called *Stakeholder Dialogue*. This has particular ethics and design principles which set it apart from other participatory approaches. It involves a group of stakeholders in a face-to-face and principled negotiation process, designed for the situation and comprising a staged sequence of deliberative workshops and other activities. In our own work, to maximise the benefits of the process and the number that can be involved, our professional team train, lead and work alongside volunteer small group facilitators.

In this chapter I describe 12 keys to success related to consensus-building. I use 'tension' rather than 'conflict' because there is a gradation from low tension (parties recognise different perspectives but are willing to cooperate) to full conflict (parties are polarised, have no direct contact and the conflict is escalating). 'Sponsoring bodies' are organisations (e.g. government agencies, NGOs or partnerships) that provide the resources for a stakeholder process. Their motivation may be compliance with statutory requirements, a principled commitment to participation or, most likely, because they want well-informed decisions that have greater support and are easier to implement. 'Projects' are the people and enterprise put together to achieve a particular aim. Projects may sit within larger organisations that are the sponsoring body, or be a partnership between organisations and be the sponsoring body.

Conflicts in Conservation: Navigating Towards Solutions, ed. S. M. Redpath, R. J. Gutiérrez, K. A. Wood and J. C. Young. Published by Cambridge University Press.

Table 17.1 *Levels of influence that stakeholders can have and examples of when each level of influence is suitable.*

Information giving	Information gathering	Consultation	Shared decision-making
Purpose from sponsoring body's perspective			
• To raise awareness	• To develop own understanding to make a decision	• To be open to influence when making a decision	• To make decisions with others not for them
Examples of when this level of influence is suitable			
• The organisation is mandated to make the decision • The situation is not contentious • In an emergency	• Information will be freely given • It is clear how the information is to be used • There are good levels of trust and/or low risk of action being rejected	• Reasonable levels of trust • Sponsoring organisation has final say • Responsibility to take action rests with sponsor	• When situation complex • Responsibility to implement action needs to be shared • There are divergent views and power differences

Inclusion, deliberation and levels of influence

Key one: increase stakeholder influence

There are numerous models describing levels of stakeholder influence, but we either find these to be unclear in practice or they suggest that one end of the continuum is ethically better than another. For example, Arnstein's (1969) 'Ladder of Participation' has manipulation at one end and full citizen control at the other. While this is interesting, few environmental situations could be passed to full citizen control because they involve organisations with statutory responsibilities and the need for specialist science. The model we use instead assumes that each level of influence may be suitable depending on circumstances (Table 17.1).

In any participatory process, it is vital that the sponsoring body, facilitators and participants have as much clarity as possible about what those involved can and cannot influence. If a person enters a process in good faith, and commits time, effort and energy but ends up with no real influence, they will be disillusioned with that process and possibly engagement more generally. It will also damage social capital and trust, which has long-term consequences.

Not only is it important to have clarity over the level of influence before the process starts, but also the follow-through must match the expectation. For example, if the process is to make decisions, the decisions should be implemented. If it is consultation, organisations should tell stakeholders how their contributions were used.

The simplest way to establish the level of influence is to simply ask what happens to workshop outputs: are they the decision; or are outputs passed to a separate decision-making process as information or advice? Once this is clear the process can be designed with clarity on who will make key decisions and when.

If a process is to resolve conflict and build consensus, participants need the highest possible levels of influence. Ideally, this would be in the shared decision-making category so that no one can veto or overturn the agreements and set the resolution back. However, it may be that the negotiated agreement has to be signed off by a committee or minister, placing it functionally in the consultation category.

Key two: understand inclusion and deliberation

When considering stakeholder involvement, there is a difference between inclusion and deliberation (Studd, 2002; Table 17.2). 'Inclusion encourages breadth in decision making' (Holmes and Scoones, 2000: 31) and broadens the range of experience and knowledge involved, while deliberation occurs when 'there is sufficient and credible information for dialogue, choice and decisions and where there is space to weigh options, develop common understanding and to appreciate respective roles and responsibilities' (Brisbane Declaration, 2005: 2).

If projects include large numbers of people, there is no opportunity for deliberation so this can only be at the information-gathering level. However, if the situation is tense and complex, it is better to have an equitable balance of people from all relevant interests to facilitate in-depth deliberation and negotiation. Ideally there are sufficient resources to do both, and gather information from a larger number of people to inform the deliberations – but few public bodies and NGOs have the resources for this.

In considering the optimum level of influence, organisers also need to understand what the different levels of participation achieve. Experience suggests that organisers often assume that any kind of participation, even superficial engagement, delivers social capital and buy-in. However, support, cooperation and collective action result from the relationships, understanding and trust that develop in a well-designed, face-to-face deliberative process.

Key three: carry out systematic stakeholder identification

Once the facilitators have assessed the context, the level of influence, and the optimum mix of inclusion and deliberation, work can begin on identifying stakeholders. In academic research it may be necessary to invest resources and use

Table 17.2 *The four levels of influence in relation to the extent they can be deliberative or inclusive, the number of people that can be involved, and the related benefits of interaction and outcome.*

	Information giving	Information gathering	Consultation	Shared decision-making
Deliberation, inclusion and number of people that can be involved				
Numbers of people involved		1000s	Potentially 100s	Upper limit approx. 60 in core deliberative process
Inclusion		Most inclusive	Variable	Least inclusive
Deliberation	No deliberation	Least deliberation	Variable	Most deliberation
The benefits of the interaction and outcome generated by each level of influence				
Social capital built Knowledge exchanged Mutual understanding Decisions better informed Solutions integrated Consensus Cooperative and collective action	Delivers none of the list of benefits	Delivers fewest benefits	Variable benefits result (depends on process)	Benefits maximised

different methods to ensure the stakeholder list is comprehensive and backed by evidence such as network analysis (Reed *et al.*, 2009). Outside a research context, projects rarely have sufficient resources for this task.

Ideally, the time given to stakeholder identification should be proportionate to the level of tension among the parties and the level of influence the group will have on the outcome. In straightforward situations, we facilitate a session for project partners to generate a list of possible stakeholders based on the subject, issues, opportunities and knowledge needed. If resources allow, we cross check this against sectors (e.g. public, private, research, voluntary) and functions (e.g. users, residents, regulators). When working at a local level, we may also look for inclusion based on ethnicity, age, socio-economic group and location (residents, visitors).

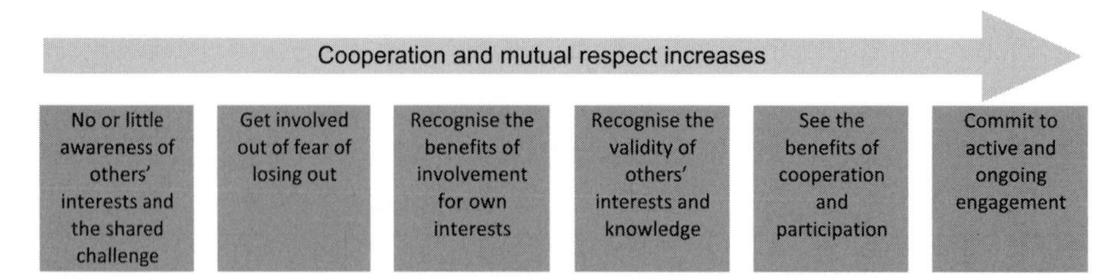

Figure 17.1 Shift in stakeholders' attitude and behaviour through a deliberative dialogue.

Where there is existing conflict, greater rigour is needed. This is because it is vital that the mix, diversity and balance of people in the core negotiation is widely accepted as equitable, balanced and legitimate. To achieve this, we use a process where the first phase is consensus building among a wide group of people about the breadth and balance of interests who should have a seat in the negotiations, and the second phase is identifying the individuals.

Designing effective processes

Key four: allow sufficient time

Projects need to allow sufficient time for all key activities. This includes scoping the context, designing the process, identifying stakeholders and then designing and preparing for workshops. In multi-workshop processes, time between workshops needs to allow for processing outputs, drafting text and carrying out any wider engagement needed to support the core deliberative dialogue.

Crucially, people also need time to build trust and change behaviour (Fig. 17.1). This includes building 'social capital' (trust and good will) and 'social learning' (understanding other perspectives and developing new shared perspectives).

Designed stakeholder processes that enable this shift have taken up to 18 months depending on the complexity, level of tension and purpose. An example is a process designed to help people reach agreement on the integrated management of a coastal and marine Natura 2000 site (see Box 13). This was a three-workshop consensus process spread over nine months. It was designed to meet all 12 principles of the Ecosystem Approach (Convention on Biodiversity) and helped multiple interests work together to reach agreement on management measures for over 50 different human uses.

Key five: design, design, design

Best practice is to design the whole consensus-building process before it starts. From reviewing other processes, we observe that ad hoc workshops usually fail to deliver the benefits claimed for participation. For best practice, design takes place at three levels: the overall process (including the number of workshops and what happens between them), the structure of each workshop, and each session within a workshop. A well-designed process will:

- have a clear sequencing and flow, with the process designed to work as a cohesive and integrated whole through the key stages of negotiation;
- have a designed sequence of questions and activities planned before, during and after each workshop;
- map the information and decision paths in the process during and between different activities; and
- use a range of facilitation skills and techniques to help people move from positional or adversarial (win/lose) negotiation behaviour to principled or cooperative negotiation behaviour (i.e. actively seeking win/win).

When considering the overall process design, different methods are suited to different levels of influence and require increasing design and facilitation competence the greater the influence and deliberation involved (Table 17.3).

Participation without design and planning at the three levels (process, workshops and sessions) leads to confusion about levels of influence (Key one) and can lead to negative consequences and even disempower stakeholders. In one example we reviewed (which has to remain anonymous), there were many sector-based workshops and people involved. This appeared inclusive, but the extent of negotiation and deliberation was not clear to those involved. The project officer led the individual workshops, but had not designed them as part of an integrated process so they were disconnected over time and between sectors. Stakeholders were not brought together to learn from and negotiate with each other face to face. This meant the workshops functioned at the 'information-gathering' level of engagement with the project officer alone reconciling different perspectives in written documents and making decisions about priorities. This lack of genuine consensus between interests, and lack of real integration and reconciliation, meant the project was in danger of failing at any stage with collateral damage for ongoing trust and action.

There is literature on critiques of participation (reviewed in NRC, 2008); most criticisms are addressed by good design. Poor or no design results in little clarity about who can influence what, the key stages of negotiation and/or appropriate stakeholder identification. A particular concern is that poorly designed participation can still appear to provide the opportunity for influence, but in reality offers little. Worse, it can consume people's time and effort so that they either do not realise they need to act outside the process to influence change or they do not have the resources to do that. In this context, the participation is actually disempowering people and undermining participatory democracy. However, good practice ethics and a skilled practitioner will address these concerns in the preparation, design, or mitigation of these risks.

Table 17.3 *Methods of engagement suited to the different levels of influence and the relative importance of both design and designer/facilitator competence.*

	Information giving	Information gathering	Consultation	Shared decision-making
Examples of methods appropriate at each level of influence				
Interactive	• Public meeting • Displays • Open days	• Workshops • Drop-in meetings • Semi-structured interviews	• Deliberative workshops • 1:1 meetings	• Facilitated consensus building processes (including deliberative workshops)
Reactive	• Press release • Leaflets • Newsletters • Advertisements	• Questionnaires • Interviews • Surveys • Opinion polls • Exhibition with feedback	• Written consultation • Online consultation	
Importance of design and of designer and facilitation competence				
Importance of coherent planned and sequenced process design	Low			High
Design and facilitation competence	Least			Most

Key six: use deliberative and principled negotiation and decision-making processes

All changes in environmental management require decision-making. Depending on the level of influence that stakeholders have this may or may not be within the stakeholder process itself (see Key one). However, where there is conflict, the best approach is a well-designed and facilitated consensus-building process that helps different interests negotiate through to final decisions. This enables people to change the way they negotiate from positional, adversarial and blocking tactics to cooperative behaviour (Table 17.4), which will be more likely to lead to collective action.

To facilitate a shift in behaviour, a negotiation process starts with opening discussions to share and explore information, knowledge and skills, followed by

Table 17.4 *Characteristics of adversarial and cooperative behaviour.*

Adversarial behaviour	Cooperative behaviour
Withhold information	Share information
Make threats	Ask questions
Argue from positions	Explore interest and needs
Attack the others' knowledge or them	Explore knowledge and perspectives
Defend position	Seek solutions
Work on each other	Work on the challenge
Actively seek win/lose	Actively seek win/win

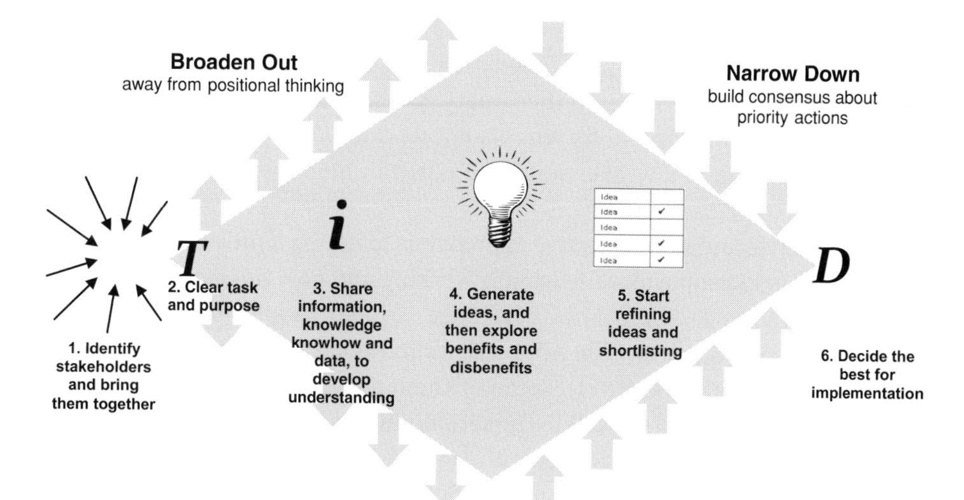

Figure 17.2 The process of broadening out, then narrowing down the discussion in a consensus-building process.

generating and exploring ideas and options, before narrowing options to ones that are mutually acceptable (Fig. 17.2).

Consensus-building does not mean everyone agrees about everything at the end of the process. Rather, alternative solutions emerge and the facilitator records the levels of support for these alternatives. The facilitator asks those who do not accept a particular solution to explain their concerns and suggest amendments. If the amendments are not acceptable to others, the facilitator will ask everyone to share responsibility to suggest modifications that increase the acceptability for all. This process might be repeated several times. If a decision deadline is approaching, the facilitator will help people build consensus about what level of consensus is acceptable to proceed. After all this has taken place, solutions are found that have broad support. Some people will actively support the solutions, while others accept them recognising that after all the hard work and creative thinking, no better alternative exists.

This process requires time. For example, a consensus-building process for the integrated management of some remote English islands had a sequence of four deliberative workshops at its heart and took 9 months (Pound, 2004a).

Key seven: shift focus from problems to strengths

The usual approach to environmental management is problem-solving (i.e. identifying the problems and working to find solutions). This can have a negative effect, causing people to feel defensive and disown the problem, blaming something or someone else. People can also feel overwhelmed with the complexity and scale of a challenge and become demotivated, feel powerless or deny a problem exists (Bandura, 2007; Sutton, 2012). Fixing problems takes resources (people, time, energy, creativity, innovation and funds). Unless there is a new injection of funds or people into the situation, resources have to be taken from what is currently working to fix what is not working. This can leave what was previously working in a weaker state and result in more problems than at the outset.

An alternative and more effective approach is to design a process that frames questions to help people identify what is working and how this can be strengthened and enhanced. Approaches developed to help people focus and build on strengths and assets include Appreciative Inquiry and Asset-based Community Development (Bushe, 2011). Some of the language and methods in these approaches do not translate well to the serious and scientific culture of environmental and conservation management. However, when designing a dialogue process, it is possible to draw on the underlying principle of helping people imagine a positive future, identify what is already working well, consider what needs to happen to enhance or strengthen current efforts and then explore what else needs to happen. This has a positive effect. People feel their contribution is identified, acknowledged and valued. The group realise good work is already being done so the challenge seems more solvable. It builds positive momentum for delivery, harnesses existing resources of time, effort, innovation and energy and works with the current momentum. In short it motivates people and plays to strengths (Yuliani *et al.*, 2008).

One example of this approach changing people's behaviour and attitudes was in a commercial fishery when we facilitated a two-workshop dialogue between fishers, regulators and environmentalists. In past encounters, fishers had felt under attack, not listened to and disrespected. As a result they resisted cooperating and there was significant tension between the different interests. At the outset we asked people to describe what a sustainable fishery would be like, identify what they were already doing to achieve that, and suggest ideas for what more they could do. This changed the whole tone of the workshops. Previously unnoticed innovations by the fishers were recognised as likely to be making a significant difference. With this, the dialogue became forward-focused and constructive. Stereotypes that each side had about the other began to break down so people worked on the shared challenge.

Quality delivery

Key eight: use skilled facilitation

We emphasise the importance of design but the quality, attitude and skills of individual facilitators can also make a significant difference (see Box 17). Key attributes include:

- knowledge, experience and skills facilitating group interaction and group process;
- providing an environment where participants can speak freely and safely;
- encouraging cooperative behaviour;
- maintaining confidentiality;
- knowing a range of facilitation techniques and tools and when to employ them;
- enabling equal opportunity, so strong voices don't dominate;
- ability to lead or be part of a team of facilitators;
- ability to handle the pressure of a live event.

The challenge is finding facilitators with these skills and a good understanding of the complexity of conservation conflicts.

Key nine: value all forms of knowledge

In a well-designed process, all types of knowledge, data, observation and opinion, including scientific, need to be shared, questioned and sifted through in a dialogue process. Good quality science and data are necessary to support deliberations and decision-making, as are other kinds of knowledge based on either stakeholders' direct experience or traditional knowledge. Experiential or traditional knowledge may be about the way things work, trends and changes, how the environment is used and who uses it, key locations for particular activities, and cultural meanings and feelings about the place. An important challenge for science-trained conservationists is to recognise the value of different forms of knowledge and different ways of relating to stakeholders (Table 17.5). Conservationists can also want to 'educate' other stakeholders. One conservationist said to me in frustration: 'you need to make them (other stakeholders) make "rational decisions in a logical way"'. Although traditional decision theory (and many conservationists) assume humans are 'rational optimisers', research now suggests that most decisions are made in our subconscious mind where heuristics are used and emotions may be the driving force (Douglas, 2011; Hammerstein

Table 17.5 *Shift in attitudes and behaviour needed in the culture of conservationists. (Adapted from Van Boven and Hesselink, 2003.)*

Change in attitude of conservationists	
From:	To:
Focus on scientific and technical knowledge	Many forms of knowledge are needed and used
Seeing other stakeholders as the problem	Realising we are all part of the problem
Seeing other stakeholders as a distraction and drain on resources	Realising they are a resource – of information, ideas and endeavour
Telling others what to do	Listening with an open mind
Pushing others to change	Working with others to agree change
Behaving as experts	Behaving as partners
Formal approaches	Informal and interactive approaches
Our ideas and solutions	The best, most workable ideas and solutions

and Stevens, 2012). While best-available science is needed to inform a discussion, too much data and information can be counterproductive and obscure the essential elements needed to make wise decisions (see Gladwell, 2005). It can also overwhelm people, causing them to lose confidence in their own wisdom and decision-making capacity.

We have found the optimum way of identifying the information needed for the deliberations is to ask stakeholders (including the scientists) to create a list for themselves. We then ask who has the information and can make it available; the group then helps to collate it. A considerable amount of information can result. When this happens we use a technique we developed called 'speed data' to enable participants to sift through and flag key information they want other participants to note. Where new information and research is needed, co-producing knowledge has promise. In this approach, scientists and resource users are partners in framing, undertaking and interpreting the research, which significantly increases acceptance of research findings.

Working with stakeholders and embedding a participation ethos

Key ten: have core values and ethics about participation

This key seems self-evident, but it is not always the case. Organisations can sponsor participation processes because they have to, not because they want to. Successful projects emphasise the importance of project teams (within sponsoring organisations) having or embedding a culture with strong core values and ethics about participation (Pound, 2009). Facilitating in a context where this does not exist is difficult. Without core values, organisations and individuals can inadvertently, or deliberately, 'use' stakeholders to their own ends, act in ways that are disrespectful of either the overall process or stakeholders, damage trust, and undermine the facilitator and the legitimacy of the process, which result in weak agreements. However, when stakeholders believe the process is legitimate, they are more likely to accept outcomes that are not their own first preference (NRC, 2008).

Key eleven: understand and be clear on different roles and responsibilities

There are three main roles in an effective process.

- The sponsoring body's role is to convene, fund and provide the secretariat for the process. They often also set the overall purpose and time frames for the discussion, and (particularly if they have statutory obligations) may set the level of influence others can have.
- The participants (which will include representatives of the sponsoring body) have a responsibility to share their knowledge and expertise, and negotiate a way forward.
- The facilitation team designs the process and function as an impartial third party, helping people discuss the topic in a staged, structured and focused way.

Difficulties arise when one exceeds their responsibilities. For example, if a facilitator gives a view on the content of the discussion, they lose their impartiality and the trust of participants. If a sponsoring body seeks to direct or control the facilitation team, and succeeds, it compromises their third-party role, professional ethics, and questions the legitimacy and fairness of the process. Clarity about roles is vital.

Key twelve: good practice participation becomes business as usual

Most organisations still do participation on a project by project basis. However, when organisations embed designed, facilitated dialogue as the way they work, there can be considerable long-term benefits.

In one situation, the level of opposition about designating a marine conservation site required a designed, consensus-building process to produce a management scheme. This was transformative and had multiple benefits for implementation (Pound, 2004b). After the main process the project continued to engage stakeholders in a variety of forums and held a facilitated workshop every other year to review progress. After six years the scheme came up for formal review and we facilitated another consensus process. The social capital established over the years was evident in the informal and friendly way people related to each other. Formal evaluations of participant satisfaction supported the view that there was good understanding and trust within the group (Pound, 2006).

For the environment and conservation sectors to work this way as standard practice requires greater understanding and skills capacity. Professional designers and facilitators are needed for the initial process, for formal progress reviews, or if things become more complex or tense. However, design and facilitation skills can be transferred to projects so they can continue to work this way when conducting smaller meetings and workshops.

The agencies, governments and NGOs that fund this approach also need to understand what constitutes good practice, as well as the role of designers and facilitators as independent third parties. This will change the way organisations fund and contract designer/facilitators, which is perhaps the most important key to success because the reality is that the quality of the participation process starts with the procurement process. Too often specifications have inadequate time frames, too few resources, and restricted scope and freedom for the third party to design a good process.

Conclusions

The benefits of consensus-building are not accidental. Keys to success include a well-designed and well-facilitated process, sufficient time and resources to work, and supportive organisations that respect other stakeholders and understand good practice and the challenges and benefits of working this way.

In terms of possible future trends, environmental organisations in the UK are beginning to think about co-production (e.g. Buddery and Shafique, 2013), and while some examples do exist they tend to be the result of serendipity rather than deliberate intent. We have been encouraging this approach for some years. From a research perspective there is more need to compare the relative merits of designed deliberative dialogue and other approaches that are not designed and/or not deliberative. Attempting such research would face significant challenges. Practitioners design each Stakeholder Dialogue to context so none are the same and each practitioner has their own design preferences. There is however a growing body of literature reviewing processes and identifying evidence that supports best practice including the value of design, deliberative processes and skilled facilitators (e.g. Warburton, 1997; Reed, 2008; Pound, 2009).

References

Arnstein, A. (1969). A ladder of citizenship participation. *J. Am. Inst. Planners*, 26, 216–233.

Bandura, A. (2007). Impeding ecological sustainability through selective moral disengagement. *Int. J. Innov. Sustain. Dev.*, 2, 8–35.

Brisbane Declaration. (2005). *The United Nations Brisbane Declaration on Community Engagement*. Brisbane, August 2005. International Conference on Engaging Communities. http://www.iap2.org.au/sitebuilder/resources/knowledge/asset/files/37/unbrisbanedeclarationcommunityengagement.pdf

Buddery, P. and Shafique, A. (2013). *Environmental Protection and Management – A Social Productivity Approach for SEPA and SNH*. London: RSA 2020 Public Services.

Bushe, G. R. (2011). Appreciative inquiry: theory and critique. In *The Routledge Companion to Organizational Change*, eds. D. Boje, B. Burnes and J. Hassard, pp. 87–103. Oxford: Routledge.

Douglas, K. (2011). Decision time: how you make up your mind. *New Sci.*, 212(2838), 38–41.

Gladwell, M. (2005). *Blink – The Power of Thinking without Thinking*. New York, NY: Little, Brown and Company.

Hammerstein, P. and Stevens, J. R. (2012). *Evolution and the Mechanisms of Decision Making*. Ernst Strüngmann Forum Report (vol. 11). Cambridge, MA: MIT Press.

Holmes, T. and Scoones, I. (2000). *Participatory Environmental Policy Processes: Experiences from North and South*. Institute for Developmental Studies Working Paper 113. Brighton: University of Sussex, Institute for Development Studies.

NRC. (2008). *Public Participation in Environmental Assessment and Decision Making*. Washington, DC: The National Academies Press.

Pound, D. (2004a). *Stakeholder Dialogue – A Good Practice Guide for Users – Including 'Making the Most of the Islands' An Example from the Isles of Scilly*. Bristol: Countryside Agency.

Pound, D. (2004b). From conflict to cooperation: Thanet Coast and Natura 2000. In *Achieving Environmental Objectives, The Role and Value of Communication, Education, Participation and Awareness (CEPA) in Conventions and Agreements in Europe*, eds. G. Martin-Mehers, D. Hamu, E. Auchincloss and W. Goldstein, pp. 122. Gland, Switzerland and Cambridge, UK: IUCN Commission on Education and Communications.

Pound, D. (2006). *The Nature of Our Coast: Helping People and Wildlife Thrive: NE Kent European Marine Site Management Scheme Review*. Peterborough: English Nature.

Pound, D. (2009). *Adopting Effective Stakeholder Engagement Processes to Deliver Regional Marine Protected Area (MPA) Network*. Natural England Commissioned Report, Number 008. Peterborough: Natural England.

Reed, M. S. (2008). Stakeholder participation for environmental management: a literature review. *Biol. Conserv.*, 141, 2417–2431.

Reed, M. S., *et al.* (2009). Who's in and why? Stakeholder analysis as a prerequisite for sustainable natural resource management. *J. Environ. Manage.*, 90, 1933–1949.

Studd, K. (2002). An introduction to deliberative methods of stakeholder and public participation. *English Nature Research Reports* 474. Peterborough: English Nature.

Sutton, R., Douglas, K. and Murphy, A. (2012). *Engaging Communities in Climate Change and Adaptation Measures. A Review of Relevant Psychological Science*. University of Kent: CC2150 InterReg Project.

Van Boven, G. and Hesselink, F. (2003). *Conservation Results by Managing Change, The Role of Communication, Education and Public Awareness*. Gland, Switzerland: IUCN Commission on Education and Communication.

Warburton, D. (1997). *Participatory Action in the Countryside – A Literature Review*. Cheltenham: Countryside Commission.

Yuliani, E. L., Adnan, H. and Indriatmoko, Y. (2008). *The Use of Appreciative Inquiry as a Tool for Enhancing Adaptive Capacity in Natural Resources Management*. Cheltenham: IASC Conference.

Box 17

Access management by local consensus: mediated negotiations in the UK Peak District

Roger Sidaway
Independent Facilitator, Edinburgh, Scotland

The long history to allow public access on privately owned, uncultivated land in Britain is core to the dispute over access to moorland in the Peak District National Park in Northern England, in particular the controversy over the perceived risk to wildlife from public access. Here I describe deliberations of the Peak District Access Consultative Group (PDACG) in the National Park that I facilitated (see also Sidaway, 2005).

Specific provisions in the National Parks and Access to the Countryside Act 1949 enable local authorities to enter into access agreements with landowners. The Peak District was the first English national park to be designated in 1951 and the national park authority has been assiduous in negotiating access agreements. Under the terms of an agreement, landowners and their tenants receive compensation for access and are able to close their moors to the public for up to 14 days a year for grouse hunting. Bye-laws regulating public access are enforced by Park personnel. A review of access agreements promoted the concept of limiting public access to 'corridors' and advocated research into the effects of access on the breeding populations of upland waders (Peak Park Joint Planning Board, 1992). However, the supposition that recreational disturbance was harming waders was controversial. The Ramblers' Association, an advocacy group for access, applied political pressure on the Board to increase access to all moorland areas, while conservation interests advocated stricter restrictions on access. In 1992, the Board consulted approximately 50 organisations on a draft Access Strategy which proposed establishing the PDACG, whereby principal interests would be represented and would address the controversy over recreational disturbance to waders.

A small task force (group) of nine was established with three members each representing landowners, hikers and conservation interests. There was resistance to the group being chaired by a member or officer of the Board, as the Board was not perceived to be neutral. However, it was agreed that the Board would provide logistical support and officers would attend PDACG meetings. The group would be led by a neutral facilitator and hold six-monthly meetings under a set of agreed procedures.

Following a meeting in June 1993, the PDACG met six times between September 1993 and June 1994. During the initial meetings, members made presentations on their organisations' interests about access to moorlands. Although their objectives differed, they identified mutual interests in conserving and managing moorlands within the national park. A list of considerations to include in local Access Management Plans was compiled, which the group applied during a pilot planning exercise.

The group then collaboratively prepared and agreed to a report in 1994, which was endorsed by all organisations. It advocated the following management strategy to the Board:

(a) accept a proposed moorland code of behaviour and establish a viable network of wildlife sanctuary areas for certain species and conservation management measures for others, and

(b) undertake an annual programme of monitoring (Peak District Access Consultative Group, 1994).

However, the Board in November 1994 welcomed the report but recognised that there was still much to be done. Most members of the group felt that they had a better understanding of others' points of view and that the initial approach of identifying common ground among stakeholders was more worthwhile than dwelling on conflicts that were only too familiar. The strengths of this mediation were its inclusiveness, fairness, and enabling of personal relationships to develop. The apparent main weakness was the long timescale.

Many procedural details were important. The size of the group was considered to be about right, but probably at the top limit to operate this kind of procedure. The composition of the group was seen to be fair, but there needed to be clearer arrangements about how other interested organisations could become involved. The group had opted for evening meetings, but these frequently over-ran time limits. With hindsight, most members agreed that full day or weekend meetings would have been more productive. The employment of an independent facilitator was seen as necessary at this stage, and it was the pilot planning exercise that grounded and focused an otherwise abstract discussion and gave some confidence that a collaborative approach (using mediation) could probably resolve most practical problems.

If the timescale appears extended, it should be seen in the context of a conflict occurring over 70 years. Yet final agreement was contingent on further work being done to resolve outstanding technical issues, such as the scale of wildlife sanctuary areas and on resources being obtained to monitor and evaluate a wide range of factors. Success gained at each stage was likely to reinforce confidence in collaborative planning as a way of resolving problems.

The Board did not act on the recommendations of the PDACG and with hindsight the failure to obtain, in advance, a firm commitment to do so was a fundamental weakness of the exercise. Various reasons were given for inaction, which certainly contributed to scepticism on the part of the Ramblers, at least, on the value of voluntary negotiations.

However, the Board subsequently decided to prepare a management plan using consensus building for the 542 ha North Lees estate which it owns and that includes Stanage Edge, an important site for rock climbing, as well as farmland and moorland of high recreational value. The Stanage Forum was established in 2000 with open membership and a nominated 18-member steering group whose meetings were independently facilitated. The group met 22 times, reporting back to the forum at significant stages in the plan's preparation. The plan was publicly launched at the forum's fourth meeting in September 2002.

References

Peak District Access Consultative Group (1994). *Report of the Peak District Access Consultative Group to the Peak Park Joint Planning Board*. Bakewell, Derbyshire: PPJPB.

Peak Park Joint Planning Board (1992). *Strategy for Access to Open Country: A Balanced Approach*. Bakewell, Derbyshire: PPJPB.

Sidaway, R. (2005). *Resolving Environmental Disputes*. London: Earthscan.

CHAPTER EIGHTEEN

Conservation conflict transformation: the missing link in conservation

FRANCINE MADDEN
Human–Wildlife Conflict Collaboration
and
BRIAN McQUINN
Human–Wildlife Conflict Collaboration and Oxford University

Conservation conflicts are an increasing threat to many species of wildlife around the world (Madden, 2004; Michalski *et al.*, 2006). As we have seen earlier in this book, conservation conflicts often serve as proxies for underlying social conflicts, including struggles over group recognition, empowerment, identity and status (Coate and Rosati, 1988; Burton, 1990; Satterfield, 2002; Madden, 2004; Dickman, 2010; Madden and McQuinn, 2014). Such complexity undermines the receptivity of diverse stakeholders to find common ground that would benefit both people and wildlife. As a result, conservation goals are adversely impacted (Madden, 2004; Redpath *et al.*, 2013). Moreover, conservationists' lack of explicit capacity to transform these social conflicts further compromises the broader goals of conservation and limits their ability to find resolution and commitment on the substantive issues. Even where stakeholder engagement is acknowledged, recommended or conducted (e.g. Treves *et al.*, 2009; Barlow *et al.*, 2010, Redpath *et al.*, 2013; Box 9), such well-meaning efforts often do not address the full suite of underlying social and psychological conflicts at play, nor do they create the necessary social conditions for positive, transformative change. For instance, if the act of bringing stakeholders together to address wildlife impacts or conservation solutions does not also provide a sufficient process for genuinely improving relationships among individuals, building trust and empowering people early, increasing equitable and inclusive decision-making among stakeholders, even palatable decisions on substantive issues may ultimately be rejected by key stakeholders.

In our work we have adapted an approach to conservation from a niche within peace-building: conflict transformation (CT). At its core, CT conceptualises current disputes as opportunities to constructively change the underlying relationships, decision-making processes and social systems that can serve as a

Conflicts in Conservation: Navigating Towards Solutions, ed. S. M. Redpath, R. J. Gutiérrez, K. A. Wood and J. C. Young. Published by Cambridge University Press.

foundation for sustainable conservation action (Lederach *et al.*, 2007; Madden and McQuinn, 2014). In this sense, a CT orientation recognises conflict as a natural, and potentially constructive and creative, part of human interaction. Hence, the transformation of conflict implies that the goal is not necessarily to end conflict, but to harness its ebb and flow as a means to sustain dynamic problem-solving within a given context (Deutsch, 1973; Lederach, 2003).

Conservation conflict transformation (CCT)

We summarise the following two models as a practical orientation in CCT: the Levels of Conflict Model and the Conflict Intervention Triangle Model (Moore, 1986; Walker and Daniels, 1997; Canadian Institute for Conflict Resolution, 2000; Madden and McQuinn, 2014). The Levels of Conflict Model is an analytical tool that portrays the complexities and intensities of a conservation conflict. The Conflict Intervention Triangle Model depicts the three causes of conflict and enables a practitioner to think, conceptually, about project design. Both models help deepen understanding of conflict dynamics and provide a comprehensive, conceptual framework for considering alternative conservation responses to a conflict.

Levels of Conflict Model: an analytical model

According to this model, conflict can occur at three levels: dispute, underlying and identity-based (Canadian Institute for Conflict Resolution, 2000). We describe them as levels to indicate the increasing complexity and gravity of the conflict at each deeper level.

Dispute is the first level of conflict and is the observable expression of the conflict (see Fig. 18.1). Disputes are what people talk about and relate to the physical or material impacts. For example, people may be in dispute over the loss of crops to geese (Box 7) or loss of livestock to lions (Box 2). Every conflict contains a dispute. Conflicts can occur at the dispute level, but often the dispute is a symptom of a deeper conflict. A focus solely on settling disputes explains, in part, why conservationists are often perplexed when a problem is apparently 'solved' (Chapter 8), yet the conflict remains or even intensifies.

The underlying level of conflict is driven by a history of unresolved disputes. Thus, the settlement of past disputes – decisions, actions, projects, or interactions – may heighten emotional reaction, intensity and frustration around current disputes. Adding historical grievances infuses current disputes with additional complexity (Chapter 4). Individuals in a conflict may further mask the significance of history by focusing only on the current dispute. They may also view the current event as an opportunity to rectify past wrongs. Hence, when underlying conflict is also present in a current dispute, one can anticipate more reactions and repercussions that distort the dynamics of a conflict. Unfortunately, it is nearly impossible to avoid underlying conflict in conservation because conflicts are often of long duration.

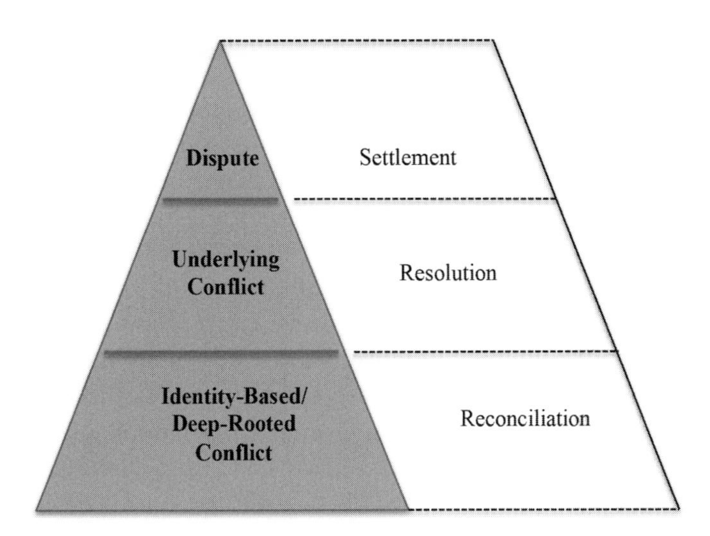

Figure 18.1 Levels of conflict (and corresponding processes). This model depicts the levels of conflict according to depth, source and likely intractability, which is potentially evident in any conflict scenario (Canadian Institute for Conflict Resolution, 2000: 73). *Source*: The Canadian Institute for Conflict Resolution (2000).

The third level of conflict is identity-based or deep-rooted conflict. Identity-based conflict is evident when one party makes assumptions or holds prejudices about another party. Individuals within parties may never have met or they may not have personal historical links, yet they may typecast their adversaries.

In conservation conflicts, as in other types of conflict, deep prejudices and disparaging assumptions may be involved, leading to a perceived threat to one's group identity or way of life. For example, US laws and conservation actions for wolves in the American Midwest are perceived by ranchers and hunters to be a direct threat to their way of life (Box 15). This perception is less about lost revenue because of wolf predation on livestock as it is about how the wolf has come to symbolise oppression by authorities seeking to protect wolves. Likewise, wildlife conservation organisations are often criticised because of a perception that they care more for wildlife than for people (Madden, 2004). Conversely, the use of lethal control to manage wolves may be perceived as an attack on the values held by animal welfare groups. The result of perceived threats on identity is that groups then focus on protecting identity, which constricts their receptivity for finding common ground and shared solutions.

From the platform of CT, underlying and deep-rooted conflicts are less about wildlife and more about people's recognition and respect, cultural and social identity, fear of a loss of control, anger over historical grievances, and trust and exclusion. A conflict that starts as a dispute may over time become an identity-level conflict, as the disputants align their identities with their positions in the dispute (Lederach, 1997).

Identity-based conflict is intense, complex and pervasive in conservation. It may involve power imbalances, non-material identity needs, history, cultural beliefs, aspirations for group recognition, respect, autonomy, or basic dignity. Displays of such conflicts may appear 'irrational' compared to the disputes being negotiated.

The processes needed to address each level of conflict are quite different. The dispute model uses the term 'settlement' to describe the relatively simple and straightforward process that is required to solve conflict at the dispute level. Disputes in civil society are often settled in courts using a rights-based system within a set of legal codes (see Chapter 8). Conservation groups often use lawsuits strategically to secure protections for endangered species and for changing government management actions. Similarly, governments use laws and law enforcement as a means to safeguard compliance. That said, compliance with a 'settlement' does not mean the underlying conflict is resolved, if underlying level conflicts exist. This assumption lies at the core of the CCT approach. If underlying and deep-rooted conflicts are not addressed, settlements are often short-lived because those involved in the conflict tend to use (or create) new circumstances to rectify historically alleged wrongs.

The Levels of Conflict Model uses the expression 'resolution' to explain the process used to solve underlying conflicts and 'reconciliation' to suggest the change in process required to transform identity-based conflicts. Many current conflict resolution, negotiation and participatory planning processes address the dispute and to some degree the underlying level of conflict, but these processes typically ignore the deeper identity-based level that may drive the creation of additional future disputes (Lederach, 2003).

In reality, most conservation conflicts involve identity-based conflict and consequently need 'reconciliation' processes to transform conflict into constructive, genuine and sustained commitment to conservation actions. Typical conflict resolution processes are designed to address the immediate dispute and to some extent the underlying conflicts, yet they do not often integrate a full treatment of the human social and psychological needs involved in identity-based conflict. In fact, research indicates that when deeply held beliefs are in question, the force of opposition can actually intensify rather than lessen when material inducements are presented as a means to settlement (Ginges *et al.*, 2007).

Whereas disputes lean toward the concrete, measureable and easily described, identity-based conflicts are typically indistinct, not material and not always expressed. Where they are expressed, they are often not fully accounted for in decision-making processes. Further complicating matters, identity-based conflicts may be expressed as dispute-level conflicts for several reasons. First, articulating the identity-level conflict as a dispute gives concrete attention and clarity to a group's concern (Rothman, 1997). Second, it is easier or more socially tolerable to speak of identity-based conflict in terms of material losses rather

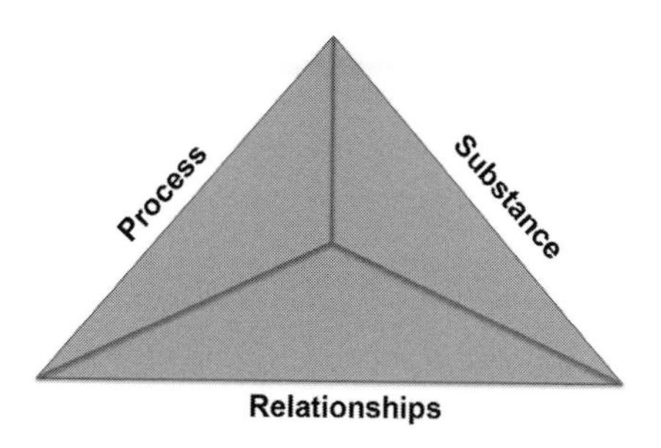

Figure 18.2 Conflict Intervention Triangle Model. This model offers a conceptual framework for understanding the sources of conflict, as well as focal points for intervention and change. *Sources:* Moore (1986) and Walker and Daniels (1997: 22).

than social or psychological ones. Finally, the goal of conservation is to protect wildlife or ecosystems, which tends to direct discussions toward the wildlife or ecosystems and away from impacts that conservation decisions may have on people (Madden, 2004; Dickman, 2010).

Many conflicts over wildlife conservation are proxies for social conflicts concerning a group's desire to meet its non-material needs. Such needs often do not seem to be related to conservation goals, so there is temptation to disregard these dynamics. However, unless these undercurrents are acknowledged and dealt with, conservation accomplishments are typically temporary or incur negative costs. The CCT framework allows a more complete examination of these deeper-rooted conflicts to better understand what often seems to be 'irrational' behaviour undermining conservation efforts (Madden and McQuinn, 2014). Only by addressing the relationships and conflict dynamics underlying conservation efforts can sustainable coexistence goals be realised. That said, the changes to process design suggested by models like this have yet to be systematically integrated into conservation planning. Indeed, decision-making and thorough analysis of social conflict dynamics are not standardised within conservation programme design (Ellis *et al.*, 2005).

The Conflict Intervention Triangle Model: designing interventions

While the Levels of Conflict Model provides insights into the depth and complexity of conflict in a given context, the Conflict Intervention Triangle Model offers a conceptual framework for understanding where action is needed to address the conflict. Process, relationships and substance are all sources of conflict, as well as focal points for intervention and change (Moore, 1986; Walker and Daniels, 1997; Madden and McQuinn, 2014). All three components must be analysed and strengthened to ensure robust conservation programmes (Fig. 18.2). Taken together, the Levels of Conflict Model provides insights into the extent to which changes are needed in the process and relationships aspects of a conflict intervention.

'Substance' includes all the material or physical concerns of people in a conflict and is, typically, a strength and focus of conservation programmes.

Substance issues relate to dispute-level conflict described previously. These might include technical or economic solutions to prevent or mitigate wildlife damage. Logically, conservationists tend to focus on substance issues in a conflict, often ignoring or discounting other important sources of conflict that are motivated by process and relationship concerns.

'Process' elements relate to decision-making power and how and by whom it is employed. Often, parties might initially approve a specific solution, but if they do not feel sufficiently included and valued in the decision-making process, they may nonetheless eventually rebuff the solution, even if it is a solution that satisfies their substantive concerns.

The CT framework suggests conservationists should re-evaluate what successful stakeholder engagement and decision-making actually require. A re-evaluation would affect a variety of traditional approaches to conflict resolution, management and planning (Sites, 1990; Ralls and Starfield, 1995; Lindstrom and Smith, 2001; Nie, 2003; Llambí *et al.*, 2005; Raik *et al.*, 2005; Coleman *et al.*, 2011). While many of these methodologies require stakeholder engagement, they often move straight to group decision-making or they impose a formulaic process upon a set of relationships that is not yet ready for this type of engagement.

A process that misses the deeper-rooted level conflict may actually aggravate the conflict. Indeed, many efforts overlook or fail to account for what is really at stake for participants and thus they perpetuate among stakeholders their lack of genuine authority over problem definition and decision-making (Innes and Booher, 1999).

Lovallo and Sibony (2010) reviewed 1048 business decisions over 5 years, and determined that 'process mattered more than analysis in determining the quality of outcomes, by a factor of six' (Lovallo and Sibony, 2010: 6). Conservationists and governments often oppose changes to decision-making processes that require giving up some decision-making control. Yet, involving other stakeholders in the decision-making process may expand the range of acceptable solutions by addressing these deeper-rooted conflict dynamics and empowering community decision-making. For example, conservationists that implemented a community anti-poaching programme in northern Mozambique realised that if the community did not support these efforts, the programme would fail. And in fact, the project area was experiencing snaring of wildlife and poaching of elephants, despite the publicly vocalised commitment by the community to stop these activities. In a moment of crisis (and opportunity), the conservationists changed their orientation to community engagement and turned over significant control over decision-making authority, all of which resulted in a deeper commitment by the community to stop snaring and poaching. In fact, for the first time in the history of the protected area, community members voluntarily turned in snares to the project (Beggs, 2012). Such anecdotal reports are reinforced by more rigorously evaluated, longer-term applications of this approach within the peace-building field (Lederach, 1997, 2003; Anderson *et al.*, 2003; Lederach *et al.*, 2007).

'Relationships' comprise the third side of the conflict intervention triangle. Indeed, the quality of a relationship or the level of respect, trust and genuine commitment that exists between stakeholders can be a source of contention. In our experience, this cause of conflict is often overlooked, evaded or treated too casually by those who superficially mark other groups as 'partners in conservation' when that relationship is still wrought with feelings and evidence of their mutual distrust. Stakeholders will undervalue or even sabotage solutions if they do not also meet underlying social and psychological needs, including those met through strong, positive relationships (Burton, 1990; Sites, 1990; Satterfield, 2002).

Limits of current approaches

The influence of human social and psychological needs as well as personal values in conflict is unmistakable, even when not identified. Even conservationists seldom scrutinise how their own values and beliefs either shape their worldview or are perceived (either implicitly or explicitly) by others involved in a conflict (Satterfield, 2002; Ellis *et al.*, 2005). As a result, non-material social and psychological needs and values are seldom explicitly included in conservation decision-making (Ellis *et al.*, 2005). Yet, solutions are likely to be more successful when such issues are fully and appropriately integrated (Lederach, 1997, 2003; Coleman and Deutsch, 2012).

Gaps in conservation capacities

Conservation professionals characteristically and understandably enter the field because of their desire to protect wildlife and wild nature, not people. While educational opportunities in the 'human dimensions' of conservation have noticeably increased over the years (Decker *et al.*, 2012), even human dimensions practitioners often lack the hands-on 'know-how' to transform social conflict into constructive opportunities for coexistence (Sample *et al.*, 1999; Bonine *et al.*, 2003; Bruskotter and Shelby, 2010). Most conservationists do not receive practical training or possess the capacity to navigate destructive social dynamics, reconcile deep-rooted conflicts, foster common ground among people who may be deeply antagonistic to conservation goals and create constructive change processes that will lead to better, more durable solutions. Even where practitioners or organisations create success in one instance, they struggle and often fail to replicate their success in other contexts (Bonine *et al.*, 2003). Where curricula include participatory planning or conflict management, the treatment of conflict often goes no deeper than an effort to identify 'common interests' that may be obscured or unachievable if more intractable social conflict exists (Lederach *et al.*, 2007; Coleman *et al.*, 2011). Such processes may not only prevent the potential for a reconciliation of conflict, but also may fuel hostilities (Sites, 1990; Ginges *et al.*, 2007).

Conflict transformation versus resolution: advancing the field

Views differ as to whether conflict transformation as a practice belongs within the spectrum of conflict resolution approaches that range from dispute settlement at one end to conflict transformation at the other (Ramsbotham *et al.*, 2011: 9), or whether CT offers a distinct undertaking from that of traditional conflict resolution approaches (Francis, 2002; Lederach, 2003). Either way, both sides of the debate recognise that conflict transformation engages conflict 'at the deepest level' (Ramsbotham *et al.*, 2011). It is when practitioners consciously and consistently consider this deeper level that we see the greatest need for change in our field. As a result, we are inclined to make the distinction between conflict transformation and other forms of conflict resolution, management and planning.

According to Lederach (2003), conflict transformation has some similarities to conflict resolution, but it differs in several key respects. Conflict resolution typically addresses the immediate issue at hand and, in some situations, may engage limited components of underlying conflict, while too often not considering the more complex identity-level conflict. Yet, this more complex, deep-rooted conflict is prevalent in wildlife conservation conflicts and may be the ultimate source of intractability when dealing with species like wolves in the American Midwest (Naughton-Treves *et al.*, 2003; Box 15). Conflict transformation offers a different vision and evolution in approach: it conceptualises immediate disputes as opportunities to analyse and transform the underlying relationships and social systems that have the ability to adversely impact conservation efforts (Lederach *et al.*, 2007). Conflict transformation is a process that is used to recognise and address the immediate needs of situations, while simultaneously allowing incorporation of the underlying structures within a system that generate new conflicts tomorrow. Thus, because both conflict resolution and conflict transformation operate on a mid- to long-term time scale, we consider CT to be a more appropriate framework in which to address conservation conflict and its long-term goals. Moreover, we feel that CT more consciously upholds the recognition that conflict has the potential to be an indispensable and valued aspect of human resourcefulness and innovation (Deutsch, 1973).

Conflict transformation is a process that allows people to address the current, visible problems in conservation while concurrently allowing employment of processes that focus on both the substance and the relationship aspects of the conflict, all with the goal of enabling long-term constructive change in the social system. The deeper, less-visible origins of conservation conflict are the energy force driving the creation of what is visible and tangible on the surface. Indeed, genuine, positive, sustainable change is impossible until the underlying structures are constructively transformed.

Conclusion

Given our experience adapting conflict transformation to conservation and building capacity among diverse stakeholders, we argue that a significant limitation of current conservation approaches is a lack of widespread capacity and integration of conservation conflict transformation. Extensive anecdotal evidence suggests this approach is successful, even where traditional methods have failed, but we recognise the urgent need to test this approach with independent data.

The goal of conservation conflict transformation is to attain sustainable positive coexistence, not merely temporary solutions or short-term compliance. CT aims to reconcile negative relationships between conflicting stakeholders and transform the enabling environment, impacting the viability of conservation efforts. CT empowers stakeholders in conservation to work collaboratively through constructive change processes to foster durable and equitable conditions for coexistence among parties in conflict. Moreover, CT views coexistence as continuously evolving and developing quality of relationships. The continual engagement that sustains constructive and positive relationships and decision-making processes will enable more effective and creative adaptation of conservation programmes to changes in the social and ecological systems. To that end, a thorough analysis of conflict dynamics is an essential first step to avoid unintended consequences and foster social conditions that support decision-making towards sustainable conservation.

We suggest that by using CCT, conservation may begin to reorient its own understanding of what it means to achieve coexistence. In that effort, we may begin to transform the underlying structures of conflict and support positive changes in attitudes, relationships and behaviours, leading to constructive social change and sustainable coexistence. Indeed, the future of wildlife conservation depends on diverse people working collaboratively and cohesively in an atmosphere of mutual respect, trust, tolerance and commitment.

Acknowledgements

We would like to thank the over 500 conservation leaders and practitioners, farmers, ranchers, government officials, NGO practitioners, development workers and other conservation stakeholders since 2008 that have participated in the Human–Wildlife Conflict Collaboration's (HWCC) CT capacity building and conflict intervention efforts. These diverse individuals and groups have implemented CT in innovative and impactful ways to successfully transform conflict in their projects and their communities.

References

Anderson, M. B., Olson, L. and Doughty, K. (2003). *Confronting War: Critical Lessons for Peace Practitioners*. Cambridge, MA: The Collaborative for Development Action.

Barlow, A. C. D., Greenwood, C. J., Ahmad, I. U. and Smith, J. L. D. (2010). Use of an action-selection framework for human–carnivore conflict in the Bangladesh Sundarbans. *Conserv. Biol.*, 24, 1338–1347.

Beggs, C. (2012). Wildlife Conflict Management: In Practice in Mozambique. http://wildnet.org/sites/default/files/WCN_Newsletter_Fall12.pdf

Bonine, K., Reid, J. and Dalzen, R. (2003). Training and education for tropical conservation. *Conserv. Biol.*, 17, 1209–1218.

Bruskotter, J. T. and Shelby, L. B. (2010). Human dimensions of large carnivore conservation and management. *Hum. Dimens. Wildl.*, 15, 311–314.

Burton, J. (1990). *Conflict: Basic Human Needs*. New York, NY: St. Martin's Press.

Canadian Institute for Conflict Resolution. (2000). *Becoming a Third-Party Neutral: Resource Guide*. Ottawa, Ontario: Ridgewood Foundation for Community-Based Conflict Resolution.

Coate, R. A. and Rosati, J. A. (1988). *The Power of Human Needs in World Society*. Boulder, CO: Lynne Rienner Publishers.

Coleman, P. T. and Deutsch, M. (2012). Psychological components of sustainable peace: an introduction. In *Psychological Components of Sustainable Peace*, eds. P. T. Coleman and M. Deutsch, pp. 1–14. New York, NY: Springer.

Coleman, P. T., Vallacher, R., Bartoli, A., Nowak, A. and Bui-Wrzosinska, L. (2011). Navigating the landscape of conflict: applications of dynamical systems theory to addressing protracted conflict. In *The Non-Linearity of Peace Processes: Theory and Practice of Systemic Conflict Transformation*, eds. D. Körppen, N. Ropers and H. J. Gießmann, pp. 39–56. Opladen /Farmington Hills: Barbara Budrich Verlag.

Decker, D. J., Riley, S. J. and Siemer, W. F. (2012). Human dimensions of wildlife management. In *Human Dimensions of Wildlife Management*, second edition, eds. D. J. Decker, S. J. Riley and W. F. Siemer, pp. 3–14. Baltimore, MD: John Hopkins Press.

Deutsch, M. (1973). *The Resolution of Conflict*. New Haven, CT: Yale University Press.

Dickman, A. J. (2010). Complexities of conflict: the importance of considering social factors for effectively resolving human–wildlife conflict. *Anim. Conserv.*, 13, 458–466.

Ellis, C., Koziell, I., McQuinn, B. and Stein, J. (2005). Approaching the table: transforming conservation-community conflict into opportunity. In *Beyond the Arch: Community and Conservation in Greater Yellowstone and East Africa. 7th Biennial Scientific Conference on the Greater Yellowstone Ecosystem*, ed. A. Wondrak Biel, pp. 82–95. Mammoth Hot Springs: Yellowstone National Park.

Francis, D. (2002). *People, Peace, and Power: Conflict Transformation in Action*. London: Pluto Press.

Ginges, J., Atran, S., Medlin, D. and Shikaki, K. (2007). Sacred bounds on rational resolution of violent political conflict. *Proc. Natl Acad. Sci. USA*, 104, 7357–7360.

Innes, J. E. and Booher, D. E. (1999). Consensus building and complex adaptive systems. *J. Am. Plann. Assoc.*, 65, 412–423.

Lederach, J. P. (1997). *Building Peace: Sustainable Reconciliation in Divided Societies*. Washington, DC: United States Institute of Peace.

Lederach, J. P. (2003). *The Little Book of Conflict Transformation*. Intercourse, PA: Good Books.

Lederach, J. P., Neufeldt, R. and Culbertson, H. (2007). *Reflective Peacebuilding: A Planning, Monitoring and Learning Toolkit*. Notre Dame, IN: The Joan B. Kroc Institute for International Peace Studies, University of Notre Dame and Catholic Relief Services.

Lindstrom, M. J. and Smith, Z. A. (2001). *The National Environmental Policy Act: Judicial Misconstruction, Legislative Indifference, and Executive Neglect*. College Station, TX: Texas A&M University Press.

Llambí, L. D., *et al.* (2005). Participatory planning for biodiversity conservation in the High Tropical Andes: are farmers interested? *Mt. Res. Dev.*, 25, 200–205.

Lovallo, D. and Sibony, O. (2010). *The Case for Behavioral Strategy*. Boston, MA: McKinsey.

Madden, F. (2004). Creating coexistence between humans and wildlife: global perspectives on local efforts to address human–wildlife conflict. *Hum. Dimens. Wildl.*, 9, 247–257.

Madden, F. and McQuinn, B. (2014). Conservation's blind spot: a case for conflict transformation in wildlife conservation. *Biol. Cons*, 178, 97–106.

Michalski, F., Boulhosa, R. L. P., Faria, A. and Peres, C. A. (2006). Human–wildlife conflicts in a fragmented Amazonian forest landscape: determinants of large felid depredation on livestock. *Anim. Conserv.*, 9, 179–188.

Moore, C. W. (1986). *The Mediation Process: Practical Strategies for Resolving Conflict*. Hoboken, NJ: Wiley.

Naughton-Treves, L., Grossberg, R. and Treves, A. (2003). Paying for tolerance: rural citizens' attitudes toward wolf depredation and compensation. *Conserv. Biol.*, 17, 1500–1511.

Nie, M. A. (2003). *Beyond Wolves: The Politics of Wolf Recovery and Management*. Minneapolis, MN: University of Minnesota Press.

Raik, D. B., Lauber, T. B., Decker, D. J. and Brown, T. L. (2005). Managing community controversy in suburban wildlife management: adopting practices that address value differences. *Hum. Dimens. Wildl.*, 10, 109–122.

Ralls, K. and Starfield, A. M. (1995). Choosing a management strategy: two structured decision-making methods for evaluating the predictions of stochastic simulation models. *Conserv. Biol.*, 9, 175–181.

Ramsbotham O., Woodhouse, T. and Miall, H. (2011). *Contemporary Conflict Resolution*. Cambridge: Polity.

Redpath, S. M., *et al.* (2013). Understanding and managing conservation conflicts. *Trends Ecol. Evol.*, 28, 100–109.

Rothman. J. (1997). *Resolving Identity-Based Conflict: In Nations, Organizations, and Communities*. San Francisco, CA: Jossey-Bass Publishers.

Sample, V. A., Ringgold, P. C., Block, N. E. and Giltmier, J. W. (1999). Forestry education: adapting to the changing demands on professionals. *J. Forest.*, 97, 4–10.

Satterfield, T. (2002). *Anatomy of a Conflict: Identity, Knowledge and Emotion in Old-growth Forests*. Michigan: Michigan State University Press.

Sites, P. (1990). Needs as analogues of emotions. In *Conflict: Human Needs Theory*, ed. J. W. Burton, pp. 7–33. New York, NY: St. Martin's Press.

Treves, A., Wallace, R. B. and White, S. (2009). Participatory planning of interventions to mitigate human–wildlife conflict. *Conserv. Biol.*, 23, 1577–1587.

Walker, G. and Daniels, S. (1997). Foundations of natural resource conflict: conflict theory and public policy. In *Conflict Management and Public Participation in Land Management*, eds. B. Solberg and S. Miina, pp. 13–36. Joensuu, Finland: European Forest Institute.

Box 18

Conservation conflicts from livestock depredation and human attacks by tigers in India

Mysore D. Madhusudan
Nature Conservation Foundation, 3076/5, 4th Cross, Gokulam Park, Mysore 570002, India

Over the last century, the tiger *Panthera tigris* has declined in global distribution and abundance. Over half of the world's 3600 remaining tigers are believed to occur in India. Despite receiving the highest protection under Indian law, tigers and their habitats continue to come under intense anthropogenic pressures. They share their range with over 1.2 billion people and 0.5 billion livestock, many of whom are still forest-dependent yet live in a region with a fast-growing economy hungry for natural resources. One of the main consequences of such overlap between tigers and people is increased interaction between them, which results in conservation conflicts.

The chief aspects of this conflict are the costs imposed via livestock depredation and attacks on humans. When tigers are killed in reprisal for such losses, conflict occurs between those interested in tiger conservation and those negatively affected by tigers.

A key conflict management issue is that reliable information on people–tiger interactions has proven difficult to obtain. Although this occurs for many reasons, available research suggests considerable spatio-temporal variation in the occurrence and intensity of these interactions. Another linked issue is when patterns of interaction are poorly documented, understanding the processes that underlie interactions becomes difficult. While ecology offers insight into the 'ultimate' drivers of these interactions (Madhusudan and Mishra, 2003), understanding various 'proximate' factors driving interactions remains a vital knowledge gap, especially for management. Instead, there is much conjecture about the mechanisms of conflict. This often leads to contradictory claims about its causes from various parties involved and disagreements on the appropriate management actions. For example, when a tiger occurs in a human-dominated area, the belief that tigers are obligate forest species that invariably avoid contact with people is grounds for delaying intervention, whereas the notion that they can adapt to human-dominated areas may necessitate decisive interventions. Thus, generating reliable knowledge can improve opportunities for conflict management.

There is also a need to understand economic and livelihood consequences of tiger depredation. Researchers have estimated material and monetary losses from tiger depredation, and also evaluated the effectiveness of compensation schemes to ameliorate conflict (Madhusudan, 2003; Karanth *et al.*, 2013). The transition from rural production systems to market-linked economies also poses challenges to conflict resolution. For example, people seem to perceive losses more negatively when their production is intended for markets rather than for subsistence (Owen, 2013).

While ecological and economic knowledge is important, it is not uncommon to encounter complex cultural and religious traditions that enable communities across India that experience serious losses of livestock, or even loss of human lives, not only to perceive no negative interaction with tigers, but also to venerate them (e.g. Jalais, 2011). People in rural communities are often willing to coexist with large carnivores (Athreya *et al.*, 2013). While such apparent tolerance is not uniform or assured, it creates 'cultural spaces' for species such as the tiger that are as important to secure as 'ecological spaces'. To use economic measures such as compensation schemes without eroding intrinsic cultural or ethical motivations of people to coexist with tigers is a peculiar challenge for Indian conservation.

India's public institutions also play a vital role in management of tiger–human interactions. Forest departments enforce India's traditionally preservationist wildlife laws and have often imposed restrictions on access and use of forest resources by local communities. This has exacerbated conflict between communities who want forest access and the government. Judicial rulings that wildlife is government property (Ganjapure, 2012) can dislodge the view that conflict resolution is a shared responsibility and, instead, deepen conservation conflicts.

Yet, despite the ecological, economic and cultural complexity, India's approach to this conservation conflict is rather uniform. It advocates approaches ranging from relocation of human settlements from tiger reserves to the lethal control of tigers, especially those involved in persistent attacks on humans, in human-dominated landscapes (Karanth and Gopal, 2005). While such approaches may be necessary, they are certainly not sufficient, especially given newer challenges and opportunities. For example, market economies risk eroding cultural tolerance of tigers, but they also bring new instruments such as livestock insurance programmes that could better compensate livestock losses. Although governments may alienate rural people through exercise of authority, they can also be a positive force, such as when the Madhya Pradesh government provided a legal guarantee (MP Public Services Guarantee Act, 2010) for time-bound disbursement of compensations for livestock killed by tigers. Reducing tiger–human interactions and associated conflicts, after all, is as much a challenge to creative ability as it is a test of management capability.

References

Athreya, V., Odden, M., Linnell, J. D. C., Krishnaswamy, J. and Karanth, U. (2013). Big cats in our backyards: persistence of large carnivores in a human dominated landscape in India. *PLoS ONE*, 8, e57872.

Ganjapure, V. (2012). Wild animals government property: High Court. *Times of India*, Nagpur edition, 13 April.

Jalais, A. (2011). *Forest of Tigers: People, Politics and Environment in the Sundarbans*. London: Routledge.

Karanth, K. K., Gopalaswamy, A. M., Prasad, P. K. and Dasgupta, S. (2013). Patterns of human–wildlife conflicts and compensation: insights from Western Ghats protected areas. *Biol. Conserv.* 16, 175–185.

Karanth, K. U. and Gopal, R. (2005). An ecology-based policy framework for human–tiger coexistence in India. In *People and Wildlife: Conflict or Coexistence*?, eds. R. Woodroffe, S. Thirgood and A. Rabinowitz, pp. 373–387. Cambridge: Cambridge University Press.

Madhusudan, M. D. (2003). Living amidst large wildlife: livestock and crop depredation by large mammals in the interior villages of Bhadra Tiger Reserve, south India. *Environ. Manage.*, 31, 466–475.

Madhusudan, M. D. and Mishra, C. (2003). Why big, fierce animals are threatened: conserving large mammals in densely populated landscapes. In *Battles Over Nature: Science and the Politics of Conservation*, eds. V. Saberwal and M. Rangarajan, pp. 31–55. New Delhi: Permanent Black.

Owen, N. (2013). *Conservation, Conflict and Costs: Living with Large Mammals in the Nilgiri Biosphere Reserve, India*. Leeds: University of Leeds.

CHAPTER NINETEEN

Legislated collaboration in a conservation conflict: a case study of the Quincy Library Group in California, USA

R. J. GUTIÉRREZ
University of Minnesota

ANTONY S. CHENG
Colorado State University

DENNIS R. BECKER
University of Minnesota

SCOTT CASHEN
California

DAVID GANZ
United States Aid for International Development

JOHN GUNN
Spatial Informatics Group

MICHAEL LIQUORI
Soundwatershed

AMY MERRILL
Stillwater Sciences

D. S. SAAH
Spatial Informatics Group

and

WILLIAM PRICE
Pinchot Institute for Conservation

Nearly 258 million ha (28%) of the United States is publicly owned land that is managed by federal government agencies. For example, the US Department of Agriculture's Forest Service (USFS) manages over 77 million ha of national forests and grasslands for the benefit of the American public. Given its legal directive to manage multiple uses, it is not surprising that conflicts arise among stakeholders over how this land should be used (Lansky, 1992). The USFS has much

Conflicts in Conservation: Navigating Towards Solutions, ed. S. M. Redpath, R. J. Gutiérrez, K. A. Wood and J. C. Young. Published by Cambridge University Press.

discretion in how land is managed, yet must often balance conflicting values of public use and benefit (Nie, 2004). As national priorities, social preferences and public awareness of national forest goods, services and values have changed over time, USFS managers have faced increased pressure to balance consumptive uses with the need for environmental protection. Competing stakeholder demands coupled with increased environmental risks (wildfires, tree diseases and insect epidemics) have resulted in an escalating conservation conflict that is manifested in administrative appeals, lawsuits and a growing distrust of the agency.

Over time, the USFS has embraced new directions and management paradigms to reduce conflict. Some of these have been ecosystem management, adaptive management and now collaborative management (e.g. Holling, 1978; Maser, 1988; Franklin, 1992; Boyce and Haney, 1997; Wondolleck and Yaffee, 2000; Brown *et al.*, 2004). These approaches reflect changing societal values, political pressures and new scientific information.

A persistent conflict has been the logging of trees in national forests and related impacts on forest ecosystems (Lansky, 1992). The USFS' timber sale programme has supported jobs and community stability through economic development. Logging has also been a mechanism to reduce the risk of wildfire by reducing tree density (fuel for fires) and vertical stand diversity ('ladder' fuels; North *et al.*, 2009). However, logging can also negatively affect forest integrity, watershed quality, wildlife, aesthetic and spiritual values of forests (Satterfield, 2002; North *et al.*, 2009). Differences of opinion also exist within the USFS about the appropriate role and level of logging to achieve multiple use and ecosystem management goals. Thus, a potentially intractable situation revolves around logging.

By the 1980s, conflicts among forest preservationists, the forest products industry, government land managers and others over the management of national forests in the Sierra Nevada and Cascade ranges of northern California had become so severe that most timber management activities proposed by the USFS resulted in intense political and legal battles. Sharing a common belief that USFS timber management practices were not satisfying anyone, community leaders, environmental activists and timber industry representatives in and around Quincy, California, USA began meeting in the town library to discuss their concerns (Fig. 19.1). In 1992, this coalition of individuals officially became known as the Quincy Library Group, and in 1993 the group presented a 'Community Stability Proposal' (hereafter Community Stability Plan) to the USFS as a bottom-up approach to management of local forests (Marston, 2001). The intent of the Community Stability Plan was to increase timber production, while simultaneously promoting forest health and the economic stability of local communities. Expected benefits of the plan included jobs, reduced risk of catastrophic wildfires, habitat conservation and watershed restoration. As part of its planning process, the USFS considered but rejected the Community

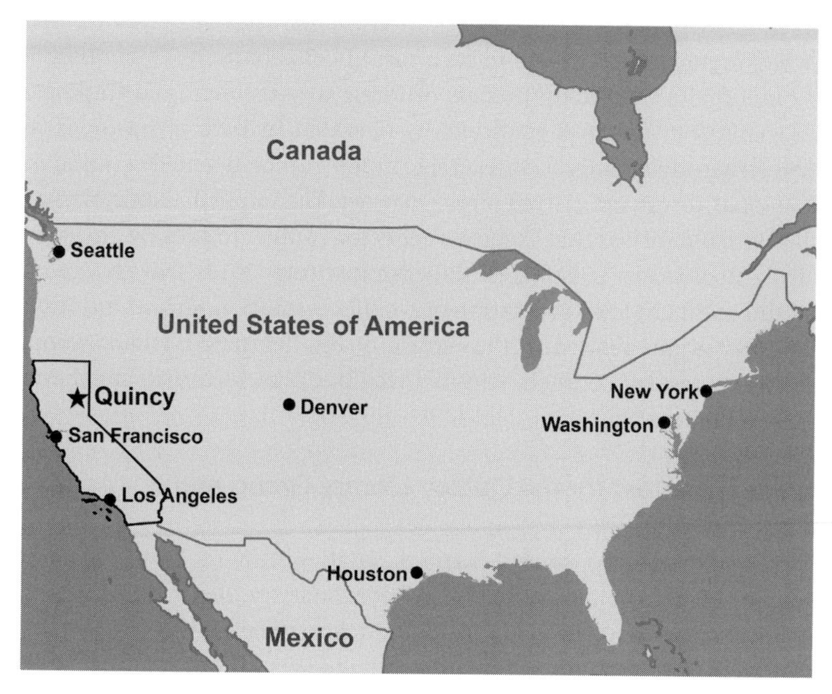

Figure 19.1 Map of the United States showing location of the Quincy Library Group project area in northeastern California.

Stability Plan. Regardless, the Quincy Library Group and its Community Stability Plan were hailed as models for resolving long-standing conflict over national forest logging (Fitzsimmons, 2012), although they were not without their critics (Blumberg, 1999; Coggins, 1999; McCloskey, 1999).

Following the rejection of their plan, the Quincy Library Group sought help from the US Congress, which passed legislation that required the USFS to implement the Community Stability Plan. The Herger–Feinstein Quincy Library Group Forest Recovery Act of 1998 (hereafter, The Act) authorised an initial five-year programme of work for the USFS to conduct a pilot project to test the management actions outlined in the Community Stability Plan on not less than 16,187 ha annually. Management actions included: (1) creating 'fuel breaks' for wildfire protection; (2) allowing individual tree selection harvest (logging individual trees to promote regeneration or growth of remaining trees and to generate timber); and (3) allowing group selection (logging small [~1–2 ha] areas of trees to promote regeneration of trees and to generate timber).

The Act essentially mandated collaboration between the USFS and the Quincy Library Group. The legislation also established an independent science panel to review the success of The Act. Because the collaboration and implementation of the Community Stability Plan would probably not have occurred without The

Act, we, as this independent science panel, interpreted, without prejudice, the legislative requirements of The Act to be a forced collaboration. Thus, the US Congress supported a bottom-up process of forest management and imposed it on an organisation that has traditionally operated in both a top-down (a hierarchical structure) and decentralised (i.e. much authority and discretion is vested in regional, forest and district levels) manner. The implicit assumption of this legislative mandate was that Congress expected conflict to subside. In 2013, we provided a final report to Congress (Pinchot Institute, 2013). Our charge in this review did not include an evaluation of conflict resolution, so we here offer a series of observations related to the nature of this legislated collaboration and its ability to resolve conflicts within three national forests in northern California.

Socio-political context for the Quincy Library Group and Community Stability Plan

The Quincy Library Group started with a small group of local citizens, 'The Friends of Plumas Wilderness', concerned about the negative effects of logging on local national forests and a local politician concerned about community development (Marston, 2001). The political and social situation related to logging in Quincy at the time was acrimonious because of declining old-growth forests, the extensive use of clear-cut logging and the application of herbicides to control 'undesirable' vegetation. At the time, similar situations were common throughout the Pacific Northwest (Satterfield, 2002). The Quincy Library Group's work was unique because it focused on a region's national forests spanning more than 619,000 ha in northeastern California. The Quincy Library Group sought to accommodate what had become opposing positions – timber harvesting versus environmental protection – by creating a compromise strategy (the Community Stability Plan) for managing the area's forests. The Quincy Library Group did not expect USFS' management approach to solve the problems of job loss and local economic decline, wildfire abatement, conservation of habitat and protection of watersheds because their management approach was not sufficiently aggressive (Marston, 2001).

Attempts to resolve local environmental conflicts are inextricably linked to processes occurring at different levels within organisations and at other institutions altogether. These multiple, nested venues of conflict management interact and can conflate one another (Wyborn and Bixler, 2013). Indeed, while the environmentalists and logging interests were at an impasse in Quincy, other broader regional land management plans and policies were being developed for the California spotted owl *Strix occidentalis occidentalis* and national forests in the Sierra Nevada (Box 19). These broader plans would ultimately shape USFS management direction, which heightened interest by the Quincy Library Group to incorporate local stakeholders into USFS planning. In the case of the spotted owl, it was known that logging had led to declines of the owl elsewhere so the USFS commissioned a scientifically credible management plan for the owl population in the Sierra Nevada (Owl Plan; Verner *et al.*, 1992). In the Quincy Library Group's case, the potential implementation of the Owl Plan was feared to be harmful to the timber industry. This motivated one local politician to contact discretely the director of the Friends of the Plumas Wilderness to find a common solution. Neither party approved of the Owl Plan, albeit for different reasons, and saw an opportunity in cooperation to circumvent the Owl Plan to achieve a better result for their respective interests.

The initial meetings between the local politician and the leader of the 'Friends' environmental group laid the foundation for the compromise. In the first public meeting involving a broad range of stakeholders, held at the Quincy Library, there was general consensus among participants that a new way forward was preferable to the existing conflict. A core group became the Quincy Library Group. After this meeting, there was an extremely rapid pace of activity that led to the development of the Community Stability Plan within eight months, and it was immediately submitted to the USFS. The Quincy Library Group presented the Community Stability Plan to the public after it had submitted it to the USFS, suggesting limited broader public input to the Community Stability Plan. Thus, a perceived mutual threat, the Owl Plan, motivated opponents to draw together (Marston, 2001).

Did legislated collaboration result in amelioration of the conflict?

The political and ideological dynamics within the Quincy Library Group were predictably volatile and resulted in the eventual departure of some local conservation proponents. Aside from reported personality conflicts (Marston, 2001), there also was a division over the content of the Community Stability Plan and the perceived influence of certain parties. Specifically, two of the three leaders represented economic interests; one was the former local politician and the other a representative of a prominent local timber company – Sierra Pacific Industries, a private company operating large sawmills in California and a potent political and economic force in the region. It was no secret that the major environmental, non-governmental organisations in the region disdained Sierra Pacific and saw the Quincy Library Group as a cynical front for Sierra Pacific's desire to sustain high timber harvesting on local national forests. In an expression of the 'fixed pie' bias in conflict and negotiation theories (Bazerman, 1983), many environmentalists were convinced that if Sierra Pacific Industries supported Quincy Library Group agreements, it must necessarily be bad for the environment.

While subsequent Quincy Library Group legislation initially created an opportunity for ameliorating conflict, it did so only for those who remained engaged in the process. Stakeholders who harboured concerns about the Quincy Library

Group, logging and forest management left the process and became active in appealing USFS projects. Disengaged local environmentalists expressed concerns not only about the influence of the timber industry within Quincy Library Group in general and Sierra Pacific Industries in particular, but also they perceived the use of group selection harvest as a subterfuge for logging old-growth trees. They were also concerned about the effects of such large-scale actions on the land. Passage of the legislation actually intensified the conflict because it drew in over 100 regional, state and national groups in opposition. National environmental groups considered the legislation a dangerous precedent of piecemeal legislation designed to favour local interests on land that was owned by all Americans (McCloskey, 1999).

Although the legislation initially funded a 5-year pilot project, the work ultimately continued for 13 years (2000–2012), making it the largest and longest such 'community forest' landscape project attempted in the United States. Alignment of political interests, recurrence of large catastrophic wildfires and the desire to reduce large wildfire events resulted in approximately $350 million being authorised by Congress. However, the problem of The Act being nested within other political, administrative and legal venues addressing national forest management became apparent when the USFS developed a region-wide forest management strategy called the Sierra Nevada Forest Plan (commonly called the Sierra Nevada Framework) that was announced in 2001 and revised in 2004. The Sierra Nevada Framework established region-wide guidelines for national forest management activities throughout the $> 63{,}000\ km^2$ Sierra Nevada such as limitations on the size of trees allowed for harvesting. As the USFS attempted to implement The Act, projects were appealed on the basis of their departure from the Sierra Framework standards, making it difficult to distinguish conflicts associated with The Act from conflicts over the Sierra Framework. They were different venues embroiled in the same general conflict. Nonetheless, over time, The Act provided consistent funding to implement USFS projects, and it focused pressure and attention on the USFS to fix issues that had been unresolved for several decades.

What barriers to conflict resolution were most influential in this legislated collaboration?

One immediate barrier was the 'marginalisation' of stakeholders, either by exclusion from the initial Quincy Library Group or by stakeholders withdrawing from the process. Alienation led to the formation of a constituency that resisted implementation and, more importantly, was able to attract outside support for their resistance. These outside groups lobbied Congress and filed appeals and lawsuits during implementation of The Act. This marginalisation also appeared to result in a hardening of rhetorical positions by proponents and opponents of commercial logging. The environmentalists who withdrew felt their interests could be better represented by working outside the structure of The Act and initially were successful in stopping several major projects. They essentially changed the way the USFS approached environmental planning by either significantly delaying projects or causing project revision to appease their interests. The legislation, however, required the Quincy Library Group and the USFS to continue working together to resolve issues. Eventually, scientists and prominent environmental interests co-authored conceptual approaches for managing Sierra Nevada forests (North *et al.*, 2009; North, 2012). These approaches, ironically, provided an opportunity for disenfranchised groups to negotiate independently with the USFS on implementation of specific projects. Thus, the schemes that emerged as a result of the Quincy Library Group process were used as a mediation platform by outside stakeholders who appealed projects. This might not have occurred without the long-term appropriations that resulted in continued confrontation and renegotiation. The two reports containing conceptual approaches represented watershed moments that changed the dynamic of The Act's implementation. Until this time, the Quincy Library Group via the Community Stability Plan exerted significant influence and pressure on the USFS to implement forest prescriptions pursuant to the law. The authors of the technical reports questioned the wisdom of those legal prescriptions and offered science-based alternatives that allowed timber harvesting, yet attempted to achieve the ecological objectives many environmentalists were most concerned about. One can argue that The Act, rather than ameliorating conflict on a broad scale, allowed intensification of the conflict because of marginalisation and focus of a disproportionate amount of attention and resources on a single programme. However, one can also argue that, at the local scale, the marginalisation was mitigated over time by the gradual willingness of the USFS and environmental groups to negotiate over the implementation of projects.

Another barrier was the inconsistent and sometimes weak support by USFS leadership at the district and regional levels. This was a systemic issue born from the change in the USFS' historical stewardship role to one more subject to outside influence, conflicting legislation and planning mandates created by the Sierra Nevada Framework, and the challenge of integrating such a programme into ongoing tasks and budgeting processes. Leadership was resistant because their authority and discretion were partially usurped by the legislation that provided local, external interests with some influence over management actions. Weak support also affected efforts to manage and monitor projects at different landscape scales. Such support by leadership would have required profound changes in thinking and practice relative to past practices.

Compounding the inconsistent support by USFS leadership, another apparent barrier was, from our perspective, that the exact roles of the implementation team, other than as accounting and coordination mechanisms, were unclear.

Because 'line officers' (forest managers) held ultimate authority for implementing projects, the implementation team did not have clear authority and mandate to implement projects. Thus, there was variability among line officers in their creativity, technical ability and willingness to implement projects. The lack of authority also created an atmosphere of uncertainty about decision-making and limited the range of actions adopted by the USFS personnel, particularly regarding monitoring (Pinchot Institute, 2013). This was further exacerbated by changing management strategies resulting from the Sierra Nevada Framework. Such uncertainty led to conflict between the USFS and both the Quincy Library Group and outside stakeholders about the pace and scale of projects.

Two final barriers related to the bureaucratic structure of the USFS. First, there was high turnover in leadership because of promotion, transfers and retirement. Many forest supervisors and district rangers served on the three national forests over the 13 years of implementation. With such shifting of leaders it was hard to keep perspective, focus and commitment for a single programme, even if it was central, by legislation, to the mission of these national forests. Second, the lack of an efficient process for appeals hampered progress, often preventing projects from being realised. Opponents of the USFS bogged down the agency with legal and administrative challenges, and because of its bureaucratic structure these legal challenges resulted in management uncertainty, excessive documentation and more unnecessary, complex bureaucracy.

Discussion

The Act initiated a pilot project that was intended to be a national demonstration of landscape-scale forest management that could simultaneously improve community economic stability, reduce the size and severity of wildfire, protect California spotted owls and improve the condition of water resources. It introduced a new model of forest management and direction that has since been replicated to various degrees. Indeed, the 2001 USFS Collaborative Forest Restoration Program in New Mexico and later the 2010 Collaborative Forest Landscape Restoration Program, authorised under the Federal Landscape Restoration Act of 2009, now span more than 4.4 million ha of public and private land across 33 national forests in 13 states. Lessons from this pilot project provided insight to ongoing landscape restoration efforts in which collaboration is legislated. However, place-based legislative solutions for conflict resolution also will risk being perceived as exclusive to one local community and area. This perception manifested itself in the large coalition of organisations that opposed the Quincy legislation.

The amelioration of conflict was mixed, based on our review. On the one hand, The Act and its process gave members of the Quincy Library Group a reason to persist and remain relevant. In this way, the legislation sustained one collaboration while ossifying conflict between the Quincy Library Group and other local and regional environmental organisations. On the other hand, the disengagement of influential environmental groups and individuals from the process, regardless of the reasons, led to conflict that continues today, although with some progress and unexpected consequences noted. This conflict was manifested not only in media and personal exchanges but also by legal and process challenges to specific projects. There were 20 appeals among the 417 forest management projects proposed over 13 years, of which 6 were litigated (Pinchot Institute, 2013). Many more projects were terminated before being offered, delayed to bolster scientific justification, or were simply not attempted for fear of litigation. Substantial time and money were spent by all parties involved, particularly the USFS, defending instead of implementing decisions. This also had the ancillary effect of creating an atmosphere of uncertainty among USFS staff who felt decisions impeded their ability to implement projects as specified in The Act. Appeals, lawsuits and institutional fear were evidence prima facie that substantial conflict remained and detracted from full implementation of The Act. However, the threat of appeals and lawsuits also had the, perhaps unintended, effect of forcing USFS project planners to scale down proposed projects and to consult more closely with environmentalists who remained outside The Act's process. This upset the balance of power initially possessed by the Quincy Library Group, but marked the beginning of new efforts to collaborate on USFS project planning. Finally, USFS leadership was not provided a mechanism for ownership that, concomitantly, did not threaten their historic procedures, which made the USFS more problematic as a reliable vehicle for delivery and contributed to unsatisfactory results in relation to the large monetary investment. Thus, while conflict was partially reduced locally, it remained an important undercurrent throughout the 13-year project.

To the degree that the project fostered constructive dialogue among antagonists, it was successful. However, the initiating Quincy Library Group clearly failed to recognise (or did not care) that the scale of conflict went well beyond the local area (e.g. national forest policy). The failure to consider broader constituencies impeded the project's effectiveness to resolve conflict. In this case, the forests and management processes that were at the root of conflict between local people having different goals were also of interest to the entire nation. The disengagement of local members of the environmental community, and more importantly, the failure to engage constituents who had a vested interest in the broader outcome, was undermining to the success of The Act. Indeed, a common theme among conflict examples in this book that have been at least partially successful was that a concerted effort was made to engage as broad a constituency of stakeholders as possible (e.g. Box 6; Box 7; Box 9; Box 20). The Quincy Library Group appeared to try to transform what was really a national issue into a local one. Yet this has not been an uncommon direction taken by local interests, and, in the case of The Act, it was even made a cause célèbre for co-management of public lands by local interests. The net result was a limited constituency and broad resistance.

The Quincy Library Group process was not a conflict resolution process per se; rather, it was a community project designed to achieve particular objectives of forest management and economic stability that would reduce local conflict. Therefore, it can be evaluated with respect to conflict resolution. Redpath *et al.* (2013) provided a theoretical roadmap to conflict resolution that depicts two separate but interacting processes that could lead to various win–lose outcomes. Legislated (mandated) collaboration meant that one side was forced to collaborate with another side – top-down process. However, The Act essentially resulted in a hybrid conflict approach; it was top-down by virtue of legislation, but bottom-up (grassroots) by instigation. Although collaboration and communication eventually improved among stakeholders (e.g. outside environmental groups with USFS), The Act was not immediately embraced by USFS personnel. Moreover, its acceptance was uneven among forest supervisors over the course of the project. In addition, there were disenfranchised local and external area stakeholders who were outside the process, so they can be perceived as losers in this process. Those that remained engaged can be perceived as winners because their persistence allowed them influence with the USFS. Thus, the legislated collaboration created winners and losers as a result of process. It was also possible that environmentalists who left the initial dialogue felt less constrained in how they engaged the USFS than if they were bound by congressional legislation. Interestingly, the USFS was a partial 'winner' in that they received a significant infusion of funds from the federal government to facilitate implementation of the project even if they felt their management prerogative was usurped. It could also be argued that the resistance by environmentalists impeded the implementation of The Act such that the local community was a loser in terms of economic activity or that environmental restoration was impeded. However, it was unclear what would have happened in the absence of the project or if the project had been fully implemented because there was a huge economic decline unfolding on both national and world scales. We believe that the initial failure of agreement among stakeholders likely resulted in the perception that the legislation was an imposition favourable to one side rather than the support mechanism it was intended to be.

Perhaps the most important aspect of The Act that had the potential to appease those groups against the project was the legislated requirement for scientific assessment of environmental effects (i.e. monitoring). The importance of monitoring for building trust among stakeholders and allaying fears about negative project impacts cannot be overstated because monitoring directly answers the concerns of stakeholders (Ralph and Poole, 2003). Monitoring also formed the basis for adaptive management by the USFS (*sensu* Holling,

1978). Finally, monitoring could have created the neutral foundation upon which trust could be built among stakeholders. Monitoring and minimising negative environmental impacts were prominently in The Act; yet monitoring accomplishments were uneven at best and sometimes a critical failure (Pinchot Institute, 2013). Failure occurred because the monitoring programme lacked sufficient input or guidance from either senior scientists or administrators at critical times, individuals or groups failed to complete critical objectives, and when failures occurred there was apparently no accountability levied by administrators. In addition, little authority was conferred on those charged with organising monitoring. This ultimately reduced the potential for conflict resolution and undermined the credibility of the USFS to conduct adaptive management.

Redpath *et al.* (2013) discussed the option of marginalising recalcitrant stakeholders in a conflict. In essence, this happened when some environmental interests left the Quincy Library Group. We do not attribute fault to this fracturing of the group to anyone, but rather, we suggest modifying Redpath *et al.*'s (2013) suggestion that proceeding without stakeholders might only work if the stakeholders only have marginal power or influence. In this case, marginalisation of powerful local interests by the Quincy Library Group in combination with its failing to recognise scales of influence resulted in a much stronger resistance than might have been present had greater patience and commitment to local resolution been exercised during the formation of the Quincy Library Group. The Quincy Library Group conflict had multiple dimensions and stakeholders (local, regional, national; federal management; public ownership of land) that were either not recognised or ignored, which invited frustration and unanticipated protraction of conflict. It also appeared that the Quincy Library Group positioned itself as the dominant force by external coordination with the media and legislative offices. Inclusivity or the attempt to be inclusive has been a theme that runs central to successful resolution of conservation conflicts; but inclusivity was incomplete with Quincy Library Group.

In summary, the Quincy Library Group was born of frustration and led to a limited partnership among local interests who achieved a degree of local cooperation with disproportionate political influence as indicated by Congressional support and funding. It was laudable for members of Quincy Library Group to take the bold step to engage in personal dialogue with their adversaries. It forced the cooperation of a large federal agency as a result of its ability to influence the passage of legislation to support their cause and through time achieved some level of genuine cooperation within Quincy and with the USFS. Yet, the many project appeals and court litigation suggested that the legislated cooperation did not resolve conflict from other outside perspectives. Although it was clear that these appeals had their own benefit by motivating greater coordination between USFS and antagonistic stakeholders about specific project proposals, it was the publication of key conceptual documents by scientists (e.g. foresters, fire ecologists, wildlife biologists) that allowed a way forward in negotiations by disaffected parties by creating a mutually acceptable framework, based on scientific principles, to facilitate discussions and find common ground.

Acknowledgement

We thank the Quincy Library Group, community members and USFS staff for information and helpful discussions. In particular, we thank Colin Dillingham for serving as liaison to the Pinchot review team and to the Pinchot Institute for Conservation for administrative support.

References

Bazerman, M. (1983). Negotiator judgment: a critical look at the rationality assumption. *Am. Behav. Sci.*, 27, 211–228.

Blumberg, L. (1999). Preserving the public trust: public lands management must reflect both local and national priorities. *Forum Appl. Res. Public Pol.*, 14.

Boyce, M. S. and Haney, A. (1997). *Ecosystem Management*. New Haven, CT: Yale University Press.

Brown, R. T., Agee, J. K. and Franklin, J. F. (2004). Forest restoration: principles in the place of context. *Conserv. Biol.*, 18, 903–912.

Coggins, G. C. (1999). Regulating federal natural resources: a summary case against devolved collaboration. *Ecol. Law Q.*, 25, 602–610.

Fitzsimmons, A. K. (2012). *Reforming Federal Land Management: Cutting the Gordian Knot*. Lanham, MD: Rowman & Littlefield.

Franklin, J. F. (1992). Scientific bases for new perspectives in forests and streams. In *Watershed Management: Balancing Sustainability and Environmental Change*, ed. R. J. Naiman, pp. 25–72. New York, NY: Springer.

Holling, C. S. (1978). *Adaptive Environmental Assessment and Management*. New York, NY: John Wiley & Sons.

Lansky, M. (1992). *Beyond the Beauty Strip: Saving What's Left of Our Forests*. Gardiner, ME: Tilbury House.

Marston, E. (2001). The Quincy Library Group: a divisive attempt at peace. In *Across the Great Divide: Explorations in Collaborative Conservation and the American West*, eds. P. Brick, D. Snow and S. Van de Wetering, pp. 79–90. Washington, DC: Island Press.

Maser, C. (1988). *The Redesigned Forest*. San Pedro, CA: R. & E. Miles.

McCloskey, M. (1999). Local communities and the management of public forests. *Ecol. Law Q.*, 25, 624–629.

Nie, M. (2004). Statutory detail and administrative discretion in public lands governance: arguments and alternatives. *J. Environ. Law Litig.*, 19, 223–291.

North, M. (2012). *Managing Sierra Nevada Forests*. General Technical Report GTR-PSW-237. Albany, CA: Pacific Southwest Station.

North, M., Stine, P., O'Hara, K., Zielinski, W. and Stephens, S. (2009). *An Ecosystem Management Strategy for Sierran Mixed-conifer Forests*. General Technical Report GTR-PSW-220. Albany, CA: Pacific Southwest Station.

Pinchot Institute. (2013). *Independent Science Panel Report: Herger–Feinstein Quincy Library Group Forest Recovery Act*. Washington, DC: Gifford Pinchot Institute for Conservation.

Ralph, S.C. and Poole, G. C. (2003). Putting monitoring first: designing accountable ecosystem restoration and management plans. In *Restoration of Puget Sound Rivers*, eds. D. R. Montgomery, S. Bolton, D. B. Booth and L. Wall, pp. 222–242. Seattle, WA: University of Washington Press.

Redpath, S. M., *et al.* (2013). Understanding and managing conservation conflicts. *Trends Ecol. Evol.*, 28, 100–109.

Satterfield, T. (2002). *Anatomy of a Conflict: Identity, Knowledge, and Emotion in Old-Growth Forests.* Vancouver: University of British Columbia Press.

Verner, J., McKelvey, K. S., Noon, B. R., Gutiérrez, R. J., Gould, Jr., G. I. and Beck, T. W. (1992). *The California spotted owl: a technical assessment of its current status.* General Technical Report. GTR-PSW-133. Albany, CA: Pacific Southwest Station.

Wondolleck, J. M., and Yaffee, S. L. (2000). *Making Collaboration Work: Lessons from Innovation in Natural Resource Management.* Covelo, CA: Island Press.

Wyborn, C. and Bixler, R. P. (2013). Collaboration and nested environmental governance: scale dependency, scale framing, and cross-scale interactions in collaborative conservation. *J. Environ. Manage.*, 123, 58–67.

Box 19

The spotted owl and conservation of old-growth forests in western North America

R. J. Gutiérrez
Department of Fisheries, Wildlife, and Conservation Biology, University of Minnesota, St. Paul, Minnesota 55108, USA

The conservation of the spotted owl *Strix occidentalis* has been a nearly four-decade-long conflict between people who wish to see old-growth forests preserved for aesthetic, spiritual and ecological reasons and those who view ancient trees as valuable commodities. Thus, it embodies much of the root causes of conservation conflicts (Redpath *et al.*, 2013). Superficially, it is a clash of values about exploitation of natural resources and the interests of local versus external stakeholders. However, the conflict runs deeper because with the passage of the US Endangered Species Act in 1973 (ESA) the power that was traditionally held by influential economic and political groups was reduced greatly because designating the owl as endangered changed the balance of power. In addition, the ESA allows designation of 'critical habitat' for endangered species (USDC, 1973) and this was viewed as an opportunity to use the owl as a surrogate for protecting ancient forests (USDC, 1973). To counter this tactic, the timber industry used the threats of job loss and reduction in community stability to gain social and political support to increase logging on public land (HFQLG, 1998). Each major group has 'grassroots' elements to support its broader agenda.

The owl was a little-known species before the ESA, but attained national and international consciousness about what was both good and bad about conservation in general and about the ESA in particular. Stakeholders developed both grassroots and external support; communities became divided, advocates travelled far to become active in conflict and various forms of social protest were engendered (Satterfield, 2002). With large coalitions forming, resource managers were forced to develop conservation plans to protect both owls and old forests (Yaffee, 1994). The sheer number of parties in conflict ensured the interest of the American Congress. With legal and political (ideological) issues at stake, scientists were asked to provide information about the owl to either 'prove or disprove' either its dependence on ancient forests or its population status (stable or declining). Scientists responded by gathering the largest information base for an endangered species in North America and perhaps the world (Gutiérrez *et al.*, 1995). There were 20,000–30,000 owls in the early stages of the conflict, but some populations have declined $> 50\%$. Initially, some stakeholders doubted the owl was endangered because of these numbers, but research showed the owl had declined precipitously in many areas (e.g. Forsman *et al.*, 2011). Moreover, the public accepted an important scientific conclusion – it was not the number of owls that was important but their rate of decline. Research extended beyond science to law, social science and economics. However, even researchers came into conflict among themselves when their work associations, their findings and their views about forest and wildlife management differed from others (e.g. see Rosenberg *et al.*, 2012).

The complexity and magnitude of the conflict, its longevity, the economic stakes (billions of dollars and thousands of jobs), the vitriol among parties in conflict and the attraction of the courts and politicians made this conflict precedent-setting at US national level because the legal strength of the ESA became more apparent than previously recognised; species that are endangered cannot be allowed to become extinct for any reason except if it is in the nation's interest. Almost lost in this morass of conflict was the need to balance owl and old-forest conservation with logging. Here, trust has been a major issue, as the public did not trust either natural resource agencies or the timber industry to adhere to agreements; the reverse was also true. Thus, intransigence of ideological position and trust has been a major impediment to conflict resolution. Science was not sufficient to quell this conflict so new initiatives emerged to resolve conflicts (e.g. Marston, 2001; Box 3 in Redpath *et al.*, 2013). It remains to be seen whether these initiatives can provide impetus for resolution of this conflict. However, it is clear that the structural impediments for resolution, including money, loss of historical power and ideology, will continue to influence persistence and the outcome of this conflict. Moreover, conservation initiatives often constrain land management so there will be continued resistance to conservation by land managers and forestry interests (Collins *et al.*, 2010).

References

Collins, B. M., Stephens, S. L., Moghaddas, J. J. and Battles, J. (2010). Challenges and approaches in planning fuel treatments across fire-excluded forested landscapes. *J. Forest.*, 108, 24–31.

Forsman, E. D., *et al.* (2011). *Population Demography of Northern Spotted Owls*. Studies in Avian Biology No. 40. Berkeley, CA: University of California Press.

Gutiérrez, R. J., Franklin, A. B. and LaHaye, W. S. (1995). Spotted owl (*Strix occidentalis*). In *The Birds of North America No. 179: Life Histories for the 21st Century*, eds. A. Poole and F. Gill. Washington, DC: The Philadelphia Academy of Sciences and The American Ornithologists' Union.

HFQLG (1998). Herger–Feinstein Quincy Library Group Forest Recovery Act (PL 105–277, div. A, Section 101(e) [title IV, section 401], 21 October 1998), 112 United States Statutes at Large, pp. 2681-231–2681-305.

Marston, E. (2001). The Quincy Library Group: a divisive attempt at peace. In *Across the Great Divide: Explorations in Collaborative Conservation and the American West*, eds. P. Brick, D. Snow and S. Van de Wetering, pp. 79–90. Washington, DC: Island Press.

Redpath, S. M., *et al.* (2013). Understanding and managing conservation conflicts. *Trends Ecol. Evol.*, 28, 100–109.

Rosenberg, K. K., Vesely, D. G. and Gervais, J. A. (2012). Maximizing endangered species research. *Science*, 337, 799.

Satterfield, T. (2002). *Anatomy of a Conflict: Identity, Knowledge, and Emotion in Old-Growth Forests*. Vancouver: University of British Columbia Press.

USDC (U.S. Dept. of Commerce) (1973). Endangered Species Act of 1973. National Oceanic and Atmospheric Administration, National Marine Fisheries Service Enforcement Division, Washington, DC.

Yaffee, S. L. (1994). *The Wisdom of the Spotted Owl: Policy Lessons for a New Century*. Washington, DC: Island Press.

CHAPTER TWENTY

Finding a way out of conservation conflicts

STEPHEN M. REDPATH
University of Aberdeen
R. J. GUTIÉRREZ
University of Minnesota
KEVIN A. WOOD
Bournemouth University & Centre for Ecology & Hydrology
and
JULIETTE C. YOUNG
Centre for Ecology & Hydrology Edinburgh

The world is undergoing rapid change from increasing human pressure. The scale and intensity of this change are deeply worrying from a conservation perspective. For example, we see severe threats to species, habitat and ecosystems from poaching (Maisels *et al.*, 2013), the illegal use of poison (Ogada, 2014), overharvesting (Pinsky and Palumbi, 2014) and agricultural expansion (Laurance *et al.*, 2014). In this book we have focused on how those who represent conservation arguments (conservationists) can respond to these types of challenges. These conservation conflicts arise because one side is passionate about the need to conserve biological diversity, whether for moral, intrinsic or anthropocentric reasons, and the other side may be more focused on different objectives related to human livelihoods and well-being. That is not to say that those arguing for human livelihoods do not recognise the need to conserve biodiversity, and vice versa, but each side may question the relative importance of the arguments, or the specific objectives, or the methods used to achieve those objectives. What is clear is that conservationists are antagonists in these conflicts, and this realisation is important because in order to navigate a path out of destructive conflict, conservationists will need to recognise their role in these issues, address the roots of the problem and be clear about their objectives and about how they engage with the other parties (Redpath *et al.*, 2014).

Throughout the book, we have presented a range of richly complex and multilayered examples. Each has its own idiosyncrasies, but together they expose general principles and highlight what is needed to map and manage conservation conflicts. In this final chapter we build on these perspectives and draw

Conflicts in Conservation: Navigating Towards Solutions, ed. S. M. Redpath, R. J. Gutiérrez, K. A. Wood and J. C. Young. Published by Cambridge University Press.

out the principles and steps towards collaborative conflict management. While we recognise that conflicts may be a force for good (Coser, 1956), the conflicts presented here are more often damaging and costly both to humans and biodiversity. So our aim is not to eliminate conflict, which would be futile, but rather to consider how we should manage it positively when it arises to minimise its negative consequences. We start by considering the process of mapping conflict and the input required from natural and social sciences and humanities.

Mapping conflict

Identifying the problem

Once a conflict has arisen, the first step is to identify the relevant stakeholders (Reed *et al.*, 2009) and to understand their different interpretations and perceptions about the conflict. Such understanding is vital, otherwise any subsequent conflict management process cannot begin or may be flawed and lead to ineffective outcomes (Young *et al.*, 2013a). Stakeholders on either side may have very different views of what the problem is and indeed whether there actually is a conflict. For example, conservationists often see impacts by stakeholders on biodiversity, or vice versa, rather than a conflict between their own world view and that of others. In part this is because they see the imperative of biodiversity conservation and so do not see themselves as stakeholders. This in turn can hinder the search for solutions.

We also need to understand what motivates the conflict, building on the typology of conflict (Chapter 1). To what extent is the conflict driven by perceived injustices, differences in beliefs and values, or by disagreements about information or impact? There are many examples in this book of conflicts that are seemingly over one issue, but are in fact over several very different issues, such as political ideology and socio-economic identity (e.g. see Boxes 2 and 10). A deeper understanding of the motivations for conflict will help guide appropriate mitigation and conflict management.

Identifying values and positions

We need to understand the values and positions held by stakeholders. In Chapter 2, Holland argues that 'especially if ideals or values are involved then the possibility of resolution must always be problematic because ideals and values are precisely the kinds of thing that do not allow negotiation or compromise'. We therefore need to be able to distinguish between those deeply held values which will be non-negotiable and the positions and interests that parties hold, which may be more flexible.

There are two other related aspects. First, we need to remember that people have multiple values and identities and that there are risks associated with pigeonholing them under one set of values. In Chapter 10, Dower points out that 'those who hunt as a sport may be in alliance with a group wanting to preserve an area of countryside in order to stop certain development whereas on most other issues they would be diametrically opposed'. Related to this is the fact that as well as focusing on the differences among stakeholders, as parties often do in conflict, we also need to understand the values that are common among parties (see Chapter 10). In this way we can highlight shared understanding and values and thereby create space for discussion of shared solutions.

Identifying the scale of the conflict

Conflicts invariably occur across multiple social, geographical and temporal scales (Chapter 15). It is important, therefore, to be clear about how people represent an issue in terms of a particular scale, as this will influence the way in which the issue is managed (Young *et al.*, 2013b). A key step here is to involve all relevant stakeholders in this framing. Despite the current focus (at least in theory) on scale-adapted governance (Newig and Fritsch, 2009; Buizer *et al.*, 2011; Kok and Veldkamp, 2011), and experience from some disciplines and processes that are more used to thinking at larger scales (see Chapter 12; Box 7), there is little evidence of situations where relevant stakeholders have jointly framed the scale of a conservation conflict, although see Young *et al.* (2013b) and Box 4 for one example. There is, however, evidence of mis-framing conflicts. In Chapter 19, the issue of scale was ignored in the Quincy Library Group process, which resulted in a focus on the local issue when it actually was also a national issue. This reflects again that 'local' conflicts may be linked to wider scales and patterns of political relationships and of biodiversity use (Meadowcroft, 2002). So, while thinking at the local scale may be easier from a practical perspective, developing processes to encourage the identification of the largest adaptable scale may in the long term lead to more sustainable and effective conservation conflict management processes (Box 1).

Multiple disciplines and knowledge

As discussed throughout this book, conservation conflicts are complex. We need insight from relevant social sciences and economics and from humanities as well as from ecological science (e.g. Boxes 11 and 13). The specific disciplines required are likely to vary with the specific nature of the conflict, although it seems inevitable from the examples in this book that most conservation conflicts will have dimensions that require input to understand the ecological interactions, the positions, goals, values and relations of the humans involved, the politics, the history, the legal framework and the ethical arguments to help guide subsequent conflict management.

The knowledge that scientists create from research is only part of the relevant knowledge needed to resolve conflicts. Other forms, such as indigenous knowledge or that derived from stakeholders' experiences of working in a conflict, are also valid and relevant. There is good reason to suspect that these forms of knowledge provide useful insight of great benefit to conservation (Fazey *et al.*, 2006; Fraser *et al.*, 2006; Pilgrim and Pretty, 2010; Adams and Sandbrook, 2013). However, they can also be challenging for quantitatively trained natural scientists to accept, and difficult to combine with traditional scientific data, although a framework for their integration has been suggested (Raymond *et al.*, 2010). As Adams and Sandbrook (2013) point out, the effort to combine knowledge is likely to be worthwhile, as there is a strong and urgent argument to make a subtle move away from 'evidence-based' conservation to 'evidence-informed' conservation. We suggest that this may be particularly urgent in conflict situations, where a shared understanding of the evidence among scientists, stakeholders and policy makers is so important.

Legislation

Legislation is a central element in conflicts that can act as a motivation for action, but can also act as a position to hide behind and therefore a barrier to the search for solutions. The mapping of conflicts needs to include consideration of the relevant legislation and an acknowledgment of the boundaries to possible solutions that this sets. That said, strict, inflexible legislation can lead to disenfranchised stakeholders, sometimes to the detriment of conservation goals (Woodroffe *et al.*, 2005). As Heydon *et al.* (2011) point out, 'failure to adequately address these challenges and make provisions for the future will lead to increased conflict... and, ultimately, to the reversal in the favourable public attitudes, the growth of which we have witnessed over the past half century. This is not an outcome that any of us wish to see'. A more flexible approach is therefore likely to reduce conflict and aid the search for shared solutions.

Parties are more likely to accept laws prohibiting activities when members of their own group have been involved in the development of a law that has benefits for the majority, and that people can live with. Where members of a group have had no involvement in a law that concerns them, there is a danger that this group will perceive the law as being imposed upon them by outside forces, and hence may react antagonistically towards it. For example, Box 12 illustrates the contrast between a relatively ineffective law against lead shot in England passed without much support from the shooting community, and a similar law in Denmark passed with support from the shooting community, which has proven to be much more effective than the former, with better compliance.

Conflict management

Even if we have a good grasp of the underlying causes of conflict, a clear idea of the positions and values of the relevant stakeholders at appropriate scales and understand the ecological, economic and social impacts, that does not mean that we are any nearer to resolving the problem. Conflict situations are invariably emotionally charged and difficult to navigate. People on at least one side may feel aggrieved and frustrated, so parties may simply not be prepared to engage with those representing opposing views. For example, conservationists may not engage with those involved in illegal activity, feeling that their interests are better served by using law enforcement and media campaigns (Box 3). In some cases this adversarial approach may indeed lead to one party winning, and achieving their objectives. However, as such approaches do not deal with the underlying social conflict, they are likely to lead to the 'losing' stakeholders having a strong sense of grievance (Chapters 8 and 13), so that the conflict in some form is likely to re-emerge elsewhere. The alternative approach is for parties in conflict to come together to recognise the conflict as a shared problem and therefore seek shared solutions through engagement. In the above example this would mean conservationists and those involved in illegal behaviour engaging in dialogue to find solutions that lead to the end of illegal activity and that both sides find acceptable.

As discussed in Chapter 1, this brings us to an important question: under what conditions is it better to fight or engage for conservation objectives? Will the long-term interests of conservation be best served through the use of legislation and enforcement or by tackling the underlying conflict and engaging with other parties to reduce conflict and maximise the meeting of interests? In some cases the activity may be so damaging to the interests of conservation that enforcement is perceived as the only logical alternative, irrespective of the costs outlined above. However, if the aim is to reduce conflict, we know that appropriate dialogue and collaboration lead to better relationships, reduced conflict and improved social outcomes (Beierle and Konisky, 2001; Ansell and Gash, 2008; Jones-Walters and Cil, 2011). We would therefore hypothesise that the long-term benefits to conservation will be enhanced through collaborative approaches, as the relationships and understanding between conservationists and other parties will be improved, allowing the space for shared solutions. Currently, however, there is little available evidence to allow us to test this hypothesis (Young *et al.*, 2013a).

Game theory provides some insight into these problems (Chapter 14; Colyvan and Regan, 2011). The conditions under which parties would be expected to fight or cooperate in the search for solutions will be dependent on factors such as altering the relative costs and benefits involved and distinguishing the non-negotiable values from the more readily negotiable interests (Colyvan and Regan, 2011; Ramsbotham *et al.*, 2011). In many cases, the move from an adversarial to a collaborative approach to conflict management will require a change or catalyst. Sometimes, an aggrieved party will encourage like-minded individuals to begin dialogue or change practices (see Box 12), or a government minister or agency might intervene following political or legislative pressure (e.g. Boxes 4, 9, 12, 13 and 14). Sometimes a local champion will emerge, who can help drive the process forward (Young *et al.*, 2012).

Designing the process

Of course, even when parties are prepared to engage in a collaborative process, that is just the starting point. The design of the process to fit the situation is crucial. In some cases the appropriate process will involve the key stakeholders, sometimes with independent facilitators, sometimes with government and their agencies and sometimes with relevant scientists. Participatory and deliberative processes to try and resolve conflict are discussed in detail by Reed and Sidoli del Ceno (Chapter 16), Pound (Chapter 17) and Madden and McQuinn (Chapter 18) and in Box 17. Each outlines good practice principles. In Chapter 16, the focus is on identifying and involving the right people; ensuring the right atmosphere; and making the process relevant. Pound (Chapter 17) outlines 12 keys to success relating to inclusion, deliberation and levels of influence; designing effective processes, quality delivery; working with stakeholders and embedding a participation ethos. Finally, in Chapter 18, Madden and McQuinn describe conflict transformation as a way of conceptualising conflicts as 'opportunities to analyse and transform the underlying relationships and social systems that have the ability to adversely impact conservation efforts'.

It is worth noting that even if stakeholders are prepared to enter into dialogue, that does not mean that they will subsequently be prepared to negotiate a solution. For example, in the case of raptors in the UK (Box 3) a seven-year dialogue process was established to search for shared solutions to the problem, but the process collapsed when one side withdrew because of what they perceived as a lack of progress. Without a genuine willingness from all sides to find shared solutions and the leadership in place to act on recommendations coming out of such dialogue, these processes can be hollow and ultimately damaging to conflict management.

A central element in the design of these processes is that deliberations and agreements must be perceived as fair by all parties involved. Ultimately, if stakeholders remain dissatisfied or feel an agreement is not fulfilled, the conflict will resurface. Researchers also have to consider their role in these processes, and the credibility of their science. Are they advocating one side in the conflict or are they there to present the science and data to understand and help map the conflict? The credibility of science is dependent on the impartiality of those conducting research and their ability to objectively weigh up the evidence for and against alternative standpoints. However, much of the interest in conservation conflicts has naturally come from conservation biologists, who are unlikely to be best placed to provide impartial advice. At the very least, they will be perceived as being biased and their views may be ignored or rejected. In all situations, scientists involved have to be very aware of the personal values and 'baggage' that they bring with them and to be aware how they may be perceived by those engaged in conflict.

Another important element lies in identifying the objectives. One of the complications integral to conflict is that the goals are often unclear and may even be disputed within a group of stakeholders. For example, in a situation where farmers and conservationists are in conflict over the impact of a predator, some farmers may seek to minimise the impact on their livestock, whereas others may want to reduce the national populations of the predator or even eliminate it entirely. Some of these goals may be openly expressed and others not. Similarly, some conservationists may want to allow the predator to achieve its natural density, whereas others may want to enforce the legislation or resolve the conflict. Identifying and agreeing these positions and objectives is clearly important as they will dictate the most appropriate and effective mitigation strategy.

Exploring solutions

Once the objectives have been agreed, the next step is to consider the range of solutions available to achieve those objectives. In many of the conflicts addressed in this book, at least part of the solution is likely to involve a focus on human–wildlife interactions and impacts. For example, if there is a conflict over wolves, possible solutions may include financial compensation, lethal control, zoning, non-lethal technical solutions to reduce impact, population management and improved livestock husbandry, to name but a few. Some of these may be favoured more by one side of the conflict than others (see Boxes 8 and 15). Commonly, conservationists may favour less intrusive, non-lethal techniques to reduce impact, whereas stakeholders whose livelihoods are threatened may favour lethal control. Therefore, the first step is to agree a potential solution that is acceptable to both sides, then to test the effectiveness of that solution in the field through monitoring and experimentation, thereby improving understanding, reducing the level of uncertainty and adapting the management approach accordingly. In other words, following the framework for adaptive management set out by Holling (1978) and more recently by Williams *et al.* (2009).

On the face of it this may seem simple, but there are a number of hurdles to get over. One challenge for scientists is that their data may often be at variance with stakeholders' understanding. For example, stakeholder perceptions of the impact of predators on their prey may differ greatly from the scientific data. In such situations, these stakeholders may find such evidence hard to accept as it clashes with their expectations (Wood *et al.*, 2014). One reason for this is that science often addresses narrow questions that may not be relevant to the issues that matter to the stakeholders involved in conflict. This mismatch between science and practitioners' needs can lead to tension. If scientific data are to be useful in addressing a conservation conflict, then the science cannot be carried out in isolation (e.g. Box 9). The researcher must consider what knowledge is needed by stakeholders as well as the capacity of the practitioners to act on such knowledge. For example, researchers testing alternative management options need to consider whether such options are within the capabilities of the practitioners to carry out (Wood *et al.*, 2013). Working closely with stakeholders can help ensure that potential scientific solutions are achievable and acceptable. Thus, in any conservation conflict there is a need to develop the science in collaboration with the stakeholders to ensure acceptance and to ensure the science is addressing the right questions at the right scales. Furthermore, because conflict encourages heightened scrutiny of ideas and data, the quality of science produced can benefit the conflict management process greatly (e.g. Box 19 and Chapter 12).

Another hurdle is lack of transparency and simplicity. Monitoring has to be transparent and clear to allow stakeholders to assess whether agreements are viable and meet their concerns and procedures are in compliance. For example, the individual-based models used to address the conflict between conservationists and fishermen over shellfisheries have proven too complex for most stakeholders to understand, with the result that model results are often not trusted (Box 14). This suggests two things. First, peer review by the stakeholders could be carried out to assess the monitoring. Second, a neutral third party could carry out the monitoring. In some cases, a government has a legal requirement or authority to conduct monitoring so a strictly neutral third-party monitor is not possible. When these situations occur, as they often do in the USA, the agency charged with monitoring should make the entire monitoring process and results transparent and available to stakeholders. In some cases, resolution processes come to conclusion successfully, but there is no element that assures compliance (see Box 15). When this occurs, agreements are tenuous. Monitoring informs stakeholders both when measures of success are failing and when compliance is failing. For example, if information from monitoring shows stakeholders that a measure of success is not being achieved, stakeholders can ask: is the mechanism designed to achieve the desired result flawed or is it the people charged with utilising the mechanism that are failing? If there is no monitoring of what happens post-management, or no assessment of the social and conservation outcomes of conflict management, it is difficult to learn from such processes and avoid making future mistakes. There is also the question of who undertakes monitoring or adaptive management – again, trust has to exist or be built with those monitoring in order for results to be viewed as credible by stakeholders (see Chapters 3 and 12).

Finally, uncertainty can also present a challenge by delaying processes and forestalling decisions. Uncertainty is a crucial element which is highlighted in a number of chapters and boxes (e.g. Chapters 3 and 14; Boxes 15 and 18). There are different types of uncertainty, including process uncertainty, measurement uncertainty, structural uncertainty and implementation uncertainty (Milner-Gulland and Rowcliffe, 2007). Some of these uncertainties are inherently easier to deal with than others. From both an adaptive management and a conflict management perspective, the important element is being transparent about these uncertainties and how the process will reduce the levels of uncertainty and enable the stakeholders to learn and adapt.

Principles of collaborative conflict management

We consider five general principles that are foundations for successful, collaborative conflict management.

Communication

Communication is essential not only for sharing information and ideas and building understanding and trust, but also for displaying both respect for others and a willingness to collaborate. Madden and McQuinn, together with Pound and Sidaway (Chapters 17 and 18; Box 17), elaborate on this critical dimension of dialogue among parties in conflict. If our discourse is not civil, respectful and neutral in our portrayal of conflict dynamics, it will likely lead to the worsening of relationships and undermine any process designed to resolve conflicts. Aggressive or dismissive communication will reduce the likely success of the conflict management process because it undermines respect and trust and exaggerates either preconceived notions about others or personal conflicts independent of the conservation conflict. Lambert and Whitehouse (Chapters 4 and 7) show that understanding the historical dynamics and culture of peoples involved allow us to frame our communication in a sensitive and appropriate way (see also Box 2). The tone and manner of communication projects the impression that the parties in conflict have some knowledge of the historical, cultural and political dimensions of the other parties in the conflict so that their use of language encourages dialogue rather than discourages it. Finally, it is important that communication is not dominated by certain interests. Therefore, communication requires careful consideration of who the so-called 'stakeholders' are (see inclusiveness principle below), and the need to involve and to exchange information with all interested parties, across relevant scales.

Transparency

Transparency is critical in enabling those involved in the conflict management process to understand position and goals, develop shared understanding of the available knowledge and its gaps and uncertainties and see how decisions are made (Adams *et al.*, 2003; Sarewitz, 2004; Reed, 2008; Salafsky, 2011). Transparency thereby provides the potential to make better, informed decisions. In contrast, a lack of transparency can lead to confusion over the involvement of stakeholders, differing expectations from stakeholders regarding the outcomes of the management process, mistrust, frustration and the disengagement of stakeholders.

Inclusiveness

Conflict management requires concerted consideration of who the 'relevant' stakeholders are and when and how they should be included (Boxes 2, 9 and 20; Chapters 16 and 17). These may not always be local stakeholders, but all stakeholders who are likely to be affected by the conflict and can make a difference or contribution. For example, in some conflicts, it may be essential to include the appropriate policy makers/advisers, perhaps even at the national level, as they hold considerable power (Richards *et al.*, 2004) that may be needed to enact any decisions emerging from the conflict management process. In addition, the process needs to consider when to include stakeholders. Indeed, the timing of including stakeholders in a conflict management process is important. Several authors in this book have highlighted the benefits of early inclusiveness (e.g. Box 6). Engaging stakeholders too late in a process can make people feel they have no say in the process, and can lead to processes being seen as tokenistic. Finally, there needs to be serious consideration of how to engage stakeholders (Young *et al.*, 2014). Such processes need to be 'selected and tailored to the decision-making context, considering the objectives, types of participants and appropriate level of engagement' (Reed, 2008: 2424). For example, some stakeholders may not be used to or comfortable with large meetings. Tailoring the way in which stakeholders are involved in the process according to the needs and preferences of stakeholders will require attention.

Influence

Stakeholders need to understand the level of influence they can have in a conflict management process. Advocates of the deliberative model of democracy claim that such a perspective can 'increase the quality of democratic judgements' (Warren, 1996: 46) and has 'transformative potential', as the process of discussing issues with people with often conflicting views can enable people to gain new information and rethink their own positions (Young, 2000). Such deliberative processes can also allow 'those with no or a weak voice to exert influence on decision-making outcomes' (Collins and Burgess, 1999: 1–2). As opposed to the individual interest-based approach of the representative model, the deliberative democracy model advocates a 'relative common good arising out of the free deliberation and negotiation among organised interested groups' (Beierle, 1998: 2). Professionals, such as mediators or facilitators, have a potentially important role to play in this model, acting as 'teachers and interpreters' (Fischer,

2004: 21), enabling citizens to better understand complex issues and make informed political decisions. Although the level of influence that stakeholders may have will obviously vary depending on the different contexts in which conflicts are embedded, it is essential for stakeholders to know the boundaries within which they can operate. In Chapter 18, Pound emphasises the importance of being as clear as possible about what engaged stakeholders can and can't influence. Failure to make boundaries clear may lead to disillusionment and be damaging to social capital and trust. Frank (Box 8) provides some examples of different levels of influence, and concludes that conflict management initiatives were more successful when stakeholders felt they had been listened to and had had an influence on the outcome. As outlined by Pound in Chapter 18, 'if a process is to resolve conflict and build consensus, participants need to deliberate in depth and negotiate over the way forward with high levels of influence'.

Trust

Trust is at the heart of resolution – if there is no trust between stakeholders then it must be built, through techniques such as mediated dialogue (Chapters 16, 17, 18, and Boxes 12 and 17). There is clear evidence that dialogue improves trust between stakeholders (see, for example, Boxes 1, 4, 5, 11 and 13). Public participation literature has emphasised the need for independent expert mediators or facilitators (Chess and Purcell, 1999; Rowe and Frewer, 2000; Reed, 2008). Results from recent studies (e.g. Young *et al.*, 2014), however, suggest the opposite. In the case study described in Boxes 4 and 13, the facilitator was not independent, but a local 'champion' working for a District Salmon Fishery Board. As such, while independent facilitators may improve levels of stakeholder involvement, there is a need to acknowledge and understand the potential advantages, such as trust and knowledge, and disadvantages, such as bias, of an internal stakeholder taking on this role. The advantages of an internal stakeholder may, for example, be particularly apparent in the context of natural resource management where conflicts are common and where stakeholder groups are often disenfranchised. Another linked principle, therefore, is to remember that however robust a process may be, personality (also of the third party) has much bearing on the outcome of conflict management discussions.

Conclusions

There is a clear and urgent need to find ways of meeting the objectives of biodiversity conservation and enhanced human livelihoods and well-being. Increasingly these two elements impinge on each other and we have to develop effective ways of dealing with both aspects in emerging conflicts. We argue that we need a diversity of approaches to improve our understanding of the complexity inherent in conflict; complexity that we can ill afford to ignore. We also argue that we need to find ways of supporting and strengthening collaborative conflict management. This requires involving the right people, having a transparent and inclusive process with clear, agreed objectives and working to improve shared understanding and trust.

Our ability to reduce the negative consequences of these conflicts will, we argue, be dependent on recognising the problems as shared. We ultimately want to find solutions that enable us to conserve biodiversity in just and socially acceptable ways. This means recognising the humanity on both sides and supporting the processes that enable this to happen. This will require leadership and boldness from all involved to move positions, be open to other world views and be prepared to argue, debate and negotiate towards solutions that are mutually beneficial and of long-term benefit to conservation.

References

Adams, W. M. and Sandbrook, C. (2013). Conservation, evidence and policy. *Oryx*, 47, 329–335.

Adams, W. M., *et al.* (2003). Managing tragedies: understanding conflict over common pool resources. *Science*, 302, 1915–1916.

Ansell, C. and Gash, A. (2008). Collaborative governance in theory and practice. *J. Public Adm. Res. Theory*, 18, 571.

Beierle, T. C. (1998). Public participation in environmental decisions: an evaluation framework using social goals. Discussion paper 99-06. Washington, DC: Resources for the Future.

Beierle, T. C. and Konisky, D. M. (2001). What are we gaining from stakeholder involvement? Observations from environmental planning in the Great Lakes. *Environ. Plann. C: Govern. Pol.*, 19, 515–527.

Buizer, M., Arts, B. and Kok, K. (2011). Governance, scale and the environment: the importance of recognizing knowledge claims in transdisciplinary areas. *Ecol. Soc.*, 16, 1–18.

Chess, C. and Purcell, K. (1999). Public participation and the environment: do we know what works? *Environ. Sci. Technol.*, 33, 2685–2692.

Collins, K. and Burgess, J. (1999). Summary of the London seminar. In: *Deliberative and Inclusionary Processes: A Report from Two Seminars*, CSERGE Working Paper PA 99-06, eds. T. O'Riordan, J. Burgess and B. Szerszynski.

Colyvan, M.J. and Regan, H.M. (2011). The conservation game. *Biol. Conserv.*, 144, 1246–1253.

Coser, L. (1956). *The Function of Social Conflict*. New York, NY: Free Press.

Courchamp, F., Hoffmann, B. D., Russell, J. C., Leclerc, C. and Bellard, C. (2014). Climate change, sea-level rise, and conservation: keeping island biodiversity afloat. *Trends Ecol. Evol.* http://dx.doi.org/10.1016/j.tree.2014.01.001

Fazey, I., Fazey, J., Salisbury, J., Lindenmayer, D. B. and Dovers, S. (2006). The nature and role of experiential knowledge for environmental conservation. *Environ. Conserv.*, 33, 1–10.

Fischer, F. (2004). Professional expertise in deliberative democracy: facilitating participatory inquiry. *The Good Soc.*, 13, 21–27.

Fraser, D. J., Coon, T., Prince, M. R., Dion, R. and Bernatchez, L. (2006). Integrating traditional and evolutionary knowledge in biodiversity conservation: a population level case study. *Ecol. Soc.*, 11, 4.

Heydon, M. J., Wilson, C. J. and Tew, T. (2011). Wildlife conflict resolution: a review of problems, solutions and regulation in England. *Wildl. Res.*, 37, 731–748.

Holling, C. S. (1978). *Adaptive Environmental Assessment and Management*. Chichester: Wiley.

Jones-Walters, L. and Cil, A. (2011). Biodiversity and stakeholder participation. *J. Nat. Conserv.*, 19, 327–329.

Kok, K. and Veldkamp, T. A. (2011). Scale and governance: conceptual considerations and practical implications. *Ecol. Soc.*, 16, 1–10.

Laurance, W. F., Sayer, J. and Cassman, K. G. (2014). Agricultural expansion and its impacts on tropical nature. *Trends Ecol. Evol.*, 29, 107–116.

Maisels, F., *et al.* (2013). Devastating decline of forest elephants in Central Africa. *PLoS One*, 8(3), e59469.

Meadowcroft, J. (2002). Politics and scale: some implications for environmental governance. *Landscape Urban Plan.*, 61, 169–179.

Milner-Gulland, E. J. and Rowcliffe, M. J. (2007). *Conservation and Sustainable Use: A Handbook of Techniques*. Oxford: Oxford University Press.

Newig, J. and Fritsch, O. (2009). Environmental governance: participatory, multi-level – and effective? *Env. Pol. Gov.*, 19, 197–214.

Ogada, D. L. (2014). The power of poison: pesticide poisoning of Africa's wildlife. *Ann. N Y Acad. Sci.* DOI: 10.1111/nyas. 12405.

Pilgrim, S. E. and Pretty, J. (2010). *Nature and Culture*. London: Earthscan.

Pinsky, M. L. and Palumbi, S. R. (2014). Meta-analysis reveals lower genetic diversity in overfished populations. *Molec. Ecol.*, 23, 29–39.

Ramsbotham, O., *et al.* (eds). (2011). *Contemporary Conflict Resolution*. Third edition. Cambridge: Polity Press.

Raymond, C. M., Fazey, I., Reed, M. S., Stringer, L. C., Robinson, G. M. and Evely, A. C. (2010). Integrating local and scientific knowledge for environmental management. *J. Environ. Manage.*, 91, 1766–1777.

Redpath, S. M., Bhatia, S. and Young, J. (2014). Tilting at wildlife: reconsidering human–wildlife conflict. *Oryx*, doi: 1001017/S0030605314000799.

Reed, M. S. (2008). Stakeholder participation for environmental management: a literature review. *Biol. Conserv.*, 141, 2417–2431.

Reed, M. S., *et al.* (2009). Who's in and why? A typology of stakeholder analysis methods for natural resource management. *J. Environ. Manage.*, 90, 1933–1949.

Richards, C., Sherlock, K. and Carter, C. (2004). *Practical Approaches to Participation*. SERP Policy Brief No. 1. Aberdeen: Macaulay Institute.

Rowe, G. and Frewer, L. J. (2000). Public participation methods: a framework for evaluation. *Sci. Technol. Human Val.*, 25, 3–29.

Salafsky, N. (2011). Integrating development with conservation: a means to a conservation end, or a mean end to conservation? *Biol. Conserv.*, 144, 973–978.

Sarewitz, D. (2004). How science makes environmental controversies worse. *Environ. Sci. Pol.*, 7, 385–403.

Warren, M. (1996). Deliberative democracy and authority. *Am. Polit. Sci. Rev.*, 90, 46–60.

Williams, B. K., Szaro, R. C. and Shapiro, C. D. (2009). *Adaptive Management: The U.S. Department of the Interior Technical Guide*. Washington, DC: Adaptive Management Working Group, U.S. Department of the Interior.

Wood, K. A., Stillman, R. A., Daunt, F. and O'Hare, M. T. (2013). Evaluating the effects of population management on a herbivore grazing conflict. *PLoS ONE*, 8, e56287.

Wood, K. A., Stillman, R. A., Daunt, F. and O'Hare, M. T. (2014). Chalk streams and grazing mute swans. *Br. Wildl.*, 25, 171–176.

Woodroffe, R., *et al.* (2005). *People and Wildlife: Conflict or Coexistence?* Cambridge: Cambridge University Press.

Young, I. M. (2000). *Inclusion and Democracy.* Oxford: Oxford University Press.

Young, J. C. and Marzano, M. (2012). Embodied interdisciplinarity: what is the role of polymaths in environmental research? *Environ. Conserv.*, 37, 373–375.

Young, J. C., Butler, J. R. A., Jordan, A. and Watt, A. D. (2012). Less government intervention in biodiversity management: risks and opportunities. *Biodivers. Conserv.*, 21, 1095–1100.

Young, J. C., *et al.* (2013a). Does stakeholder involvement really benefit biodiversity conservation? *Biol. Conserv.*, 158, 359–370.

Young, J. C., Jordan, A., Searle, K. R., Butler, A. and Simmons, P. (2013b). Framing scale in participatory biodiversity management may contribute to more sustainable solutions. *Conserv. Lett.*, 6, 333–340.

Young, J. C., *et al.* (2014). Improving science-policy dialogue to meet the challenges of biodiversity conservation: having conversations rather than talking at one-another. *Biodivers. Conserv.*, 23, 387–404.

Box 20

Baboons and conflict on the Cape Peninsula, South Africa

M. Justin O'Riain

Department of Biological Science, University of Cape Town, Private Bag X3, Rondebosch, 7701, South Africa

The Cape Peninsula baboon population (ca. 500 individuals in 16 troops, ranging across 250 km^2) is isolated by Cape Town's surrounding infrastructure. Being adaptive generalists, baboons have both survived the transformation of their habitat and thrived because of access to human-derived food, elimination of predators and protective management.

Conflict over baboons has occurred since the early 1600s when baboons first sought human-associated foods in gardens used to grow produce. With increasing human population troops have lost access to preferred foraging areas and hence the potential for interaction is greater (Hoffman and O'Riain, 2012).

Management authorities have attempted to reduce human–baboon interaction by educating residents, baboon-proofing waste and using 'baboon field rangers' who attempt to reduce contact between humans and baboons using aversive methods. When methods fail, a baboon may be euthanised in accordance with a 'raiding baboon protocol', derived from scientific and public participation.

Raiding behaviour and the removal of habitual raiders fuels human–human conflict on the Cape Peninsula. Managers argue that they have a legal mandate to protect the health, safety and well-being of residents by euthanising baboons when aversion fails. Opposition to 'anthropocentric management' is largely based on the perception by animal rights groups that euthanasia threatens population viability and the rights of the baboon. Raiding is seen as a natural consequence of the baboon's loss of preferred habitat and the incentives provided by human negligence (e.g. leaving doors and windows open and poor waste management). For the most part these 'baboon activists' promote the ideas that baboons share space with humans and the two primates can coexist peacefully despite the obvious health (Ravasi *et al.*, 2012) and welfare costs to humans and baboons. Moreover, they state that humans should be punished for creating opportunity for baboons through neglect.

Research began in the 1960s on the foraging ecology of wild troops living within remote areas of the Peninsula. By the late 1990s, increased interactions with humans resulted in a report about baboon population status and possible solutions to reduce interactions. The low numbers, highly skewed sex ratio (few adult males), high levels of human-induced injury and mortality and the possibility that the population may be genetically and/or behaviourally unique (i.e. marine foraging) were used as arguments to confer 'protected status' on the population. Baboon field rangers were used in 1999 to chase two of the worst raiding troops away from urban areas. Government authorities, consultants, members of the public and

NGOs formed the Baboon Management Team (BMT). The BMT devised management plans and a strategy document which included an option to euthanise baboons that posed risks to humans.

Despite these positive changes, conflict escalated. In 2002, University of Cape Town researchers determined that baboons had decreased; the sex ratio was still strongly female-biased, human-induced mortality was high and baboons in the two 'managed' troops were habitual raiders. The BMT, an excellent example of participatory planning in conservation conflict, was paralysed by polemics and monthly meetings were heated affairs with no consensus derived (Hurn, 2011). Proposed management strategies were not being implemented by the service provider responsible for daily management of baboons and there was no pure or applied research in progress to lead management discussions away from the largely subjective and emotive discourse.

Against this background, the University of Cape Town Baboon Research Unit was formed to provide data essential to understanding the causes, consequences and possible solutions to the conflict. It commenced with population-level assessments of space use and behaviour of troops across diverse habitats (Hoffman and O'Riain, 2012). It assessed population health and the efficacy of methods to reduce human–baboon interactions (Kaplan *et al.*, 2011). These data were disseminated to management authorities and greatly empowered the authorities' ability to make informed and defensible management decisions.

In 2010 the BMT was replaced by two new organisations, the Baboon Technical Team (BTT) and the Baboon Liaison Group (BLG). The BTT is comprised of city, provincial and national conservation authorities and has assumed full responsibility for short- and long-term management. The BLG is comprised of civic representatives from affected communities and is responsible for liaising between residents and the BTT. The two groups meet regularly to facilitate communication and interaction. Against the background of these improved decision-making structures supported by science, the City of Cape Town increased the budget for managing baboons from approximately U$10,000/annum to U$1 million, which enabled management of all 11 troops (up from 6).

Since these changes were effected, injuries and deaths suffered by baboons declined by > 50%. In 2012, noise and pain aversion methods were included in the management protocol, which together with improved management structures has resulted in unprecedented levels of conflict mitigation. Most troops are kept out of urban areas > 95% of the time with fencing attaining 100% success over a 1-year period. Together this equates to improved welfare and conservation status for baboons and the programme was rewarded with a public endorsement by the National Society for the Prevention of Cruelty to Animals (NSPCA). The alignment of national, provincial and city conservation managers, academic researchers, civilian representatives and animal welfare organisations under a single management umbrella has resulted in greatly reduced human–baboon interaction and as a direct consequence has largely defused the human–human conflict on the Peninsula.

References

Hoffman, T. S. and O'Riain, M. J. (2012). Monkey management: using spatial ecology to understand the extent and severity of human–baboon conflict in the Cape Peninsula, South Africa. *Ecol. Soc.*, 17, 13–21.

Hurn, S. (2011). Like herding cats! Managing conflict over wildlife heritage on South Africa's Cape Peninsula. *J. Ecol. Environ. Anthropol.*, 6, 39–53.

Kaplan, B., O'Riain, M. J., van Eeden, R. and King, A. (2011). A low-cost manipulation of food resources reduces spatial overlap between baboons (*Papio ursinus*) and humans in conflict. *Int. J. Primatol.*, 32, 1397–1412.

Ravasi, D. F., O'Riain, M. J., Davids, F. and Illing, N. (2012). Phylogenetic evidence that two distinct *Trichuris* genotypes infect both humans and non-human primates. *PLoS ONE*, 7, e44187.

© Catherine Young.

Index

Note: page numbers in *italics* refer to figures and tables, those in **bold** refer to boxes.